Problem Solving in Data Structures & Algorithms Using JAVA

First Edition

By Hemant Jain

Problems Solving in Data Structures & Algorithms Using JAVA
HEMANT JAIN

Copyright © 2016 by HEMANT JAIN. All Right Reserved.

No part of this publication may be reproduced, distributed, or transmitted in any form or by any means, including photocopying, recording, or other electronic or mechanical methods, or by any information storage and retrieval system without the prior written permission of the author, except in the case of very brief quotations embodied in critical reviews and certain other non-commercial uses permitted by copyright law.

ACKNOWLEDGEMENT

The author is very grateful to GOD ALMIGHTY for his grace and blessing.

Deepest gratitude for the help and support of my brother Dr. Sumant Jain. This book would not have been possible without the support and encouragement he provided.

I would like to express profound gratitude to my guide/ my friend Naveen Kaushik for his invaluable encouragement, supervision and useful suggestion throughout this book writing work. His support and continuous guidance enable me to complete my work successfully.

Last but not least, I am thankful to Anil Berry and Others who helped me directly or indirectly in completing this book.

Hemant Jain

TABLE OF CONTENTS

TABLE OF CONTENTS	5
TABLE OF CONTENTS	7
CHAPTER 0: HOW TO USE THIS BOOK	14
CHAPTER 1: INTRODUCTION - PROGRAMMING OVERVIEW	18
CHAPTER 2: ALGORITHMS ANALYSIS	48
CHAPTER 3: APPROACH TO SOLVE ALGORITHM DESIGN PROBLEMS	60
CHAPTER 4: ABSTRACT DATA TYPE & JAVA COLLECTIONS	65
CHAPTER 5: SEARCHING	89
CHAPTER 6: SORTING	118
CHAPTER 7: LINKED LIST	139
CHAPTER 8: STACK	182
CHAPTER 9: QUEUE	202
CHAPTER 10: TREE	213
CHAPTER 11: PRIORITY QUEUE	257
CHAPTER 12: HASH-TABLE	281
CHAPTER 13: GRAPHS	296
CHAPTER 14: STRING ALGORITHMS	316
CHAPTER 15: ALGORITHM DESIGN TECHNIQUES	342
CHAPTER 16: BRUTE FORCE ALGORITHM	348
CHAPTER 17: GREEDY ALGORITHM	354
CHAPTER 18: DIVIDE-AND-CONQUER, DECREASE-AND-CONQUER	363
CHAPTER 19: DYNAMIC PROGRAMMING	375
CHAPTER 20: BACKTRACKING AND BRANCH-AND-BOUND	384
CHAPTER 21: COMPLEXITY THEORY AND NP COMPLETENESS	387
CHAPTER 22: INTERVIEW STRATEGY	397
CHAPTER 23: SYSTEM DESIGN	399
APPENDIX	433
INDEX	434

TABLE OF CONTENTS

TABLE OF CONTENTS 5
TABLE OF CONTENTS 7

CHAPTER 0: HOW TO USE THIS BOOK 14
    WHAT THIS BOOK IS ABOUT 14
    PREPARATION PLANS 14
    SUMMARY 17

CHAPTER 1: INTRODUCTION - PROGRAMMING OVERVIEW 18
    INTRODUCTION 18
    FIRST JAVA PROGRAM 18
    OBJECT 19
    VARIABLE 20
    DATA TYPES 20
    PRIMITIVE/ BASIC DATA TYPES 20
    PARAMETER PASSING, CALL BY VALUE 22
    REFERENCE DATA TYPES 23
    PARAMETER PASSING, CALL BY REFERENCE 23
    KINDS OF VARIABLES 24
    METHODS 25
    ACCESS MODIFIERS 25
    INTERFACE 25
    RELATIONSHIP 26
    GENERAL PROTOTYPE OF A CLASS 27
    ABSTRACT CLASS & METHODS 27
    NESTED CLASS 30
    ENUMS 31
    CONSTANTS 31
    CONDITIONS AND LOOPS 32
    ARRAY 35
    TWO DIMENSIONAL ARRAY 36
    ARRAY INTERVIEW QUESTIONS 37
    CONCEPT OF STACK 40
    SYSTEM STACK AND METHOD CALLS 40
    RECURSIVE FUNCTION 42
    EXERCISES 46

CHAPTER 2: ALGORITHMS ANALYSIS 48
    INTRODUCTION 48
    ASYMPTOTIC ANALYSIS 48
    BIG-O NOTATION 48

Omega-Ω Notation ... 49
Theta-Θ Notation ... 50
Complexity analysis of algorithms ... 50
Time Complexity Order ... 51
Deriving the Runtime Function of an Algorithm ... 52
Time Complexity Examples ... 53
Master Theorem ... 56
Modified Master theorem ... 58
Exercise ... 59

CHAPTER 3: APPROACH TO SOLVE ALGORITHM DESIGN PROBLEMS ... 60

Introduction ... 60
Constraints ... 60
Idea Generation ... 61
Complexities ... 62
Coding ... 63
Testing ... 63
Example ... 64
Summary ... 64

CHAPTER 4: ABSTRACT DATA TYPE & JAVA COLLECTIONS ... 65

Abstract data type (ADT) ... 65
Data-Structure ... 65
JAVA Collection Framework ... 66
Array ... 66
Linked List ... 68
Stack ... 70
Queue ... 71
Trees ... 73
Binary Tree ... 73
Binary Search Trees (BST) ... 73
Priority Queue (Heap) ... 76
Hash-Table ... 78
Dictionary / Symbol Table ... 82
Graphs ... 85
Graph Algorithms ... 86
Sorting Algorithms ... 87
Counting Sort ... 88
End note ... 88

CHAPTER 5: SEARCHING ... 89

Introduction ... 89
Why Searching? ... 89
Different Searching Algorithms ... 89

Linear Search – Unsorted Input	89
Linear Search – Sorted	90
Binary Search	90
String Searching Algorithms	91
Hashing and Symbol Tables	91
How sorting is useful in Selection Algorithm?	91
Problems in Searching	92
Exercise	116

CHAPTER 6: SORTING 118

Introduction	118
Type of Sorting	119
Bubble-Sort	119
Modified (improved) Bubble-Sort	121
Insertion-Sort	122
Selection-Sort	124
Merge-Sort	126
Quick-Sort	128
Quick Select	130
Bucket Sort	131
Generalized Bucket Sort	133
Heap-Sort	134
Tree Sorting	134
External Sort (External Merge-Sort)	135
Comparisons of the various sorting algorithms.	136
Selection of Best Sorting Algorithm	136
Exercise	137

CHAPTER 7: LINKED LIST 139

Introduction	139
Linked List	139
Types of Linked list	139
Singly Linked List	140
Doubly Linked List	159
Circular Linked List	169
Doubly Circular list	176
Exercise	181

CHAPTER 8: STACK 182

Introduction	182
The Stack Abstract Data Type	182
Stack using Array	184
Stack using Array (Growing-Reducing capacity implementation)	186
Stack using linked list	187

PROBLEMS IN STACK 189
PROS AND CONS OF ARRAY AND LINKED LIST IMPLEMENTATION OF STACK 200
USES OF STACK 200
EXERCISE 201

CHAPTER 9: QUEUE 202

INTRODUCTION 202
THE QUEUE ABSTRACT DATA TYPE 203
QUEUE USING ARRAY 203
QUEUE USING LINKED LIST 205
PROBLEMS IN QUEUE 208
EXERCISE 211

CHAPTER 10: TREE 213

INTRODUCTION 213
TERMINOLOGY IN TREE 214
BINARY TREE 215
TYPES OF BINARY TREES 217
PROBLEMS IN BINARY TREE 220
BINARY SEARCH TREE (BST) 239
PROBLEMS IN BINARY SEARCH TREE (BST) 240
EXERCISE 254

CHAPTER 11: PRIORITY QUEUE 257

INTRODUCTION 257
TYPES OF HEAP 258
HEAP ADT OPERATIONS 259
OPERATION ON HEAP 260
HEAP-SORT 268
USES OF HEAP 275
PROBLEMS IN HEAP 276
EXERCISE 279

CHAPTER 12: HASH-TABLE 281

INTRODUCTION 281
HASH-TABLE 282
HASHING WITH OPEN ADDRESSING 283
HASHING WITH SEPARATE CHAINING 287
COUNT MAP 290
PROBLEMS IN HASHING 291
EXERCISE 294

CHAPTER 13: GRAPHS 296

INTRODUCTION 296
GRAPH REPRESENTATION 298

    Adjacency Matrix 298

    Adjacency List 299

    Graph traversals 301

    Depth First Traversal 301

    Breadth First Traversal 303

    Problems in Graph 305

    Directed Acyclic Graph 306

    Topological Sort 306

    Minimum Spanning Trees (MST) 307

    Shortest Path Algorithms in Graph 310

    Exercise 315

CHAPTER 14: STRING ALGORITHMS 316

    Introduction 316

    String Matching 317

    Dictionary / Symbol Table 322

    Problems in String 331

    Exercise 340

CHAPTER 15: ALGORITHM DESIGN TECHNIQUES 342

    Introduction 342

    Brute Force Algorithm 342

    Greedy Algorithm 343

    Divide-and-Conquer, Decrease-and-Conquer 343

    Dynamic Programming 344

    Reduction / Transform-and-Conquer 345

    Backtracking 346

    Branch-and-bound 347

    A* Algorithm 347

    Conclusion 347

CHAPTER 16: BRUTE FORCE ALGORITHM 348

    Introduction 348

    Problems in Brute Force Algorithm 348

    Conclusion 353

CHAPTER 17: GREEDY ALGORITHM 354

    Introduction 354

    Problems on Greedy Algorithm 354

CHAPTER 18: DIVIDE-AND-CONQUER, DECREASE-AND-CONQUER 363

    Introduction 363

    General Divide-and-Conquer Recurrence 364

    Master Theorem 364

    Problems on Divide-and-Conquer Algorithm 366

CHAPTER 19: DYNAMIC PROGRAMMING 375
Introduction 375
Problems on Dynamic programming Algorithm 376

CHAPTER 20: BACKTRACKING AND BRANCH-AND-BOUND 384
Introduction 384
Problems on Backtracking Algorithm 384

CHAPTER 21: COMPLEXITY THEORY AND NP COMPLETENESS 387
Introduction 387
Decision problem 387
Complexity Classes 387
Class P problems 387
Class NP problems 388
Class co-NP 391
NP–Hard: 391
NP–Complete Problems 392
Reduction 392
End Note 396

CHAPTER 22: INTERVIEW STRATEGY 397
Introduction 397
Resume 397
Nontechnical questions 397
Technical questions 398

CHAPTER 23: SYSTEM DESIGN 399
System Design 399
System Design Process 399
Scalability Theory 400
Design simplified Facebook 405
Design a shortening service like Bitly 411
Stock Query Server 413
Design a basic search engine Database 414
Duplicate integer in millions of documents 416
Zomato 419
YouTube 420
Design IRCTC 421
Alarm Clock 422
Design for Elevator of a building 423
Valet parking system 425
OO design for a McDonalds shop 425
Object oriented design for a Restaurant 427
Object oriented design for a Library system 428

Suggest a shortest path 429
Exercise 430
APPENDIX **433**
Appendix A 433
INDEX **434**

Chapter 0: How to Use this Book

What this book is about

This book is about usage of data structures and algorithms in computer programming. Data structures are the ways in which data is arranged in computers memory. Algorithms are set of instructions to solve some problem by manipulating these data structures.

Designing an efficient algorithm to solve a computer science problem is a skill of Computer programmer. This is the skill which tech companies like Google, Amazon, Microsoft, Facebook and many others are looking for in an interview. Once we are comfortable with a programming language, the next step is to learn how to write efficient algorithms.

This book assumes that you are a JAVA language developer. You are not an expert in JAVA language, but you are well familiar with concepts of references, functions, arrays and recursion. At the start of this book, we will be revising the JAVA language fundamentals that will be used throughout this book. We will be looking into some of the problems in arrays and recursion too.

Then in the coming chapter we will be looking into Complexity Analysis. Followed by the various data structures and their algorithms. Will look into a Linked-List, Stack, Queue, Trees, Heap, Hash-Table and Graphs. We will also be looking into Sorting, Searching techniques.

And we will be looking into algorithm analysis of various algorithm techniques, we will be looking into Brute-Force algorithms, Greedy algorithms, Divide and Conquer algorithms, Dynamic Programming, Reduction and Back-Tracking.

In the end, we will be looking into System Design that will give a systematic approach to solve the design problems in an Interview.

Preparation Plans

Given the limited time you have before your next interview, it is important to have a solid preparation plan. The preparation plan depends upon the time and which companies you are planning to target. Below are the three-preparation plan for 1 Month, 3 Month and 5 Month durations.

1 Month Preparation Plans

Below is a list of topics and approximate time user need to take to finish these topics. These are the most important chapters that must to be prepared before appearing for an interview.

This plan should be used when you have a small time before an interview. These chapters cover 90% of data structures and algorithm interview questions. In this plan since we are reading about the various ADT and JAVA collections in chapter 4 so we can use these datatype easily without knowing the internal details how they are implemented.

Chapter 24 is for system design, you must read this chapter if you are three or more years of experienced. Anyway, reading this chapter will give the reader a broader perspective of various designs.

Time	Chapters	Explanation
Week 1	Chapter 1: Introduction - Programming Overview Chapter 2: Algorithms Analysis Chapter 3: Approach To Solve Algorithm Design Problems Chapter 4: Abstract Data Type & JAVA Collections	You will get a basic understanding of how to find complexity of a solution. You will know how to handle new problems. You will read about a variety of datatypes and their uses.
Week 2	Chapter 5: Searching Chapter 6: Sorting Chapter 14: String Algorithms	Searching, Sorting and String algorithm consists of a major portion of the interviews.
Week 3	Chapter 7: Linked List Chapter 8: Stack Chapter 9: Queue	Linked list, Stack and Queue are one of the favorites in an interview.
Week 4	Chapter 10: Tree Chapter 23: Interview Strategy Chapter 24: System Design	This portion you will read about Trees and System Design. You are good to go for interviews. Best of luck.

3 Month Preparation Plan

This plan should be used when you have some time to prepare for an interview. This preparation plan includes nearly everything in this book except various algorithm techniques. Algorithm problems that are based on dynamic programming divide & conquer etc. Which are asked in vary specific companies like Google, Facebook, etc. Therefore, until you are planning to face interview with them you can park these topics for some time and focus on the rest of the topics.

Again, same thing here with system design problems, the more experience you are, the more important this chapter becomes. However, if you are a fresher from college, then also you should read this chapter.

Time	Chapters	Explanation
Week 1	Chapter 1: Introduction - Programming Overview Chapter 2: Algorithms Analysis Chapter 3: Approach To Solve Algorithm Design Problems Chapter 4: Abstract Data Type & JAVA Collections	You will get a basic understanding of how to find complexity of a solution. You will know how to handle new problems. You will read about a variety of datatypes and their uses.

Week 2 & Week 3	Chapter 5: Searching Chapter 6: Sorting Chapter 14: String Algorithms	Searching, sorting and string algorithm consists of a major portion of the interviews.
Week 4 & Week 5	Chapter 7: Linked List Chapter 8: Stack Chapter 9: Queue	Linked list, Stack and Queue are one of the favorites in an interview.
Week 6 & Week 7	Chapter 10: Tree Chapter 11: Heap	This portion you will read about trees and heap data structures.
Week 8 & Week 9	Chapter 12: Hash-Table Chapter 13: Graphs	Hash-Table is used throughout this book in various places, but now it's time to understand how Hash-Table are actually implemented. Graphs are used to propose a solution many real life problems.
Week 10	Chapter 23: Interview Strategy Chapter 24: System Design	Interview strategy and system design chapter are the final chapters of this course.
Week 11 & Week 12	Revision of the chapters listed above.	At this time, you need to revise all the chapters that we have seen in this book. Whatever is left needs to be completed and the exercise that may be left needing to be solved in this period of time.

5 Month Preparation Plan

In this preparation plan is made on top of 3-month plan. In this plan, the students should look for algorithm design chapters. In addition, in the rest of the time they need to practice more and more from www.topcoder.com and other resources. If you are targeting google, Facebook, etc., Then it is highly recommended to join topcoder and practice as much as possible.

Time	Chapters	Explanation
Week 1 Week 2	Chapter 1: Introduction - Programming Overview Chapter 2: Algorithms Analysis Chapter 3: Approach To Solve Algorithm Design Problems Chapter 4: Abstract Data Type & JAVA Collections	You will get a basic understanding of how to find complexity of a solution. You will know how to handle new problems. You will read about a variety of datatypes and their uses.
Week 3 Week 4 Week 5	Chapter 5: Searching Chapter 6: Sorting Chapter 14: String Algorithms	Searching, sorting and string algorithm consists of a major portion of the interviews.
Week 6 Week 7 Week 8	Chapter 7: Linked List Chapter 8: Stack Chapter 9: Queue	Linked list, Stack and Queue are one of the favorites in an interview.
Week 9 Week 10	Chapter 10: Tree Chapter 11: Heap	This portion you will read about trees and priority queue.

Week 11 Week 12	Chapter 12: Hash-Table Chapter 13: Graphs	Hash-Table is used throughout this book in various places, but now it's time to understand how Hash-Table are actually implemented. Graphs are used to propose a solution many real life problems.
Week 13 Week 14 Week 15 Week 16	Chapter 15: Algorithm Design Techniques Chapter 16: Brute Force Chapter 17: Greedy Algorithm Chapter 18: Divide-And-Conquer, Decrease-And-Conquer Chapter 19: Dynamic Programming Chapter 20: Backtracking And Branch-And-Bound Chapter 21: Complexity Theory And Np Completeness	These chapters contain various algorithms types and their usage. Once the user is familiar with most of this algorithm. Then the next step is to start solving topcoder problems from https://www.topcoder.com/
Week 17 Week 18	Chapter 22: Interview Strategy Chapter 23: System Design	Interview strategy and system design chapter are the final chapters of this course.
Week 19 Week 20	Revision of the chapters listed above.	At this time, you need to revise all the chapters that we have seen in this book. Whatever is left needs to be completed and the exercise that may be left needing to be solved in this period.

Summary

These are few preparation plans that can be followed to complete this book there by preparing for the interview. It is highly recommended that the user should read the problem statement first, then he should try to solve the problems by himself and then only he should look into the solution to find the approach of the book. Practicing more and more problems will increase your thinking capacity and you will be able to handle new problems in an interview. System design is a topic that is not asked much from a fresher from college, but as you gain experience its importance increase. We will recommend practicing all the problems given in this book, then solve more and more problems from online resources like www.topcoder.com, www.careercup.com, www.geekforgeek.com etc.

Chapter 1: Introduction - Programming Overview

Introduction

This chapter emphasizes on brush up of the fundamentals of the JAVA Programming language. It will talk about variables, references, classes, loops, recursion, arrays etc. We assume that the reader is familiar with the syntax of the JAVA programming language and knows the basics of Object-Orientation.

First JAVA Program

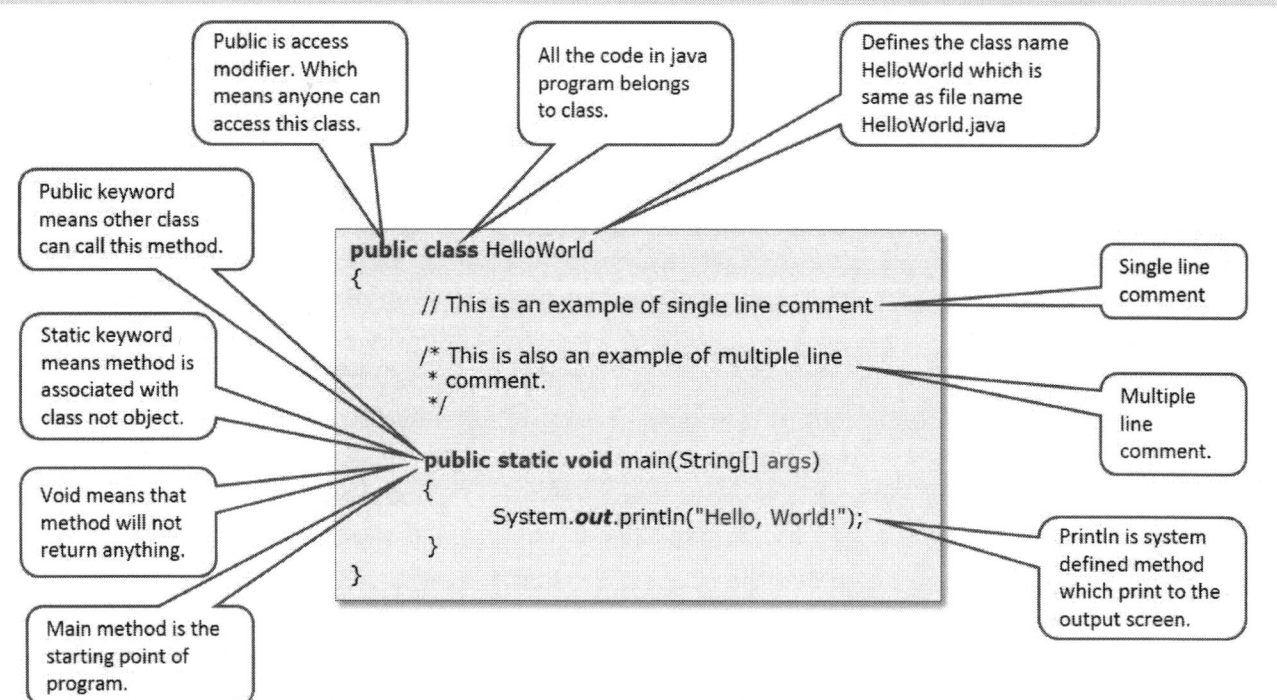

It is tradition to discuss a HelloWorld program in the start which will print the phrase "Hello, World!" to the output screen. So let us start discussing it. This is a small program but it contains many common features of all the JAVA programs.

1. This program begins with "public class HelloWorld":
    a. A class is the basic unit of encapsulation. Keyword **class** is used to create a new class in our case class HelloWorld is created. The class name is same as the name of the file, which will contain this class. Therefore, in our case the name of the file is HelloWorld.java. A class contains data and methods.
    b. The keyword **public** is an access specifier, which sets the accessibility of classes. A public class can be accessed from any class within the JAVA program.

2. Next are the comments, which are for readability of the program and are ignored by the compiler.
    a. A **single line comment** begins with two forward slash //
    b. A **multiline comment** begins with /* and ends with */.

3. "public static void main(String[] args)" :
    a. A method is a set of statements that are executed to give desire result.
    b. **main()** is a special method of a class this is the entry point of the application.
    c. The keyword **public** with the method means that it can be accessed from any class of the application.
    d. A keyword **static** in the method specifies that the method is a class method and will be called directly from the class name. It does not require any object to call this method.
    e. The **void** keyword means that this method is not going to return anything.

4. System.out.println("Hello, World!") is a system provided method which will print "Hello, World! " to the standard output.

Object

An **Object** is an entity with state and behavior. A baby, a cat, a dog, a bulb etc are all examples of objects. A baby has properties & states (name, hungry, crying, sleeping etc.) and behaviors (feed, play, etc.). Another example a bulb have two states (on, off) and the two behaviors (turn on, turn off).

Software objects are just like real world objects. They have state in the form of member variables called **fields** (isOn) and they expose behavior in the form of member functions called **methods** (turn on, turn off).

Hiding internal details (state) of the object and allowing all the actions to be performed over the objects using methods is known has **data-encapsulation**.

A **Class** is a prototype (blueprint) of objects. An object is an instance of a class. Human is a class of living being and a person named John is an instance of human class.

Example 1.1:

```java
class Bulb {
    boolean isOn=false;

    public void turnOn() {
        isOn = true;
    }

    public void turnOff() {
        isOn = false;
    }
}
```

In this example, we have a class name Bulb. It has a member variable isOn, which indicates its state that the bulb is on or off. It has two methods turnOn() and turnoff() which will change the state of the object from off to on and vice versa.

Variable

"Variables" are simply storage locations for data. For every variable, some memory is allocated. The size of this memory depends on the type of the variable.

Example 1.2:

```java
public class variableExample {
    public static void main(String[] args) {
        int var1,var2,var3; //Declaring three variables declared
        var1=100;
        var2=200;
        var3=var1+var2;
        System.out.println("Adding"+var1+"and "+var2+"will give"+var3);
    }
}
```

Analysis:
- Memory is allocated for variables var1, var2 and var3. Whenever we declare a variable (premative type variable), then memory is allocated for storing the value in the variable. In our example, 4 bytes are allocated for each of the variable.
- Value 100 is stored in variable var1 and value 200 is stored in variable var2.
- Value of var1 and var2 is added and stored in var3.
- Finally, the value of var1, var2 and var3 is printed to screen using System.out.println(); built-in method.

Data Types

The various types of variable are defined as Data Types. There are two varieties of data types available in JAVA:
1. Primitive/ Basic Data Types
2. Reference Data Types

Primitive/ Basic Data Types

Primitive data types are the basic data types, which are defined in the JAVA language. There are 8 different primitive data types - byte, short, int, long, float, double, boolean, char. All primitive values can be stored in a fixed amount of memory (between one and eight bytes)

Assignment of primitive data type is done by value. The content in primitive data type is copied to another variable.

Comparison (==) in primitive variables in done by comparing the value stored in the variable.

Parameter passing of primitive variable is done by copying the content of the variable. (i.e. pass-by-value). Since value is copied changes done on the parameter value inside a method is not reflected to the caller of the function. Returning primitive variable from a method return value stored in variable. Local variable are destroyed when return from a method.

Primitive Data Type	Explanation
Boolean	The value represents one bit of information. The allowed values are true or false.
Int	32-bit signed two's complement integer value range from -2,147,483,648 to 2,147,483,647
Char	Single 16-bit Unicode character. The value ranges from '\u0000' (or 0) to '\uffff' (or 65,535)
Byte	8-bit signed two's complement integer value range from -128 to 127
Short	16-bit signed two's complement integer. The value ranges from -32,768 to 32,767
Long	64-bit signed two's complement integer. The value ranges from -2^63 to (2^63 -1)
Float	32-bit single-precision floating-point number.
Double	64-bit double-precision floating-point number.

Example 1.3: Program demonstrating range of Primitive data type

```java
public class MinMaxValueTest {
    public static void main(String args[]) {
        byte maxByte = Byte.MAX_VALUE;
        byte minByte = Byte.MIN_VALUE;
        short maxShort = Short.MAX_VALUE;
        short minShort = Short.MIN_VALUE;
        int maxInteger = Integer.MAX_VALUE;
        int minInteger = Integer.MIN_VALUE;
        long maxLong = Long.MAX_VALUE;
        long minLong = Long.MIN_VALUE;
        float maxFloat = Float.MAX_VALUE;
        float minFloat = Float.MIN_VALUE;
        double maxDouble = Double.MAX_VALUE;
        double minDouble = Double.MIN_VALUE;

        System.out.println("Range of byte :: " + minByte + " to " +
        maxByte + ".");
        System.out.println("Range of short :: " + minShort + " to " +
        maxShort + ".");
        System.out.println("Range of integer :: " + minInteger + " to " +
        maxInteger + ".");
        System.out.println("Range of long :: " + minLong + " to " +
        maxLong + ".");
        System.out.println("Range of float :: "    + minFloat + " to " +
        maxFloat + ".");
        System.out.println("Range of double :: " + minDouble + " to " +
        maxDouble + ".");
    }
}
```

Output:
```
Range of byte :: -128 to 127.
Range of short :: -32768 to 32767.
Range of integer :: -2147483648 to 2147483647.
Range of long :: -9223372036854775808 to 9223372036854775807.
Range of float :: 1.4E-45 to 3.4028235E38.
Range of double :: 4.9E-324 to 1.7976931348623157E308.
```

Parameter passing, Call by value

Arguments can be passed from one method to other using parameters. All the basic data types when passed as parameters are by the passed-by-value. That means a separate copy is created inside the called method and the variable in the calling method remains unchanged.

Example 1.4:

```java
public static void increment(int var)
{
    var++;
}
```

```java
public static void main(String[] args)
{
    int i = 10;
    System.out.println("Value of i before increment is :  " + i);
    increment(i);
    System.out.println("Value of i before increment is :  " + i);
}
```

Output:
```
Value of i before increment is :  10
Value of i after increment is :  10
```

Analysis:
- Variable "i" is declared and the value 10 is initialized to it.
- Value of "i" is printed.
- Increment method is called. When a method is called the value of the parameter is copied into another variable of the called method. Flow of control goes to increase() function.
- Value of var is incremented by 1. But remember, it is just a copy inside the increment method.
- When the method exits, the value of "i" is still 10.

Points to remember:
1. Pass by value just creates a copy of variable.
2. Pass by value, value before and after the method call remain same.

Reference Data Types

Reference variables are used to store memory address of classes as well as arrays. Any variable other them 8 primitive data types are reference data type.

Parameter passing of a reference variable is done by copying the address of the variable. This method is called pass-by-reference. Since the object is not copied, it is shared, the changes done in a called function are also reflected to the caller of the function.

Comparison (==) in reference variable is done by comparing the address of the variable.

The default value of any reference variable is null.

Returning reference variable from a method return address of the variable. If the returned address is stored in some other variable the locally created object is not destroyed.

Parameter passing, Call by Reference

If you need to change the value of the parameter inside the method, then you should use call by reference. JAVA language by default passes by value. Therefore, to make it happen, you need to pass the address of a variable and changing the value of the variable using this address inside the called method.

Example 1.5:

```java
private static class MyInt{
            int value;
};
```

```java
public static void increment(MyInt value)
{
     (value.value)++;
}
```

```java
public static void main(String[] args)
{
      MyInt x = new MyInt();
      x.value = 10;
      System.out.println("Value of i before increment is: "+ x.value);
      increment(x);
      System.out.println("Value of i after increment is: "+ x.value);
}
```

Output:

```
Value of i before increment is :  10
Value of i after increment is :  11
```

Analysis: Object of class MyInt is passed to the method. Since the objects are passed by reference. Value change in increment() function is reflected to the original object of the caller function.

Points to remember:
1. Call by reference is implemented indirectly by passing the address of an instance of class or array to the function.

Kinds of Variables

The JAVA programming language defines three kinds of variables:
1. **Instance Variables (Non-Static)**: They are instance variables so they are unique to each instance/object of a class.
2. **Class Variables (Static):** A class variable is any field with the **static** modifier. These variables are linked with the class not with the objects of the class. There is exactly one copy of these variables regardless of how many instances of the class are created.
3. **Local Variables**: the temporary variables in a method are called local variables. The local variables are only visible to the method in which they are declared. The parameters that are passed to the methods are also local variables of the called method.

Example 1.6:

```java
class Bulb {
        //Class Variables
        private static int TotalBulbCount = 0;

        //Instance Variables
        private boolean isOn=false;

        //Constructor
        public Bulb(){
                TotalBulbCount++;
        }

        //Class Method
        public static int getBulbCount(){
                return TotalBulbCount;
        }

        //Instance Method
        public void turnOn() {
                isOn = true;
        }

        //Instance Method
        public void turnOff() {
                isOn = false;
        }

        //Instance Method
        public boolean isOnFun(){
                return isOn;
        }
}
```

Methods

There are three types of methods. Class Methods, Instance Methods and Constructors. By default, all the methods are instance methods.

Class Methods (Static): The **static** modifier is used to create class methods. Class methods with static modifier with them should be invoked with the class name without the need of creating even a single instance of the class

Instance Methods (Non-Static): These methods can only be invoked over an instance of the class.

Some points regarding Instance methods and Class methods:
1. Instance methods can access other instance methods and instance variables directly.
2. Instance methods can access class methods and variables directly.
3. Class methods can access other class methods and class variables directly.
4. Class methods cannot access instance methods and instance variables directly. To access instance variable and methods they need to create and instance (object) of class.
5. The special keyword "**this**" is valid only inside instance methods (and invalid inside class methods) as "**this**" refers to the current instance.

Constructor: It is a special kind of method, which is invoked over objects when they are created. Constructor methods have the same name as the class. Constructor method is used to initialize the various fields of the object. For the class that does not have constructors, JAVA language provides default constructors for them.

Access Modifiers

Access modifiers are used to set the visibility level to the class, variables and methods. JAVA provide four types access modifiers: default, public, protected, private.
1. **Default** modifier (No modifier needed.) has visibility to the package.
2. **Private** modifier has visibility only within its own class.
3. **Public** modifier has visibility to all the classes in the package.
4. **Protected** modifier has visibility within its own class and the subclasses its own class.

Interface

Objects define their interface as the interaction with the outside world. For example, in the bulb case switch is the interface between you and the bulb. You press the button turn on and the bulb start glowing.

Example 1.7:
```
public interface BulbInterface {
    public void turnOn();
    public void turnOff();
    public boolean isOnFun();
}
```

```java
class Bulb implements BulbInterface {

    private boolean isOn=false;

    @Override
    public void turnOn() {
        isOn = true;
    }

    @Override
    public void turnOff() {
        isOn = false;
    }

    @Override
    public boolean isOnFun() {
        return isOn;
    }
}
```

```java
public class BulbDemo {

    public static void main(String[] args) {
        Bulb b = new Bulb();
        System.out.println("bulb is on return : " + b.isOnFun());
        b.turnOn();
        System.out.println("bulb is on return : " + b.isOnFun());
    }
}
```

Analysis:
In this example, BulbInterface is the interface of Bulb class. Bulb class needs to implement all the methods of BulbInterface to implement it.

Relationship

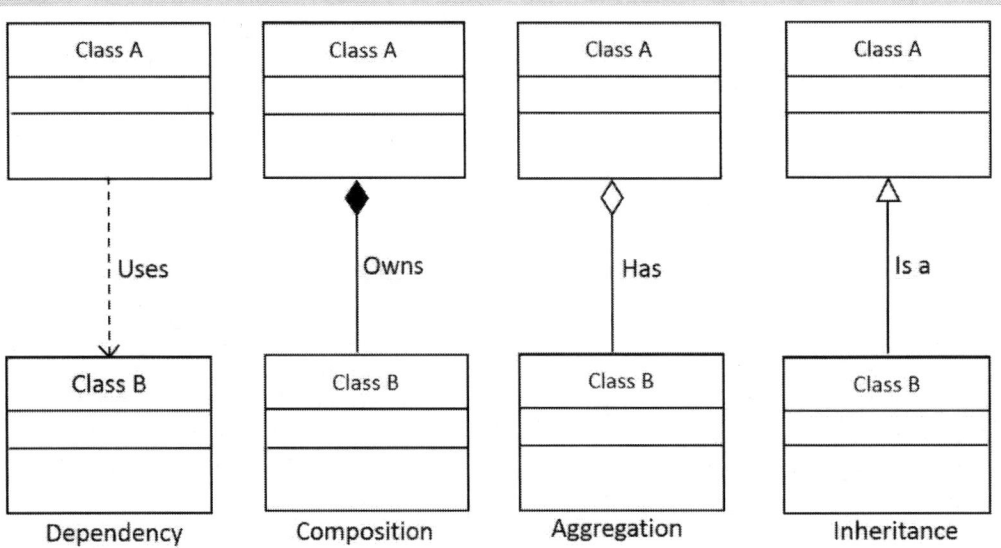

These are the various relationships that exist between two classes:

Dependency: Objects of one class use objects of another class temporarily. When a class creates an instance of another class inside its member method and use it and when the method exits, then the instance of the other class is deleted.

Association: Objects of one class work with objects of another class for some amount of time. The association is of two kinds - Aggregation and Composition.

Aggregation: One class object share a reference to objects of another class. When a class stores the reference of another class inside it. Just a reference is kept inside the class.

Composition: One class contains objects of another class. In Composition, the containing object is responsible for the creation and deletion of the contained object.

Inheritance: One class is a sub-type of another class. Inheritance is a straightforward relationship to explain the parent-child relationship.

Example 1.8:
```java
class AdvanceBulb extends Bulb {

    int intensity;

    public void setIntersity(int i) {
        intensity = i;
    }
}
```
In our example, AdvanceBulb is a sub-class of Bulb. When an object of AdvanceBulb is created, all public and protected methods of Bulb class are also accessible.

General Prototype of a Class

Example 1.9:
```java
class ChildClass extends ParentClass implements SomeInterface {
    // fields
    // methods
}
```
A ChildClass inherited from ParentClass and implements SomeInterface.

Abstract Class & Methods

An abstract method is a method which does not have a definition. Such methods are declared with abstract keyword.
A class which has at least one abstract method need to be declared as abstract. We cannot create objects of an abstract class. (A class which does not have any abstract method can also be declared as abstract to prevent its object creation.).

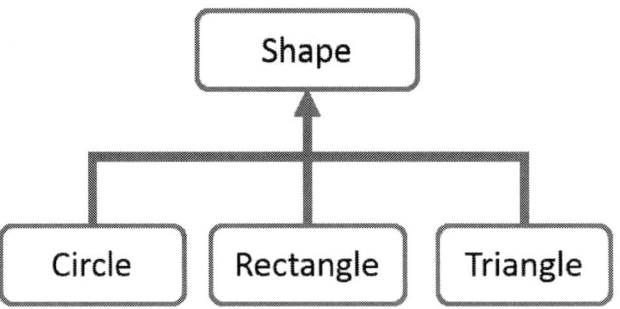

Example 1.10:

```
//Abstract Class
public abstract class Shape {

    //Abstract Method
    public abstract double area();

    //Abstract Method
    public abstract double perimeter();
}
```

Shape is an abstract class. And its instance cannot be created. Those classes, which extend it, need to implement these two functions to become concrete class whose instances can be created.

```
public class Circle extends Shape {
    private double radius;
    public Circle() {
        this(1);
    }
    public Circle(double r) {
        radius = r;
    }
    public void setRadius(double r) {
        radius = r;
    }
    @Override
    public double area() {
        // Area = πr^2
        return Math.PI * Math.pow(radius, 2);
    }
    @Override
    public double perimeter() {
        // Perimeter = 2πr
        return 2 * Math.PI * radius;
    }
}
```

Circle is a class which extends shape class and implement area() and parimeter() methods.

```java
public class Rectangle extends Shape {

    private double width, length;

    public Rectangle() {
        this(1,1);
    }

    public Rectangle(double w, double l) {
        width = w;
        length = l;
    }

    public void setWidth(double w) {
        width = w;
    }

    public void setLength(double l) {
        length = l;
    }

    @Override
    public double area() {
        // Area = width * length
        return width * length;
    }

    @Override
    public double perimeter() {
        // Perimeter = 2(width + length)
        return 2*(width + length);
    }
}
```

Same as Circle class, Rectangle class also extends Shape class and implements its abstract functions.

```java
public class ShapeDemo {

    public static void main(String[] args) {
        double width = 2, length = 3;
        Shape rectangle = new Rectangle(width, length);
        System.out.println("Rectangle width: " + width
                + " and length: " + length
                + " Area: " + rectangle.area()
                + " Perimeter: " + rectangle.perimeter());

        double radius = 10;
        Shape circle = new Circle(radius);
        System.out.println("Circle radius: " + radius
                + " Area: " + circle.area()
                + " Perimeter: " + circle.perimeter());
    }
}
```

Shape demo creates an instance of the Rectangle and the Circle class and assign it to a reference of type Shape. Finally area() and perimeter() functions are called over instance of Rectangle and Circle class.

Nested Class

A class within another class is called a nested class.

Compelling reasons for using nested classes include the following:
1. **It is a way of logically grouping classes that are only used in one place**: If a class is useful to only one other class, then it is logical to embed it in that class and keep the two together. Nesting such "helper classes" makes their package more streamlined.
2. **It increases encapsulation**: Consider two top-level classes, A and B, where B needs access to members of A that would otherwise be declared `private`. By hiding class B within class A, A's members can be declared private and B can access them. In addition, B itself can be hidden from the outside world.
3. **It can lead to more readable and maintainable code**: Nesting small classes within top-level classes places the code closer to where it is used.

A nested class has an independent set of modifiers from the outer class. Visibility modifiers (public, private and protected) effect whether the nested class definition is accessible beyond the outer class definition. For example, a private nested class can be used by the outer class, but by no other classes.

Example 1.11: Demonstrating Nested Class

```
public class OuterClass {

    class NestedClass {
        // NestedClass fields and methods.
    }
    // OuterClass fields and methods.
}
```

Let us take example of LinkedList and Tree class. Both the linked list and tree have nodes to store new element. Both the linked list and tree have their nodes different, so it is best to declare their corresponding nodes class inside their own class to prevent name conflict and increase encapsulation.

Example 1.12:

```
public class LinkedList {
    private static class Node {
        private int value;
        private Node next;
        // Nested Class Node other fields and methods.
    }
    private Node head;
    // Outer Class LinkedList other fields and methods.
}
```

```
public class Tree {
    private static class Node {
        private int value;
        private Node lChild;
        private Node rChild;
        // Nested Class Node other fields and methods.
    }

    private Node root;
    // Outer Class Tree other fields and methods.
}
```

Enums

Enums restrict a variable to have one of the few predefined values.

Example 1.13:

```
class Bulb {
    //Enums
    enum BulbSize{ SMALL, MEDIUM, LARGE }
    BulbSize size;
    //Other bulb class fields and methods.
}
```

```
public class BulbDemo {
    public static void main(String[] args) {
        Bulb b = new Bulb();
        b.size = Bulb.BulbSize.MEDIUM ;
        System.out.println("Bulb Size :" + b.size);
    }
}
```

In the above code, we made some change to Bulb class. It has one more field size and size is of type enum BulbSize. And the allowed values of the size are SMALL, MEDIUM and LARGE.

Constants

Constants are defined by using **static** modifier in combination with the **final** modifier. The final modifier indicates that the value of this field cannot be changed. The static modifier indicates that there will be only one instance of the variable and is a class variable.

For example, the following variable declaration defines a constant named PI, whose value approximates pi.

```
static final double  PI = 3.141592653589793;
```

Another example, in this we had created a constant of string type.

```
static final String text = "Hello, World!";
```

Conditions and Loops

IF Condition

If condition consists of a Boolean condition followed by one or more statements. It allows you to take different paths of logic, depending on a given Boolean condition.

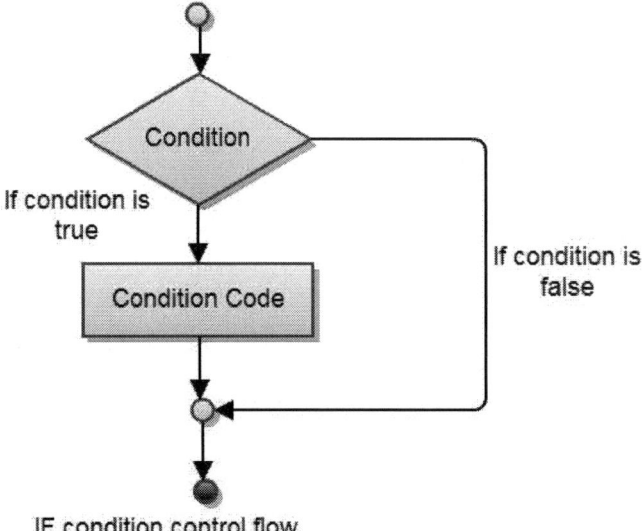

IF condition control flow

```
if (boolean_expression)
{
        // statements
}
```

If statement can be followed by else statements and an optional else statement which is executed when the Boolean condition is false.

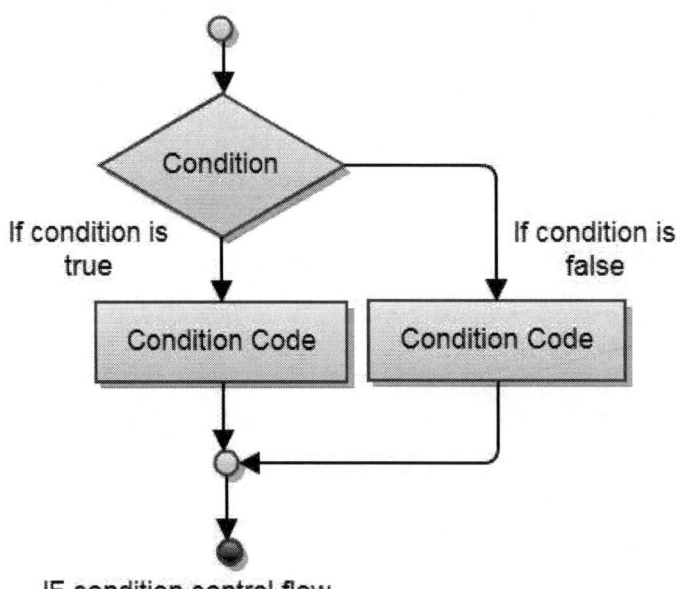

IF condition control flow

```
if(boolean_expression)
{
        // if condition statements boolean condition true
}
else
{
        // else condition statements, boolean condition false
}
```

While Loop

A while-loop is used to repeatedly execute some block of code as long as a given condition is true.

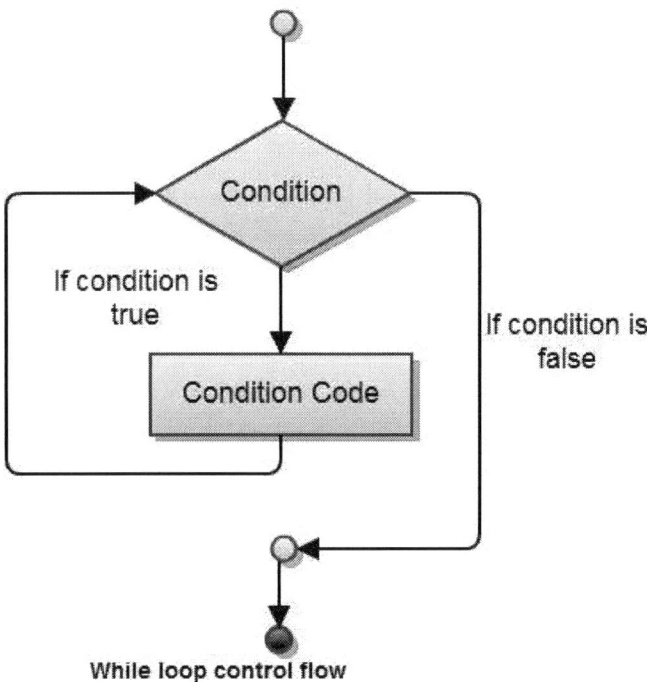

While loop control flow

Example 1.14:

```
public static void main(String[] args) {

        int[] numbers = { 1, 2, 3, 4, 5, 6, 7, 8, 9, 10};
        int sum = 0;
        int i = 0;
        while(i < numbers.length)
        {
                sum += numbers[i];
                i++;
        }
        System.out.println("Sum is :: " + sum);
}
```

Analysis:
All the variables stored in array are added to sum variabale one by on in a while loop.

Do..While Loop

A do..while-loop is similar to while-loop, but the only difference is that the conditional code is executed before the test condition. do..while-loop is used where you want to execute some conditional code at least once.

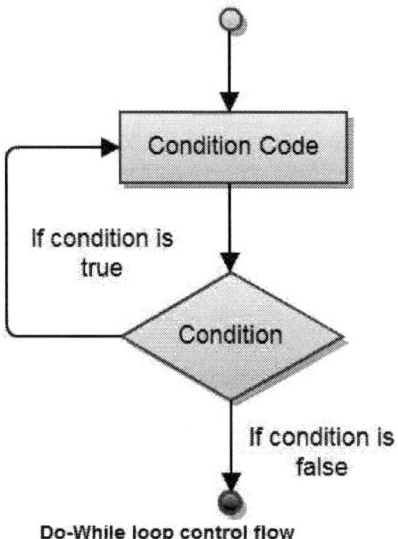

Do-While loop control flow

For Loop

For loop is just another loop in which initialization, condition check and increment are bundled together.

```
for ( initialization; condition; increment )
{
        statements
}
```

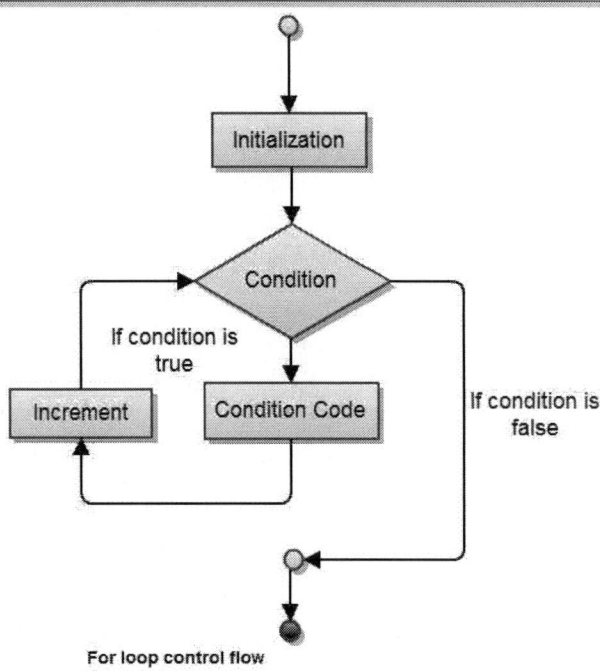

For loop control flow

Example 1.15:

```java
public static void main(String[] args) {

    int[] numbers = { 1, 2, 3, 4, 5, 6, 7, 8, 9, 10};
    int sum = 0;
    for(int i = 0; i < numbers.length; i++)
        sum += numbers[i];

    System.out.println("Sum is :: " + sum);
}
```

ForEach Loop

For loop works well with basic types, but it does not handle collections objects well. For this, an alternate syntax foreach loop is provided. JAVA does not have keyword foreach it uses for keyword.

```
for ( declaration : collection / array )
{
    statements
}
```

Example 1.16:

```java
public static void main(String[] args) {
    int[] numbers = { 1, 2, 3, 4, 5, 6, 7, 8, 9, 10};
    int sum = 0;
    for(int n : numbers)
        sum += n;
    System.out.println("Sum is :: " + sum);
}
```

Array

Arrays are the most basic data structures used to store information. An array is a data structure used to store multiple data elements of the same data type. All the data is stored sequentially. The value stored at any index can be accessed in constant time.

Example 1.17:

```java
public class Introduction {
    public static void main(String[] args) {
        arrayExample();
    }
    public static void arrayExample() {
        int[] arr = new int[10];
        for (int i = 0; i < 10; i++)
        {
            arr[i] = i;
        }
        printArray(arr,10);
    }
}
```

Analysis: Defines an array of *integer* arr. The array is of size 10 - which means that it can store 10 integers inside it. Array elements are accessed using the subscript operator []. Lowest subscript is 0 and highest subscript is (size of array – 1). Value 0 to 9 is stored in the array at index 0 to 9.
Array and its size are passed to printArray() method.

Example 1.18:

```
public static void printArray(int arr[], int count)
{
        System.out.println("Values stored in array are : ");
        for (int i = 0; i < count; i++)
        {
                System.out.println(" " + arr[i]);
        }
}
```

Analysis:
- Array variable arr and its variable count are passed as arguments to printArray() method.
- Finally array values are printed to screen using the System.*out*.println() method in a for loop.

Point to Remember:
1. Array index always starts from 0 index and highest index is size -1.
2. The subscript operator has highest precedence if you write arr[2]++. Then the value of arr[2] will be incremented.

Two Dimensional Array

We can define two dimensional or multidimensional array. It is an array of array.

Example 1.19:

```
public class Introduction {
        public static void main(String[] args) {
                twoDArrayExample();
        }
        public static void twoDArrayExample()
        {
                int[][] arr = new int[4][2];
                int count = 0;
                for (int i = 0; i < 4; i++)
                        for (int j = 0; j < 2; j++)
                                arr[i][j] = count++;

                print2DArray(arr, 4, 2);
        }

        public static void print2DArray(int[][] arr, int row, int col) {
                for (int i = 0; i < row; i++)
                        for (int j = 0; j < col; j++)
                                System.out.println(" " + arr[i][j]);
        }
}
```

Analysis:
- An array is created with dimension 4 x 2. The array will have 4 rows and 2 columns.
- Value is assigned to the array
- Finally, the value stored in an array is printed to screen by using print2DArray() method.

Array Interview Questions

The following section will discuss the various algorithms that are applicable to arrays and will follow by list of practice problems with similar approaches.

Sum Array

Write a method that will return the sum of all the elements of the *integer* array given array and its size as an argument.

Example 1.20:
```java
public static int SumArray(int arr[]) {
    int size = arr.length;
    int total=0;
    int index=0;
    for(index=0;index<size;index++)
        total = total + arr[index];
    return total;
}
```

```java
public static void main(String[] args) {
    int[] arr = {1,2,3,4,5,6,7,8,9};
    System.out.println("sum of all the values in array:" + SumArray(arr));
}
```

Sequential Search

Example 1.21: Write a method, which will search an array for some given value.
```java
public static int SequentialSearch(int arr[], int size, int value)
{
    int i = 0;
    for(i = 0; i < size; i++) {
        if(value == arr[i] )
            return i;
    }
    return -1;
}
```
Analysis:
- Since we have no idea about the data stored in the array, or if the data is not sorted then we have to search the array in sequential manner one by one.
- If we find the value, we are looking for we return that index.
- Else, we return -1 index, as we did not found the value we are looking for.

In the above example, the data are not sorted. If the data is sorted, a binary search may be done. We examine the middle position at each step. Depending upon the data that we are searching is greater or smaller than the middle value. We will search either the left or the right portion of the array. At each step, we are eliminating half of the search space there by making this algorithm very efficient the linear search.

Binary Search

Example 1.22: Binary search in a sorted array.

```
// Binary Search Algorithm – Iterative Way
public static int BinarySearch (int arr[], int size, int value)
{
        int mid;
        int low = 0;
        int high = size-1;
        while (low <= high)
        {
                mid = low + (high-low)/2;   // To avoid the overflow
                if (arr[mid] == value)
                        return mid;
                else if (arr[mid] < value)
                        low = mid + 1;
                else
                        high = mid - 1;
        }
        return -1;
}
```

Analysis:
- Since we have data sorted in increasing / decreasing order, we can apply more efficient binary search. At each step, we reduce our search space by half.
- At each step, we compare the middle value with the value we are searching for. If mid value is equal to the value we are searching for then we return the middle index.
- If the value is smaller than the middle value, we search the left half of the array.
- If the value is grater then the middle value then we search the right half of the array.
- If we find the value we are looking for then its index is returned or -1 is returned otherwise.

Rotating an Array by K positions.

For example, an array [10,20,30,40,50,60] rotate by 2 positions to [30,40,50,60,10,20]

Example 1.23:

```
public static void rotateArray(int[] a,int n,int k)
{
        reverseArray(a,0,k-1);
        reverseArray(a,k,n-1);
        reverseArray(a,0,n-1);
}
```

```
public static void reverseArray(int[] a, int start, int end)
{
    for(int i=start,j=end;i<j;i++,j--)
    {
        int temp = a[i];
        a[i]=a[j];
        a[j]=temp;
    }
}
```

1,2,3,4,5,6,7,8,9,10   =>   5,6,7,8,9,10,1,2,3,4
1,2,3,4,5,6,7,8,9,10   =>   4,3,2,1,10,9,8,7,6,5   => 5,6,7,8,9,10,1,2,3,4

Analysis:
- Rotating array is done in two parts trick. In the first part, we first reverse elements of array first half and then second half.
- Then we reverse the whole array there by completing the whole rotation.

Find the largest sum contiguous subarray.

Given an array of positive and negative *integer* s, find a contiguous subarray whose sum (sum of elements) is maximized.

Example 1.24:

```java
public class Introduction {
    public static void main(String[] args) {
        int[] arr = {1,-2,3,4,-4,6,-14,8,2};
        System.out.println("Max sub array sum :" + maxSubArraySum(arr, 9));
    }

    public static int maxSubArraySum(int[] a, int size) {
        int maxSoFar = 0, maxEndingHere = 0;
        for (int i = 0; i < size; i++) {
            maxEndingHere = maxEndingHere + a[i];
            if (maxEndingHere < 0)
                maxEndingHere = 0;
            if (maxSoFar < maxEndingHere)
                maxSoFar = maxEndingHere;
        }
        return maxSoFar;
    }
}
```

Analysis:
- Maximum subarray in an array is found in a single scan. We keep track of global maximum sum so far and the maximum sum, which include the current element.
- When we find global maximum value so far is less than the maximum value containing current value we update the global maximum value.
- Finally return the global maximum value.

Concept of Stack

A stack is a memory in which values are stored and retrieved in "last in first out" manner. Data is added to stack using push operation and data is taken out of stack using pop operation.

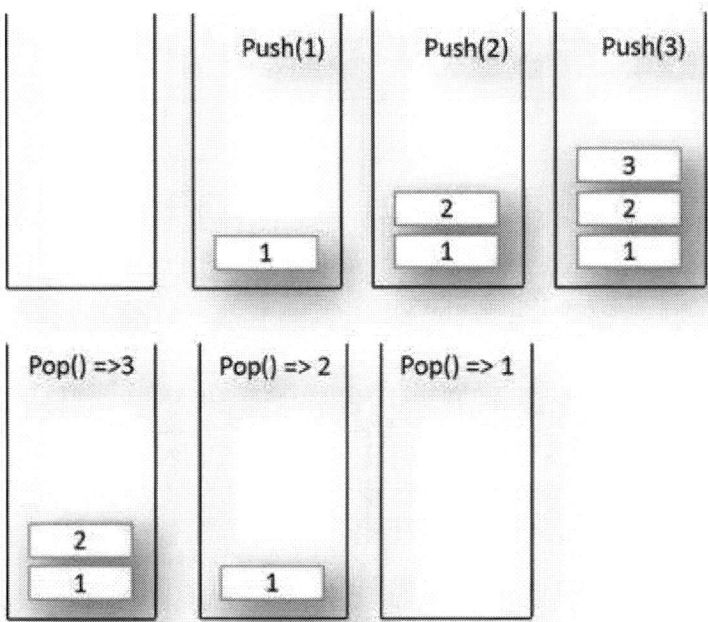

1. Initially the stack was empty. Then we have added value 1 to stack using push(1) operator.
2. Similarly, push(2) and push(3)
3. Pop operation take the top of the stack. In Stack data is added and deleted in "last in, first out" manner.
4. First pop() operation will take 3 out of the stack.
5. Similarly, other pop operation will take 2 then 1 out of the stack
6. In the end, the stack is empty when all the elements are taken out of the stack.

System stack and Method Calls

When the method is called, the current execution is stopped and the control goes to the called method. After the called method exits / returns, the execution resumes from the point at which the execution was stopped.

To get the exact point at which execution should be resumed, the address of the next instruction is stored in the stack. When the method call completes, the address at the top of the stack is taken out.

Example 1.25:

```
public static void function2()
{
        System.out.println("fun2 line 1");
}
```

```java
public static void function1()
{
    System.out.println("fun1 line 1");
    function2();
    System.out.println("fun1 line 2");
}
```

```java
public static void main(String[] args)
{
    System.out.println("main line 1");
    function1();
    System.out.println("main line 2");
}
```

Output:

```
main line 1
fun1 line 1
fun2 line 1
fun1 line 2
main line 2
```

Analysis:
- Every program starts with main() method.
- The first statement of main() will be executed. And we will print "main line 1" as output.
- function1() is called. Before control goes to function1() then next instruction that is address of next line is stored in the system stack.
- Control goes to function1() method.
- The first statement inside function1() is executed, this will print "fun1 line 1" to output.
- function2() is called from function1(). Before control goes to function2() address of the next instruction that is address of next line is added to the system stack.
- Control goes to function2() method.
- "fun2 line 1" is printed to screen.
- When function2() exits, control come back to function1(). And the program reads the next instruction from the stack, and the next line is executed. And print "fun1 line 2" to screen.
- When fun1 exits, control comes back to the main method. And program reads the next instruction from the stack and executed it and finally "main line 2" is printed to screen.

Points to remember:
1. Methods are implemented using a stack.
2. When a method is called the address of the next instruction is pushed into the stack.
3. When a method is finished the address of the execution is taken out of the stack.

Recursive Function

A recursive function is a function that calls itself, directly or indirectly.
A recursive method consists of two parts: Termination Condition and Body (which include recursive expansion).
1. Termination Condition: A recursive method always contains one or more terminating condition. A condition in which recursive method is processing a simple case and will not call itself.
2. Body (including recursive expansion): The main logic of the recursive method contained in the body of the method. It also contains the recursion expansion statement that in turn calls the method itself.

Three important properties of recursive algorithm are:
1) A recursive algorithm must have a termination condition.
2) A recursive algorithm must change its state, and move towards the termination condition.
3) A recursive algorithm must call itself.

Note: The speed of a recursive program is slower because of stack overheads. If the same task can be done using an iterative solution (loops), then we should prefer an iterative solution (loops) in place of recursion to avoid stack overhead.

Note: Without termination condition, the recursive method may run forever and will finally consume all the stack memory.

Factorial

Example 1.26: Factorial Calculation. N! = N* (N-1).... 2*1.

```java
public static int factorial(int i)
{
    /* Termination Condition */
    if(i <= 1)
        return 1;
    /* Body, Recursive Expansion */
    return i * factorial(i - 1);
}
```

Analysis: Each time method fn is calling fn-1. *Time Complexity* is **O(N)**

Print Base 10 *Integer* s

Example 1.27:

```java
public static void printInt1(int number)
{
    char digit = (char) (number % 10 + '0');
    number = number / 10;
    if (number != 0)
        printInt1(number/10);
    System.out.print(" " + digit);
}
```

Analysis:
- Each time remainder is calculated and stored its char equivalent in digit.
- If the number is greater than 10 then the number divided by 10 is passed to printInt() method.
- Number will be printed with higher order first than the lower order digits.

Time Complexity is O(N)

Print Base 16 *Integer* s

Example 1.28: Generic print to some specific base method.

```
public static void printInt2(int number, final int base)
{
    String conversion = "0123456789ABCDEF";
    char digit = (char) (number % base) ;
    number = number / base;
    if (number != 0)
        printInt2(number,base);
    System.out.print(" " + conversion.charAt(digit));
}
```

Analysis:
- Base value is provided along with the number in the function parameter.
- Remainder of the number is calculated and stored in digit.
- If the number is greater than base then, number divided by base is passed as an argument to the printInt() method recursively.
- Number will be printed with higher order first than the lower order digits.

Time Complexity is O(N)

Tower of Hanoi

The **Tower of Hanoi** (also called the **Tower of Brahma**) We are given three rods and N number of disks, initially all the disks are added to first rod (the leftmost one) in decreasing size order. The objective is to transfer the entire stack of disks from first tower to third tower (the rightmost one), moving only one disk at a time and never a larger one onto a smaller.

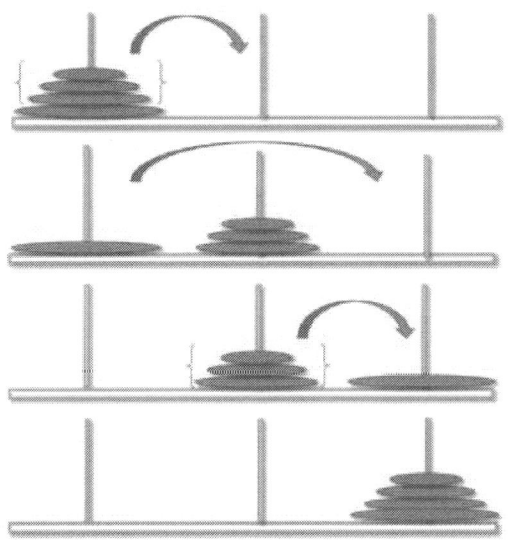

Example 1.29:

```java
public static void towerOfHanoi(int num, char src, char dst, char temp)
{
    if (num < 1)
        return;

    towerOfHanoi(num - 1, src, temp, dst);
    System.out.println("Move"+num+"disk  from peg"+src+" to peg " + dst);
    towerOfHanoi(num - 1, temp, dst, src);
}
```

```java
public static void main(String[] args)
{
    int num = 4;
    System.out.println("The sequence of moves in the Tower of Hanoi are : ");
    towerOfHanoi(num, 'A', 'C', 'B');
}
```

Analysis: TowerOfHanoi problem if we want to move N disks from source to destination, then we first move N-1 disks from source to temp, then move the lowest Nth disk from source to destination. Then will move N-1 disks from temp to destination.

Greatest common divisor (GCD)

Example 1.30:

```java
public static int GCD(int m, int n)
{
    if(m<n)
        return (GCD(n, m));
    if(m%n == 0)
        return (n);
    return(GCD(n, m%n));
}
```

Analysis: Euclid's algorithm is used to find gcd. GCD(n,m) == GCD(m, n mod m)

Fibonacci number

Example 1.31:

```java
public static int fibonacci(int n)
{
    if (n <= 1)
        return n;
    return fibonacci(n - 1) + fibonacci(n - 2);
}
```

Analysis: Fibonacci number are calculated by adding sum of the previous two number. There is an inefficiency in the solution we will look better solution in coming chapters.

All permutations of an integer array

Example 1.32:

```java
public static void printArray(int[] arr, int count) {
    System.out.print("\n Values stored in array are : ");
    for (int i = 0; i < count; i++)
    {
        System.out.print(" " + arr[i]);
    }
}
```

```java
public static void swap(int[] arr, int x, int y){
    int temp = arr[x];
    arr[x] = arr[y];
    arr[y] = temp;
    return;
}
```

```java
public static void permutation(int[] arr, int i, int length) {
    if (length == i){
        printArray(arr, length);
        return;
    }
    int j = i;
    for (j = i; j < length; j++) {
        swap(arr, i, j);
        permutation(arr, i + 1, length);
        swap(arr, i, j);
    }
    return;
}
```

```java
public static void main(String[] args)
{
    int[] arr = new int[5];
    for (int i = 0; i < 5; i++)
    {
        arr[i] = i;
    }
    permutation(arr, 0, 5);
}
```

Analysis: In permutation method at each recursive call number at index, "i" is swapped with all the numbers that are right of it. Since the number is swapped with all the numbers in its right one by one it will produce all the permutation possible.

Binary search using recursion

Example 1.33:

```
// Binary Search Algorithm – Recursive Way
public static int BinarySearchRecursive(int arr[], int low, int high, int value)
{
        int mid = low + (high-low)/2;      // To avoid the overflow
        if (arr[mid] == value)
                return mid;
        else if (arr[mid] < value)
                return BinarySearchRecursive (arr, mid + 1, high, value);
        else
                return BinarySearchRecursive (arr, low, mid - 1 , value);
}
```

Analysis: Similar iterative solution we had already seen. Now lets look into the recursive solution of the same problem in this solution also we are diving the search space into half and doing the same what we had done in the iterative solution.

Exercises

1. Find average of all the elements in an array.
2. Find the sum of all the elements of a two dimensional array.
3. Find the largest element in the array.
4. Find the smallest element in the array.
5. Find the second largest number in the array.
6. Print all the maxima's in an array. (A value is a maximum if the value before and after its index are smaller than it is or does not exist.)
   Hint:
   a) Start traversing array from the end and keep track of the max element.
   b) If we encounter an element > max, print the element and update max.

7. Print alternate elements in an array.

8. Given an array with value 0 or 1, write a program to segregate 0 on the left side and 1 on the right side.

9. Given a list of intervals, merge all overlapping intervals.
   Input: {[1, 4], [3, 6], [8, 10]}
   Output: {[1, 6], [8, 10]}

10. Write a method that will take intervals as input and takes care of overlapping intervals.

11. Reverse an array in-place. (You cannot use any additional array in other wards *Space Complexity* should be *O(1).*)

Hint: Use two variable, start and end. Start set to 0 and end set to (n-1). Increment start and decrement end. Swap the values stored at arr[start] and arr[end]. Stop when start is equal to end or start is greater than end.

12. Given an array of 0s and 1s. We need to sort it so that all the 0s are before all the 1s.
    Hint: Use two variable, start and end. Start set to 0 and end set to (n-1). Increment start and decrement end. Swap the values stored at arr[start] and arr[end] only when arr[start] == 1 and arr[end]==0. Stop when start is equal to end or start is greater than end.

13. Given an array of 0s, 1s and 2s. We need to sort it so that all the 0s are before all the 1s. And all the 1s are before 2s.
    Hint: Same as above first think 0s and 1s as one group and move all the 2s on the right side. Then do a second pass over the array to sort 0s and 1s.

14. Find the duplicate elements in an array of size n where each element is in the range 0 to n-1.

    Hint:
    Approach 1: Compare each element with all the elements of the array (using two loops) $O(n^2)$ solution
    Approach 2: Maintain a Hash-Table. Set the hash value to 1 if we encounter the element for the first time. When we same value again we can see that the hash value is already 1 so we can print that value. **O(n)** solution, but additional space is required.
    Approach 3: We will exploit the constraint "every element is in the range 0 to n-1". We can take an array arr[] of size n and set all the elements to 0. Whenever we get a value say val1. We will increment the value at arr[var1] index by 1. In the end, we can traverse the array arr and print the repeated values. Additional *Space Complexity* will be **O(n)** which will be less then Hash-Table approach.

15. Find the maximum element in a sorted and rotated array. Complexity: **O(logn)**
    Hint: Use binary search algorithm.

16. Given an array with 'n' elements & a value 'x', find two elements in the array that sums to 'x'.

    Hint:
    Approach 1: Sort the array.
    Approach 2: Using a Hash-Table.

17. Write a method to find the sum of every number in an int number. Example: input= 1984, output should be 32 (1+9+8+4).

18. Write a method to compute Sum(N) = 1+2+3+...+N.

Chapter 2: Algorithms Analysis

Introduction

Computer programmer learn by experience. We learn by seeing solved problems and solving new problems by ourselves. Studying various problem-solving techniques and by understanding how different algorithms are designed helps us to solve the next problem that is given to us. By considering a number of different algorithms, we can begin to develop pattern so that the next time a similar problem arises, we are better able to solve it.

When an interviewer asks to develop a program in an interviewer, what are the steps that an interviewee should follow? We will be taking a systematic approach to handle the problem and finally reaching to the solution.

Algorithm

An algorithm is a set of steps to accomplish a task.
An algorithm in a computer program is a set of steps applied over a set of input to produce a set of output.

Knowledge of algorithm helps us to get our desired result faster by applying the right algorithm.

The most important properties of an algorithm are:
1. Correctness: The algorithm should be correct. It should be able to process all the given inputs and provide correct output.
2. Efficiency: The algorithm should be efficient in solving problems.

Algorithmic complexity is defined as how fast a particular algorithm performs. Complexity is represented by function T (n) - time versus the input size n.

Asymptotic analysis

Asymptotic analysis is used to compare the efficiency of algorithm independently of any particular data set or programming language.

We are generally interested in the order of growth of some algorithm and not interested in the exact time required for running an algorithm. This time is also called **Asymptotic-running time.**

Big-O Notation

Definition: "f(n) is big-O of g(n)" or $f(n) = O(g(n))$, if there are two +ve constants c and n0 such that $f(n) \leq c\, g(n)$ for all $n \geq n0$,
In other words, $c\, g(n)$ is an upper bound for $f(n)$ for all $n \geq n0$

The function f(n) growth is slower than c g(n)

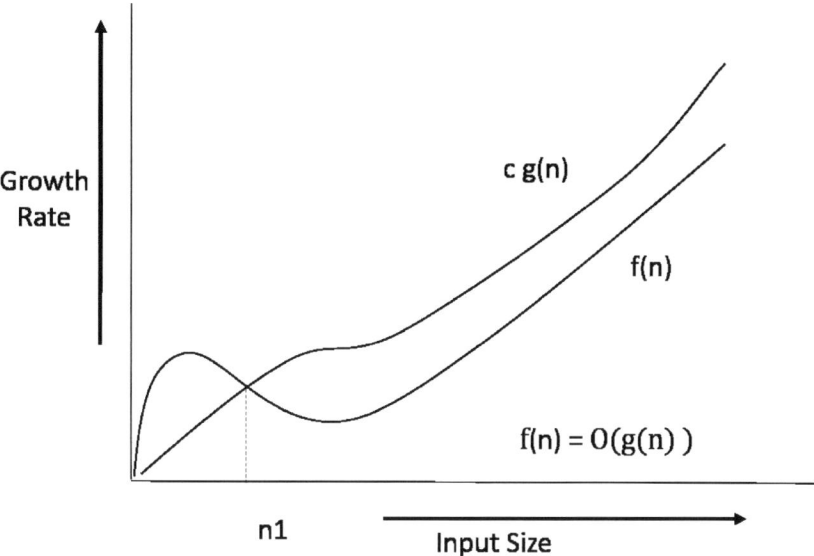

Example: $n^2 + n = O(n^2)$

Omega-Ω Notation

Definition: "f(n) is omega of g(n)." or f(n) = Ω(g(n)) if there are two +ve constants c and n0 such that
c g(n) ≤ f(n) for all n ≥ n0

In other words, c g(n) is lower bound for f(n)
Function f(n) growth is faster than c g(n)

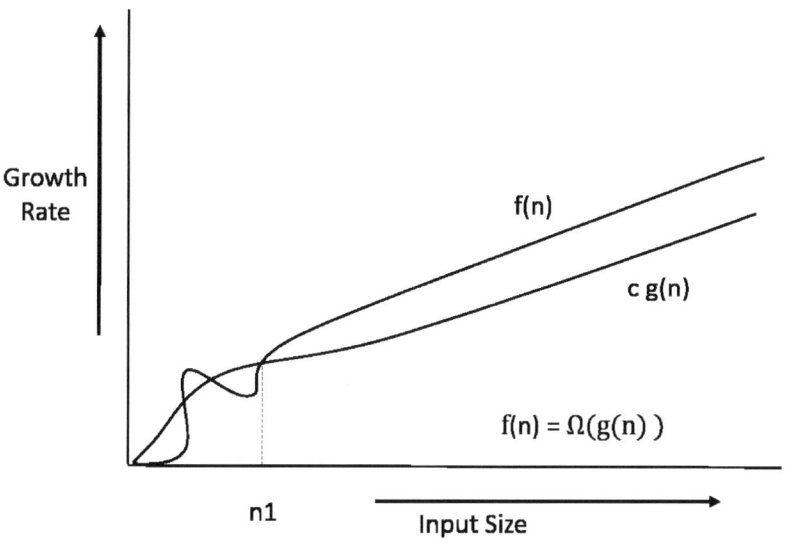

Find relationship of $f(n) = n^c$ and $g(n) = c^n$
$f(n) = \Omega\ (g(n))$

Theta-Θ Notation

Definition: "f(n) is theta of g(n)." or f(n) = Θ(g(n)) if there are three +ve constants c1, c2 and n0 such that c1 g(n) ≤ f(n) ≤ c2 g(n) for all n ≥ n0

g(n) is an asymptotically tight bound on f(n).
Function f(n) grows at the same rate as g(n).

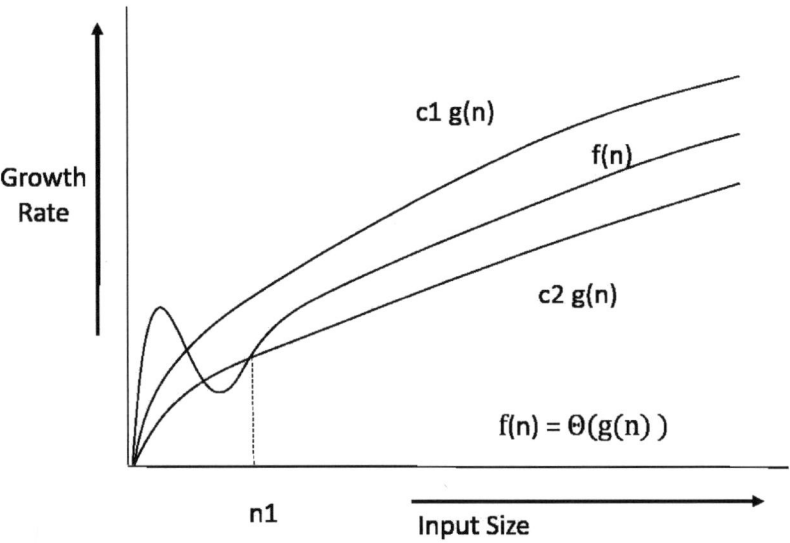

Example: $n^3 + n^2 + n = \Theta(n^2)$

Example: $n^2 + n = \Theta(n^2)$
Find relationship of $f(n) = 2n^2 + n$ and $g(n) = n^2$
$f(n) = O\ (g(n))$
$f(n) = \Theta\ (g(n))$
$f(n) = \Omega\ (g(n))$

Note:- Asymptotic Analysis is not perfect, but that is the best way available for analyzing algorithms.

For example, say there are two sorting algorithms first take $f(n) = 10000n \log n$ and $f(n) = n^2$ time. The asymptotic analysis says that the first algorithm is better (as it ignores constants) but actually for a small set of data when n is small then 10000, the second algorithm will perform better. To consider this drawback of asymptotic analysis case analysis of the algorithm is introduced.

Complexity analysis of algorithms

1) Worst Case complexity: It is the complexity of solving the problem for the worst input of size n. It provides the upper bound for the algorithm. This is the most common analysis done.
2) Average Case complexity: It is the complexity of solving the problem on an average. We calculate the time for all the possible inputs and then take an average of it.
3) Best Case complexity: It is the complexity of solving the problem for the best input of size n.

Time Complexity Order

A list of commonly occurring algorithm *Time Complexity* in increasing order:

Name	Notation	
Constant	$O(1)$	
Logarithmic	$O(\log n)$	
Linear	$O(n)$	
N-LogN	$O(n \log n)$	
Quadratic	$O(n^2)$	
Polynomial	$O(n^c)$	c is a constant & c>1
Exponential	$O(c^m)$	c is a constant & c>1
Factorial or N-power-N	$O(n!)$ or $O(n^n)$	

Constant Time: *O(1)*

An algorithm is said to run in constant time regardless of the input size.
Examples:
1. Accessing n^{th} element of an array
2. Push and pop of a stack.
3. Enqueue and remove of a queue.
4. Accessing an element of Hash-Table.
5. Bucket sort

Linear Time: *O(n)*

An algorithm is said to run in linear time if the execution time of the algorithm is directly proportional to the input size.
Examples:
1. Array operations like search element, find min, find max etc.
2. Linked list operations like traversal, find min, find max etc.

Note: when we need to see/ traverse all the nodes of a data-structure for some task then complexity is no less than *O(n)*

Logarithmic Time: *O(logn)*

An algorithm is said to run in logarithmic time if the execution time of the algorithm is proportional to the logarithm of the input size. Each step of an algorithm, a significant portion of the input is pruned out without traversing it.
Example:
1. Binary search

Note: We will read about these algorithms in this book.

N-LogN Time: $O(n\log(n))$

An algorithm is said to run in logarithmic time if the execution time of an algorithm is proportional to the product of input size and logarithm of the input size.
Example:
1. Merge-Sort
2. Quick-Sort (Average case)
3. Heap-Sort

Note: Quicksort is a special kind of algorithm to sort a list of numbers. Its worst-case complexity is $O(n^2)$ and average case complexity is $O(n \log n)$.

Quadratic Time: $O(n^2)$

An algorithm is said to run in logarithmic time if the execution time of an algorithm is proportional to the square of the input size.
Examples:
1. Bubble-Sort
2. Selection-Sort
3. Insertion-Sort

Deriving the Runtime Function of an Algorithm

Constants

Each statement takes a constant time to run. *Time Complexity* is $O(1)$

Loops

The running time of a loop is a product of running time of the statement inside a loop and number of iterations in the loop. *Time Complexity* is $O(n)$

Nested Loop

The running time of a nested loop is a product of running time of the statements inside loop multiplied by a product of the size of all the loops. *Time Complexity* is $O(n^c)$
Where c is a number of loops. For two loops, it will be $O(n^2)$

Consecutive Statements

Just add the running times of all the consecutive statements

If-Else Statement

Consider the running time of the larger of if block or else block. And ignore the other one.

Logarithmic statement

If each iteration the input size is decreases by a constant factors. $Time\ Complexity = O(\log n)$.

Time Complexity Examples

Example 1

```
int fun1(int n)
{
    int m = 0;
    for (int i = 0; i<n; i++)
        m += 1;
    return m;
}
```

Time Complexity: **O(n)**

Example 2

```
int fun2(int n)
{
    int i=0, j=0, m = 0;
    for (i = 0; i<n; i++)
        for (j = 0; j<n; j++)
            m += 1;
    return m;
}
```

Time Complexity: $O(n^2)$

Example 3

```
int fun3(int n)
{
    int i=0, j=0, m = 0;
    for (i = 0; i<n; i++)
        for (j = 0; j<i; j++)
            m += 1;
    return m;
}
```

Time Complexity: O(N+(N-1)+(N-2)+...) == O(N(N+1)/2) == $O(n^2)$

Example 4

```
int fun4(int n)
{
    int i = 0, m = 0;
    i = 1;
    while (i < n) {
        m += 1;
        i = i * 2;
    }
    return m;
}
```

Each time problem space is divided into half. Time Complexity: **O(log(n))**

Example 5

```
int fun5(int n)
{
    int i = 0, m = 0;
    i = n;
    while (i > 0) {
        m += 1;
        i = i / 2;
    }
    return m;
}
```

Same as above each time problem space is divided into half.
Time Complexity: **O(log(n))**

Example 6

```
int fun6(int n)
{
    int i = 0, j = 0, k = 0, m = 0;
    i = n;
    for (i = 0; i<n; i++)
        for (j = 0; j<n; j++)
            for (k = 0; k<n; k++)
                m += 1;
    return m;
}
```

Outer loop will run for n number of iterations. In each iteration of the outer loop, inner loop will run for n iterations of their own. Final complexity will be n*n*n.
Time Complexity: $O(n^3)$

Example 7

```
int fun7(int n)
{
    int i = 0, j = 0, k = 0, m = 0;
    i = n;
    for (i = 0; i<n; i++)
        for (j = 0; j<n; j++)
            m += 1;
    for (i = 0; i<n; i++)
        for (k = 0; k<n; k++)
            m += 1;
    return m;
}
```

These two groups of for loop are in consecutive so their complexity will add up to form the final complexity of the program.
Time Complexity: T(n) = $O(n^2) + O(n^2) = O(n^2)$

Example 8

```
int fun8(int n)
{
    int i = 0, j = 0, m = 0;
    for (i = 0; i<n; i++)
        for (j = 0; j< Math.sqrt(n); j++)
            m += 1;
    return m;
}
```

Time Complexity: $O(n * \sqrt{n}) = O(n^{3/2})$

Example 9

```
int fun9(int n)
{
    int i = 0, j = 0, m = 0;
    for ( i = n; i > 0; i /= 2)
        for ( j = 0; j < i; j++)
            m += 1;
    return m;
}
```

Each time problem space is divided into half.
Time Complexity: $O(\log(n))$

Example 10

```
int fun10(int n)
{
    int i = 0, j = 0, m = 0;
    for ( i = 0; i < n; i++)
        for ( j = i; j > 0; j--)
            m += 1;
    return m;
}
```

O(N+(N-1)+(N-2)+...) = O(N(N+1)/2) // arithmetic progression.
Time Complexity: $O(n^2)$

Example 11

```
int fun11(int n)
{
    int i = 0, j = 0, k = 0, m = 0;
    for (i = 0; i<n; i++)
        for (j = i; j<n; j++)
            for (k = j+1; k<n; k++)
                m += 1;
    return m;
}
```

Time Complexity: $O(n^3)$

Example 12

```
int fun12(int n)
{
    int i = 0, j = 0, m = 0;
    for (i = 0; i<n; i++)
        for (; j<n; j++)
            m += 1;
    return m;
}
```

Think carefully once again before finding a solution, j value is not reset at each iteration.
Time Complexity: **O(n)**

Example 13

```
int fun13(int n)
{
    int i = 1, j = 0, m = 0;
    for (i = 1; i<=n; i *= 2)
        for (j = 0; j<=i; j++)
            m += 1;
    return m;
}
```

The inner loop will run for 1, 2, 4, 8,... n times in successive iteration of the outer loop.
Time Complexity: T(n) = O(1+ 2+ 4+ ....+n/2+n) = *O(n)*

Master Theorem

The master theorem solves recurrence relations of the form:
$T(n) = a\ T(n/b) + f(n)$

Where $a \geq 1$ and $b > 1$.
"n" is the size of the problem.
"a" is a number of sub problem in the recursion.
"n/b" is the size of each sub-problem.
"f(n)" is the cost of the division of the problem into sub problem or merge of results of sub problems to get the final result.

It is possible to determine an asymptotic tight bound in these three cases:
Case 1: when $f(n) = O(n^{\log_b a - \epsilon})$ and constant $\epsilon > 1$, than the final *Time Complexity* will be:
$T(n) = \Theta(n^{\log_b a})$

Case 2: when $f(n) = \Theta(n^{\log_b a} \log^k n)$ and constant $k \geq 0$, than the final *Time Complexity* will be:
$T(n) = \Theta(n^{\log_b a} \log^{k+1} n)$

Case 3: when $f(n) = \Omega(n^{\log_b a + \epsilon})$ and constant $\epsilon > 1$, Then the final *Time Complexity* will be:
$T(n) = \Theta(f(n))$

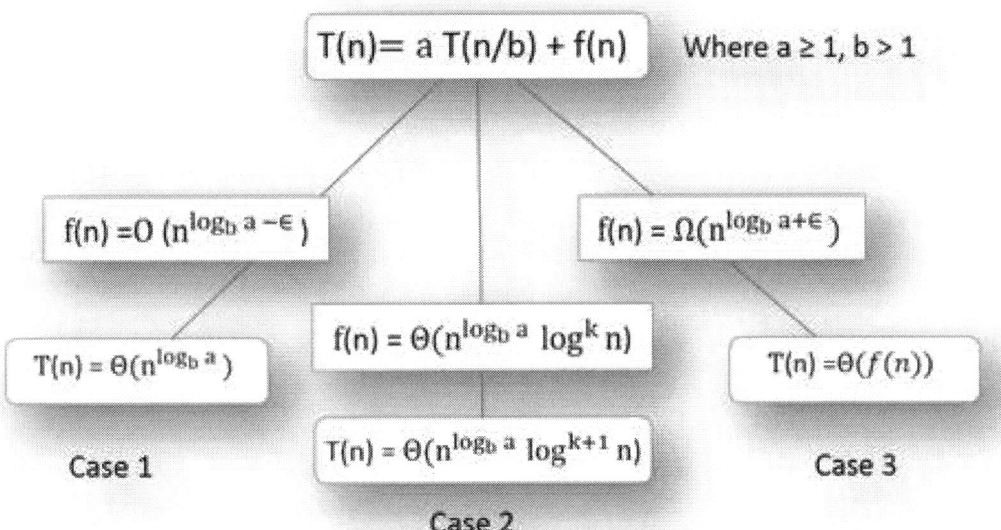

Example 14: Take an example of Merge-Sort, $T(n) = 2\,T(n/2) + n$
Sol:- $\log_b a = \log_2 2 = 1$
$f(n) = n = \Theta(n^{\log_2 2} \log^0 n)$
Case 2 applies and $T(n) = \Theta(n^{\log_2 2} \log^{0+1} n)$
$T(n) = \Theta(n \log(n))$

Example 15: Binary Search $T(n) = T(n/2) + O(1)$
Sol:- $\log_b a = \log_2 1 = 0$
$f(n) = 1 = \Theta(n^{\log_2 1} \log^0 n)$
Case 2 applies and $T(n) = \Theta(n^{\log_2 1} \log^{0+1} n)$
$T(n) = \Theta(\log(n))$

Example 16: Binary tree traversal $T(n) = 2T(n/2) + O(1)$
Sol:- $\log_b a = \log_2 2 = 1$
$f(n) = 1 = O(n^{\log_2 2 - 1})$
Case 1 applies and $T(n) = \Theta(n^{\log_2 2})$
$T(n) = \Theta(n)$

Example 17: Take an example $T(n) = 2\,T(n/2) + n^2$
Sol:- $\log_b a = \log_2 2 = 1$
$f(n) = n^2 = \Omega(n^{\log_2 2 + 1})$
Case 3 applies and $T(n) = \Theta(f(n))$
$T(n) = \Theta(n^2)$

Example 18: Take an example $T(n) = 4\,T(n/2) + n^2$
Sol:- $\log_b a = \log_2 4 = 2$
$f(n) = n^2 = \Theta(n^{\log_2 4} \log^0 n)$
Case 2 applies and $T(n) = \Theta(n^{\log_2 4} \log^{0+1} n)$
$T(n) = \Theta(n^2 \log n)$

Modified Master theorem

This is a shortcut to solving the same problem easily and fast. If the recurrence relation is in the form of T(n)= a T(n/b) + dx^s

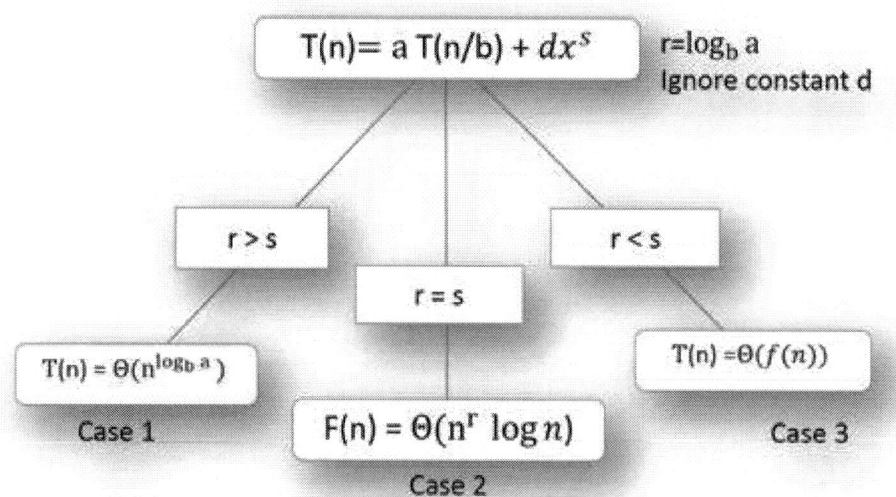

Example 19: $T(n) = 2\ T(n/2) + n^2$
Sol:- $r = log_2 2$
s = 2
Case 3: $log_2 2 < s$
T(n) =Θ(f(n)) = Θ(n^2)

Example 20: T (n) = T (n/2) + 2n
Sol:- $r = log_2 1 = 0$
s = 1
Case 3
T(n)= Θ(n)

Example 21: T (n) = 16T (n/4) + n
Sol:- r = 2
s = 1
Case 1
T(n)= Θ(n^2)

Example 22: T (n) = 2T (n/2) + n log n
Sol:- There is logn in f(n) so use master theorem shortcut will not word.
f(n) = n log(n) = Θ($n^{log_2 2} log^1 n$)
T(n) = Θ($n^{log_2 2} log^{0+1} n$) = Θ($n\ log(n)$)

Example 23: $T(n) = 2\ T(n/4) + n^{0.5}$
Sol:- $r = log_4 2 = 0.5 = $ s
Case 2: T(n) = Θ($n^{log_4 2} log^{0.5+1} n$) = Θ($n^{0.5} log^{1.5} n$)

Example 24: $T(n) = 2\,T(n/4) + n^{0.49}$
Sol:- Case 1:
T(n) = $\Theta(n^{log_4 2}) = \Theta(n^{0.5})$

Example 25: T (n) = 3T (n/3) + √n
Sol:- $r = log_3 3 = 1$
s = ½
Case 1
T(n) = $\Theta(n)$

Example 26: T (n) = 3T (n/4) + n log n
Sol:- There is logn in f(n) so see if master theorem.
f(n) = n log n = $\Omega(n^{log_4 3} \, log^1 n)$
Case 3:
T(n) = $\Theta(nlog(n))$

Example 27: T (n) = 3T (n/3) + n/2
Sol:- r=1=s
Case 2:
T(n) = $\Theta(nlog(n))$

Exercise

1. True or false
    a. $5n + 10n^2 = O(n^2)$
    b. $n \log n + 4n = O(n)$
    c. $\log(n^2) + 4 \log(\log n) = O(\log n)$
    d. $12 n^{1/2} + 3 = O(n^2)$
    e. $3^n + 11 n^2 + n^{20} = O(2^n)$

2. What is the best-case runtime complexity of searching an array?

3. What is the average-case runtime complexity of searching an array?

Chapter 3: Approach to Solve Algorithm Design Problems

Introduction

Know the theoretical knowledge of the algorithm is essential, but it is not sufficient. You need to have a systematic approach to solve a problem. Our approach is fundamentally different to solve any algorithm design question. We will follow a systematic five-step approach to reach to our solution. Master this approach and you will be better than most of the candidates in interviews.

Five steps for solving algorithm design questions are:
1. Constraints
2. Ideas Generation
3. Complexities
4. Coding
5. Testing

Constraints

Solving a technical question is not just about knowing the algorithms and designing a good software system. The interviewer wants to know you approach towards any given problem. May people make mistakes, as they do not ask clarifying questions about a given problem? They assume many things and begin working with that. There are a lot of data that is missing that you need to collect from your interviewer before beginning to solve a problem.

In this step, you will capture all the constraints about the problem. We should never try to solve a problem that is not completely defined. Interview questions are not like exam paper where all the details about a problem are well defined. In the interview, the interviewer actually expects you to ask questions and clarify the problem.

For example: When the problem statement says that write an algorithm to sort numbers.

The first information you need to capture is what king of data. Let us suppose interviewer respond with the answer *Integer.*
The second information that you need to know what is the size of data. Your algorithm differs if the input data size if 100 *integer*s or 1 billion *integers.*

Basic guideline for the Constraints for an array of numbers:
1. How many numbers of elements in the array?
2. What is the range of value in each element? What is the min and max value?
3. What is the kind of data in each element is it an *integer* or a *float*ing point?
4. Does the array contain unique data or not?

Basic guideline for the Constraints for an array of string:
1. How many numbers of elements in the array?
2. What is the length of each string? What is the min and max length?
3. Does the array contain unique data or not?

Basic guideline for the Constraints for a Graph
1. How many nodes are there in the graph?
2. How many edges are there in the graph?
3. Is it a weighted graph? What is the range of weights?
4. Is the graph directed or undirected?
5. Is there is a loop in the graph?
6. Is there negative sum loop in the graph?
7. Does the graph have self-loops?

We have already seen this in graph chapter that depending upon the constraints the algorithm applied changes and so is the complexity of the solution.

Idea Generation

We have covered a lot of theoretical knowledge in this book. It is impossible to cover all the questions as new ones are created every day. Therefore, we should know how to handle new problems. Even if you know the solution of a problem asked by the interviewer then also you need to have a discussion with the interviewer and reach to the solution. You need to analyze the problem also because the interviewer may modify a question a little bit and the approach to solve it will vary.

Well, how to solve a new problem? The solution to this problem is that you need to do a lot of practice and the more you practice the more you will be able to solve any new question, which come in front of you. When you have solved enough problems, you will be able to see a pattern in the questions and able to solve new problems easily.

Following is the strategy that you need to follow to solve an unknown problem:
1. Try to simplify the task in hand.
2. Try a few examples
3. Think of a suitable data-structure.
4. Think about similar problems you have already solved.

Try to simplify the task in hand

Let us look into the following problem: Husbands and their wives are standing in random in a line. They have been numbered for husbands H1, H2, H3 and so on. And their corresponding wives have number W1, W2, W3 and so on. You need to arrange them so that H1 will stand first, followed by W1, then H2 followed by W2 and so on.
At the first look, it looks difficult, but it is a simple problem. Try to find a relation of the final position.
$P(H_i) = i*2 - 1$, $P(W_i) = i*2$
The rest of the algorithm we are leaving you to do something like Insertion-Sort and you are done.

Try a few examples

In the same above problem if you have tried it with some example for 3 husband and wife pair then you may have reached to the same formula that we have shown in the previous section. Some time thinking some more examples try to solve the problem at hand.

Think of a suitable data-structure

For some problems, it is straight forward which data structure will be most suitable for. For example, if we have a problem finding min/max of some given value, then probably heap is the data structure we are looking for. We have seen a number of data structure throughout this book. And we have to figure out which data-structure will suite our need.

Let us look into a problem: We are given a stream of data at any time we can be asked to tell the median value of the data and maybe we can be asked to pop median data.

We can think about some sort of tree, may be balanced tree where the root is the median. Wait but it is not so easy to make sure the tree root to be a median.
A heap can give us minimum or maximum so we cannot get the desired result from it too. However, what if we use two heap one max heap and one min heap. The smaller values will go to min heap and the bigger values will go to max heap. In addition, we can keep the count of how many elements are there in the heap. The rest of the algorithm you can think yourself.

For every new problem think about the data structure, you know and may be one of them or some combination of them will solve your problem.

Think about similar problems you have already solved. Let us suppose you are given, two linked list head reference and they meet at some point and need to find the point of intersection. However, in place of the end of both the linked list to be a null reference there is a loop.

You know how to find intersection point of two intersecting linked list, you know how to find if a linked list have a loop (three-reference solution). Therefore, you can apply both of these solutions to find the solution of the problem in hand.

Complexities

Solving a problem is not just finding a correct solution. The solution should be fast and should have reasonable memory requirement. You have already read about Big-O notation in the previous chapters. You should be able to do Big-O analysis. In case you think the solution you have provided is not that optimal and there is some more efficient solution, then think again and try to figure out this information.

Most interviewers expect that you should be able to find the time and *Space Complexity* of the algorithms. You should be able to compute the time and *Space Complexity* instantly. Whenever you are solving some problem, you should find the complexity associated with it from this you would be able to choose the best solutions. In some problems there is some trade-offs between space and

Time Complexity, so you should know these trade-offs. Sometime taking some bit more space saves a lot of time and make your algorithm much faster.

Coding

At this point, you have already captured all the constraints of the problem, proposed few solutions, evaluated the complexities of the various solutions and picked the one solution to do final coding. Never ever, jump into coding before discussing constraints, Idea generation and complexity with the interviewer.

We are accustomed to coding in an IDE like visual studio. So many people struggle when asked to write code on a whiteboard or some blank sheet. Therefore, we should have a little practice to the coding on a sheet of paper. You should think before coding because there is no back button in sheet of paper. Always try to write modular code. Small functions need to be created so that the code is clean and managed. If there is a swap function so just use this function and tell the interviewer that you will write it later. Everybody knows that you can write swap code.

Testing

Once the code is written, you are not done. It is most important that you test your code with several small test cases. It shows that you understand the importance of testing. It also gives confidence to your interviewer that you are not going to write a buggy code. Once you are done with, your coding it is a good practice that you go through your code line by line with some small test case. This is just to make sure your code is working as it is supposed to work.

You should test few test cases.
Normal test cases: These are the positive test cases, which contain the most common scenario, and focus is on the working of the base logic of the code. For example, if we` are going to write some algorithm for linked list, then this may contain what will happen when a linked list with 3 or 4 nodes is given as input. These test cases you should always run in your head before saying the code is done.

Edge cases: These are the test cases, which are going to test the boundaries of the code. For the same linked list algorithm, edge cases may be how the code behaves when an empty list is passed or just one node is passed. These test cases you should always run in your head before saying the code is done. Edge cases may help to make your code more robust. Just few checks need to be added to the code to take care of the condition.

Load testing: In this kind of test, your code will be tested with a huge data. This will allow us to test if your code is slow or too much memory intensive.
Always follow these five steps never jump to coding before doing constraint analysis, idea generation, and Complexity Analysis: At least never, miss the testing phase.

Example

Let us suppose the interviewer ask you to give a best sorting algorithm.
Some interviewee will directly jump to Quick-Sort *O(nlogn)*. Oops, mistake you need to ask many questions before beginning to solve this problem.

Questions 1: What kind of data we are talking about? Are they *integer*s?
Answer: Yes, they are *integer*s.

Questions 2: How much data are we going to sort?
Answer: May be thousands.

Questions 3: What exactly is this data about?
Answer: They store a person's age

Questions 4: What kind of data-structure used to hold this data?
Answer: Data are given in the form of some array

Questions 5: Can we modify the given data-structure? In addition, many, many more…?
Answer: No, you cannot modify the data structure provided

Ok from the first answer, we will deduce that the data is *integer*. The data is not so big it just contains a few thousand entries. The third answer is interesting from this we deduce that the range of data is 1-150. Data is provided in an array. From fifths answer we deduce that we have to create our own data structure and we cannot modify the array provided. So finally, we conclude, we can just use bucket sort to sort the data. The range is just 1-150 so we need just 151-capacity integral array. Data is under thousands so we do not have to worry about data overflow and we get the solution in linear time *O(N)*.

Note: We will read sorting in the coming chapters.

Summary

At this point, you know the process of handling new problems very well. In the coming chapter we will be looking into a lot of various data structure and the problems they solve. A huge number of problems are solved in this book. However, it is recommended to first try to solve them by yourself, and then look for the solution. Always think about the complexity of the problem. In the interview interaction is the key to get problem described completely and discuss your approach with the interviewer.

Chapter 4: Abstract Data Type & Java Collections

Abstract data type (ADT)

An abstract data type (ADT) is a logical description of how we view the data and the operations that are allowed on it. ADT is defined as a user point of view of a data type. ADT concerns about the possible values of the data and what are interface exposed by it.

ADT does not concern about the actual implementation of the data structure.

For example, a user wants to store some *integers* and find a mean of it. Does not talk about how exactly it will be implemented.

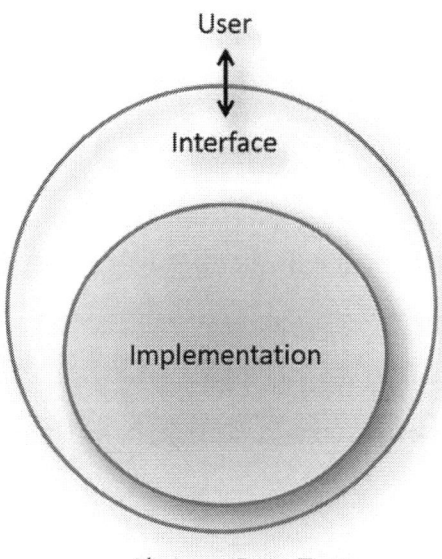

Abstract Data Type

Data-Structure

Data structures are concrete representations of data and are defined as a programmer point of view. Data-structure represents how data will be stored in memory. All data-structures have their own pros and cons. And depending upon the problem at hand, we pick a data-structure that is best suited for it.

For example, we can store data in an array, a linked-list, stack, queue, tree, etc.

JAVA Collection Framework

JAVA programming language provides a JAVA Collection Framework, which is a set of high quality, high performance & reusable data-structures and algorithms.

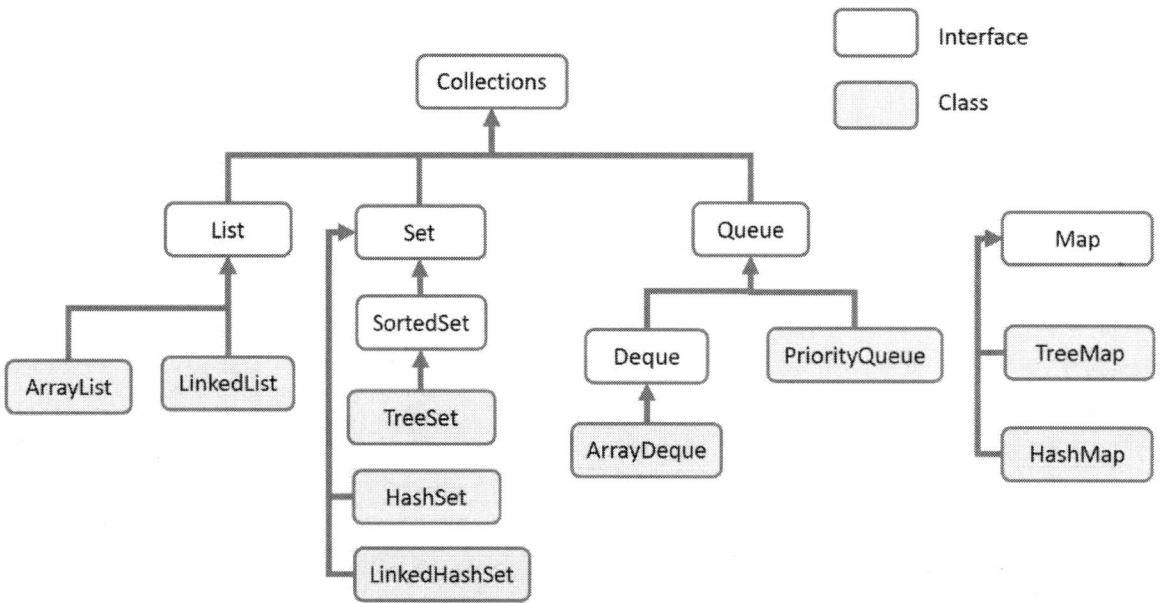

The following advantages of using a JAVA collection framework:
1. Programmers do not have to implement basic data structures and algorithms again and again. Thereby it prevents the reinvention the wheel. Thus the programmer can devote more effort in business logic.
2. The JAVA Collection Framework code is well-tested, high quality, high performance code there by increasing the quality of the programs.
3. Development cost is reduced as basic data structures and algorithms are implemented in Collections framework.
4. Easy for the review and understanding others programs as other developers also use the Collection framework. In addition, collection framework is well documented.

Array

0	1	2	3	4	5	6	7	8	9
1	2	3	4	5	6	7			

Array represents a collection of multiple elements of the same datatype. Arrays are fixed size data structure, the size of this data structure is fixed at the time of its creation. Arrays are the most common data structure used to store data.
As we cannot change the size of an array, we generally declare a large size array to handle any future data expansion. This ends up in creating a large size array, where most of the space is unused.

Note: - Arrays can store a fixed number of elements, whereas a collection stores object dynamically so there is no size restrictions it grows and shrinks automatically.

Array ADT Operations

Below is the API of array:
1. Adds an element at kth position
   Value can be stored in array at Kth position in *O(1)* constant time. We just need to store value at arr[k].
2. Reading the value stored at kth position.
   Accessing value stored a some location in array is also *O(1)* constant time. We just need to read value stored at arr[k].

Example 4.1

```java
public class ArrayDemo {
    public static void main(String[] args) {
        int[] arr = new int[10];
        for (int i = 0; i < 10; i++)
        {
            arr[i] = i;
        }
    }
}
```

JAVA standard arrays are of fixed length. Sometime we do not know how much memory we need so we create a bigger size array. There by wasting space to avoid this situation JAVA Collection framework had implemented ArrayList to solve this problem.

ArrayList implementation in JAVA Collections

ArrayList<E> in by JAVA Collections is a data structure which implements List<E> interface which means that it can have duplicate elements in it. ArrayList is an implementation as dynamic array that can grow or shrink as needed. (Internally array is used when it is full a bigger array is allocated and the old array values are copied to it.)

Example 4.2:

```java
import java.util.ArrayList;

public class ArrayListDemo {
    public static void main(String[] args) {
        ArrayList<Integer> al = new ArrayList<Integer>();
```

```
                    al.add(1); // add 1 to the end of the list
                    al.add(2); // add 2 to the end of the list
                    al.add(3); // add 3 to the end of the list
                    al.add(4); // add 4 to the end of the list
                    System.out.println("Contents of Array: " + al); // array is converted to
                                                                    // string and printed to screen

                    al.add(2,9); // 9 is added to 2nd index.
                    al.add(5,9); // 9 is added to 5th index.
                    System.out.println("Contents of Array: " + al);

                    System.out.println("Array Size: " + al.size()); // array size printed

                    System.out.println("Array IsEmpty: " + al.isEmpty());

                    al.remove(al.size() -1); // last element of array is removed.

                    System.out.println("Array Size: " + al.size());
                    al.removeAll(al); // all the elements of array are removed.

                    System.out.println("Array IsEmpty: " + al.isEmpty());
            }
    }
```

Output

```
Contents of Array: [1, 2, 3, 4]
Contents of Array: [1, 2, 9, 3, 4, 9]
Array Size: 6
Array IsEmpty: false
Array Size: 5
Array IsEmpty: true
```

Linked List

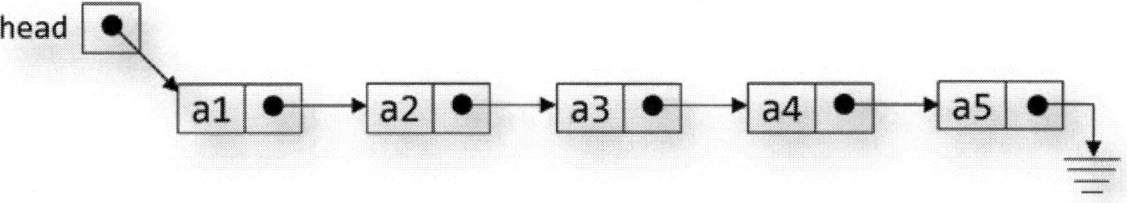

Linked lists are dynamic data structure and memory is allocated at run time. The concept of linked list is not to store data contiguously. Use links that point to the next elements.

Performance wise linked lists are slower than arrays because there is no direct access to linked list elements.

The linked list is a useful data structure when we do not know the number of elements to be stored ahead of time.

There are many flavors of linked list that you will see: linear, circular, doubly, and doubly circular.

Linked List ADT Operations

Below is the API of Linked list.

Insert(k): adds k to the start of the list
Insert an element at the start of the list. Just create a new element and move references. So that this new element becomes the new element of the list. This operation will take *O(1)* constant time.

Delete(): delete element at the start of the list
Delete an element at the start of the list. We just need to move one reference. This operation will also take *O(1)* constant time.

PrintList(): display all the elements of the list.
Start with the first element and then follow the references. This operation will take *O(N)* time.

Find(k): find the position of element with value k
Start with the first element and follow the reference until we get the value we are looking for or reach the end of the list. This operation will take *O(N)* time.
Note: binary search does not work on linked lists.

FindKth(k): find element at position k
Start from the first element and follow the links until you reach the kth element. This operation will take *O(N)* time.

IsEmpty(): check if the number of elements in the list are zero.
Just check the head reference of the list it should be Null. Which means list is empty. This operation will take *O(1)* time.

LinkedList implementation in JAVA Collections

LinkedList<E> in by JAVA Collections is a data structure which also implements List<E> interface.

Example 4.3

```java
import java.util.LinkedList;

public class LinkedListDemo {

    public LinkedListDemo() {
    }

    public static void main(String[] args) {
        LinkedList<Integer> ll = new LinkedList<Integer>();
        ll.addFirst(8); // 8 is added to the list
        ll.addLast(9); // 9 is added to last of the list.
        ll.addFirst(7); // 7 is added to first of the list.
        ll.addLast(20); // 20 is added to last of the list
        ll.addFirst(2); // 2 is added to first of list.
        ll.addLast(22); // 22 is added to last of the list.
        System.out.println("Contents of Linked List: " + ll);
```

```
            ll.removeFirst();
            ll.removeLast();

            System.out.println("Contents of Linked List: " + ll);
    }
}
```

Output:

```
Contents of Linked List: [2, 7, 8, 9, 20, 22]
Contents of Linked List: [7, 8, 9, 20]
```

Stack

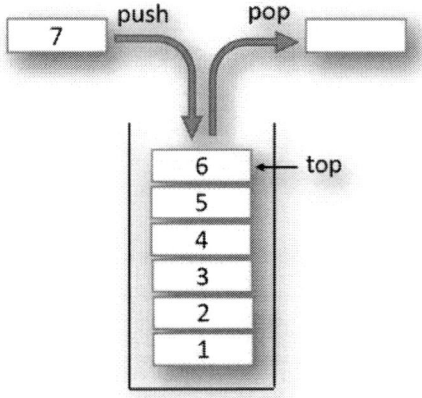

Stack is a special kind of data structure that follows Last-In-First-Out (LIFO) strategy. This means that the element that is added to stack last will be the first to be removed.

The various applications of stack are:
1. Recursion: recursive calls are preformed using system stack.
2. Postfix evaluation of expression.
3. Backtracking
4. Depth-first search of trees and graphs.
5. Converting a decimal number into a binary number etc.

Stack ADT Operations

Push(k): Adds a new item to the top of the stack

Pop(): Remove an element from the top of the stack and return its value.

Top(): Returns the value of the element at the top of the stack

Size(): Returns the number of elements in the stack

IsEmpty(): determines whether the stack is empty. It returns 1 if the stack is empty or return 0.

Note: All the above Stack operations are implemented in *O(1) Time Complexity*.

Stack implementation in JAVA Collection

Stack is implemented by calling push and pop methods of ArrayDeque<T> class.
JDK also provides Stack<T>, but we should not use this class and prefer Deque which implement collection interface.
1. First reason is that Stack<T> does not drive from Collection interface.
2. Second Stack<T> drives from Vector<T> so random access is possible so it brakes abstraction of a stack.
3. Third ArrayDeque is more efficient as compared to Stack<T>.

Example 4.4

```java
import java.util.ArrayDeque;

public class DequeStack<T> {

    private ArrayDeque<T> deque = new ArrayDeque<T>();

    public void push(T obj){
        deque.push(obj);
    }
    public T pop(){
        return deque.pop();
    }
    public T top(){
        return deque.peekLast();
    }
    public int size(){
        return deque.size();
    }
    public boolean isEmpty(){
        return deque.isEmpty();
    }
}
```

Note: - in the coming chapters we will directly use ArrayDeque object. We will not be making DequeStack wrapper class. When we use push() and pop() methods to insert and remove from a ArrayDeque we will be doing stack operations.

Queue

A queue is a First-In-First-Out (FIFO) kind of data structure. The element that is added to the queue first will be the first to be removed from the queue and so on.

Queue has the following application uses:
1. Access to shared resources (e.g., printer)
2. Multiprogramming
3. Message queue

Queue ADT Operations:

Add(): Add a new element to the back of the queue.

Remove(): remove an element from the front of the queue and return its value.

Front(): return the value of the element at the front of the queue.

Size(): returns the number of elements inside the queue.

IsEmpty(): returns 1 if the queue is empty otherwise return 0

Note: All the above Queue operations are implemented in *O(1) Time Complexity*.

Queue implementation in JAVA Collection

ArrayDeque<T> is the class implementation of doubly ended queue. If we use add(), remove() and peekFirst() it will behave as a queue. (And if we use push(), pop() and peekLast() it behave as a stack.)

Example 4.5

```java
import java.util.ArrayDeque;

public class DequeQueue<T> {
    private ArrayDeque<T> deque = new ArrayDeque<T>();

    public void add(T obj){
        deque.add(obj);
    }
    public T remove(){
        return deque.remove();
    }
    public T peek(){
        return deque.peekFirst();
    }
    public int size(){
        return deque.size();
    }
    public boolean isEmpty(){
        return deque.isEmpty();
    }
}
```

Note:- In the coming chapter we will not be making wrapper class DequeQueue<E>. We will directly use ArrayDeque<E> and will call add() and remove() methods to use it as a queue.

Trees

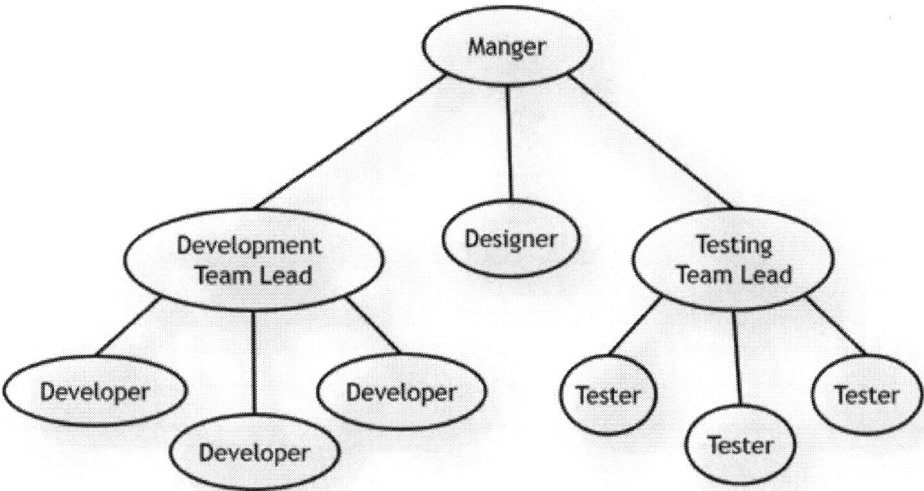

Tree is a hierarchical data structure. The top element of a tree is called the root of the tree. Except the root element, every element in a tree has a parent element, and zero or more child elements. The tree is the most useful data structure when you have hierarchical information to store.

There are many types of trees, for example, binary-tree, Red-black tree, AVL tree, etc.

Binary Tree

A binary tree is a type of tree in which each node has at most two children (0, 1 or 2) which are referred as left child and right child.

Binary Search Trees (BST)

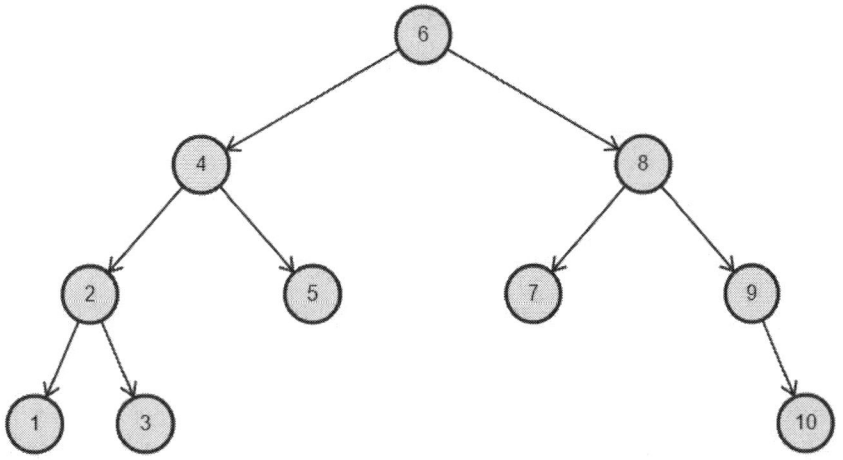

A binary search tree (BST) is a binary tree on which nodes are ordered in the following way:
1. The key in the left subtree is less than the key in its parent node.
2. The key in the right subtree is greater or equal the key in its parent node.

Binary Search Tree ADT Operations

Insert(k): Insert an element k into the tree.

Delete(k): Delete an element k from the tree.

Search(k): Search a particular value k into the tree if it is present or not.

FindMax(): Find the maximum value stored in the tree.

FindMin(): Find the minimum value stored in the tree.

The average *Time Complexity* of all the above operations on a binary search tree is O(log n), the case when the tree is balanced.
The worst-case *Time Complexity* will be **O(n)** when the tree is skewed. A binary tree is skewed when tree is not balanced.
There are two types of skewed tree.
1. Right Skewed binary tree: A binary tree in which each node is having either only a right child or no child at all.
2. Left Skewed binary tree: A binary tree in which each node is having either only a left child or no child at all.

Balanced Binary search tree

There are few binary search tree, which always keeps themselves balanced. Most important among them are Red-Black Tree (RB-Tree) and AVL tree.

The standard template library (STL) is implemented using this Red-Black Tree (RB-Tree).

TreeSet implementation in JAVA Collections

TreeSet<> is a class which implements Set<> interface which means that it store only unique elements. TreeSet<> is implemented using a red-black balanced binary search tree in JAVA Collections. Since TreeSet<> is implemented using a binary search tree its elements are stored in sequential order.

Example 4.6

```java
import java.util.TreeSet;
public class TreeSetDemo {
    public static void main(String[] args) {
        // Create a tree set.
        TreeSet<String> ts = new TreeSet<String>();
        // Add elements to the tree set.
        ts.add("India");
        ts.add("USA");
        ts.add("Brazil");
        ts.add("Canada");
```

```
            ts.add("UK");
            ts.add("China");
            ts.add("France");
            ts.add("Spain");
            ts.add("Italy");
            System.out.println(ts);
    }
}
```

Output

[Brazil, Canada, China, France, India, Italy, Spain, UK, USA]

Note:- TreeSet is implemented using a binary search tree so add, remove, and contains methods have logarithmic time complexity O(log (n)), where n is the number of elements in the set.

TreeMap implementation in JAVA Collection

A Map<> is an interface that maps keys to values. Also called a dictionary. A TreeMap<> is an implementation of Map<> and is implemented using red-black balanced binary tree so the key value pairs are stored in sorted order.

Example 4.7

```
import java.util.TreeMap;

public class TreeMapDemo {
public static void main(String[] args) {
        // create a tree map.
        TreeMap<String, Integer> tm = new TreeMap<String, Integer>();

        // Put elements into the map
        tm.put("Mason", new Integer(55));
        tm.put("Jacob", new Integer(77));
        tm.put("William", new Integer(99));
        tm.put("Alexander", new Integer(80));
        tm.put("Michael", new Integer(50));
        tm.put("Emma", new Integer(65));
        tm.put("Olivia", new Integer(77));
        tm.put("Sophia", new Integer(88));
        tm.put("Emily", new Integer(99));
        tm.put("Isabella", new Integer(100));

        System.out.println("Total number of students in class :: " + tm.size());
        for(String key : tm.keySet()){
            System.out.println(key + " score marks :" + tm.get(key));
        }
        System.out.println("Emma score present::" + tm.containsKey("Emma"));
        System.out.println("John score present:: " + tm.containsKey("John"));
        tm.remove("Emma");
        System.out.println("Emma score present::" + m.containsKey("Emma"));
    }
}
```

Output

```
Total number of students in class :: 10
Alexander score marks :80
Emily score marks :99
Emma score marks :65
Isabella score marks :100
Jacob score marks :77
Mason score marks :55
Michael score marks :50
Olivia score marks :77
Sophia score marks :88
William score marks :99
Emma present in class :: true
John present in class :: false
Emma present in class :: false
```

Priority Queue (Heap)

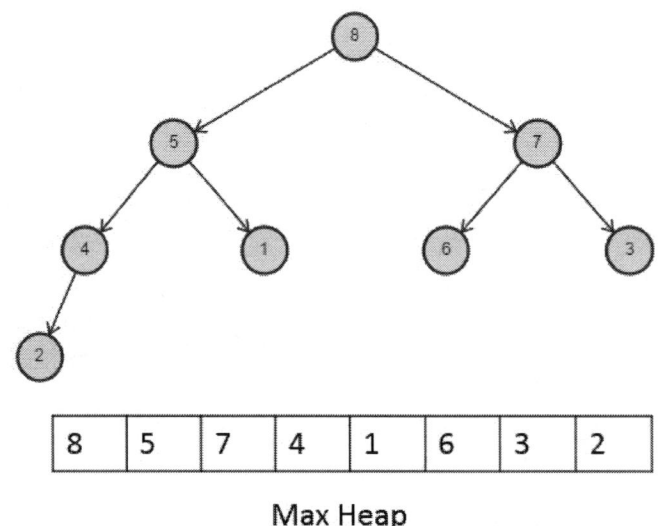

Max Heap

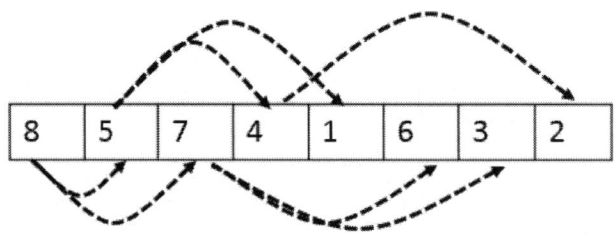

Priority queue is implemented using a binary heap data structure. In a heap, the records are stored in an array so that each key is larger than its two children keys. Each node in the heap follows the same rule that the parent value is greater than its two children.

There are two types of the heap data structure:
1. Max heap: each node should be greater than or equal to each of its children.
2. Min heap: each node should be smaller than or equal to each of its children.

A heap is a useful data structure when you want to get max/min one by one from data. Heap-Sort uses max heap to sort data in increasing/decreasing order.

Heap ADT Operations

Insert() - Adding a new element to the heap. The *Time Complexity* of this operation is O(log(n))

remove() - Extracting max for max heap case (or min for min heap case). The *Time Complexity* of this operation is O(log(n))

Heapify() – To convert a list of numbers in an array into a heap. This operation has a *Time Complexity* **O(n)**

PriorityQueue implementation in JAVA Collection

Min heap implementation of Priority Queue

Example 4.8

```java
import java.util.PriorityQueue;

public class PriorityQueueDemo {
    public static void main(String[] args) {

        PriorityQueue<Integer> pq = new PriorityQueue<Integer>();
        int[] arr = {1,2,10,8,7,3,4,6,5,9};

        for(int i: arr){
            pq.add(i);
        }

        System.out.println("Printing Priority Queue Heap : " + pq);

        System.out.print("remove elements of Priority Queue ::");
        while(pq.isEmpty() == false){
            System.out.print(" " + pq.remove());
        }
    }
}
```

Output

```
Printing Priority Queue Heap : [1, 2, 3, 5, 7, 10, 4, 8, 6, 9]
Dequeue elements of Priority Queue :: 1 2 3 4 5 6 7 8 9 10
```

Max heap implementation of Priority Queue
We just need to change collection order to make max heap from PriorityQueue<> collection.

Example 4.9

PriorityQueue<Integer> pq = **new** PriorityQueue<Integer>(Collections.*reverseOrder*());

Output

Printing Priority Queue Heap : [10, 9, 4, 6, 8, 2, 3, 1, 5, 7]
Dequeue elements of Priority Queue :: 10 9 8 7 6 5 4 3 2 1

Hash-Table

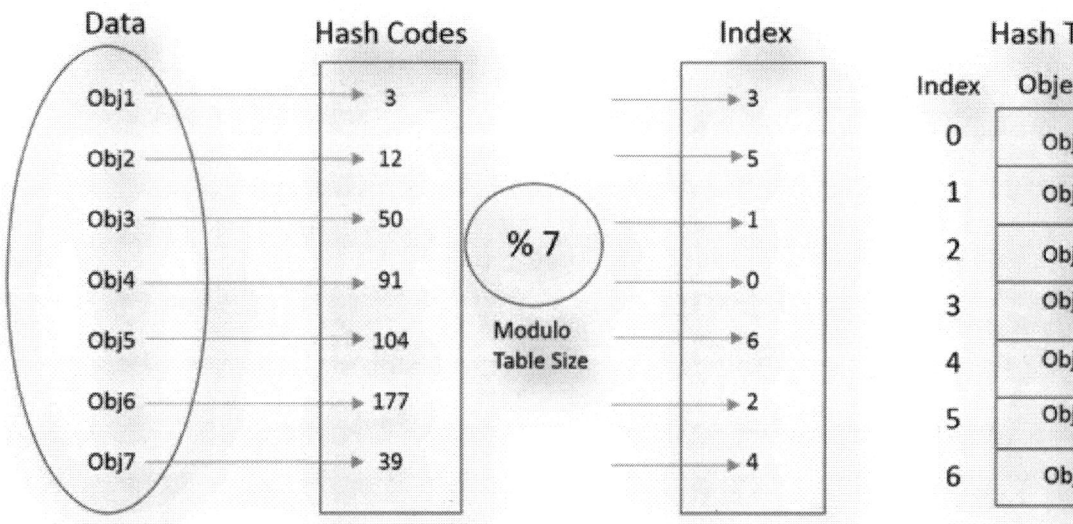

A Hash-Table is a data structure that maps keys to values. Each position of the Hash-Table is called a slot. The Hash-Table uses a hash function to calculate an index of an array of slots. We use the Hash-Table when the number of keys actually stored is small relatively to the number of possible keys.

The process of storing objects using a hash function is as follows:
1. Create an array of size M to store objects, this array is called Hash-Table.
2. Find a hash code of an object by passing it through the hash function.
3. Take module of hash code by the size of Hash-Table to get the index of the table where objects will be stored.
4. Finally store these objects in the designated index.

The process of searching objects in Hash-Table using a hash function is as follows:
1. Find a hash code of the object we are searching for by passing it through the hash function.
2. Take module of hash code by the size of Hash-Table to get the index of the table where objects are stored.
3. Finally, retrieve the object from the designated index.

Hash-Table Abstract Data Type (ADT)

ADT of Hash-Table contains the following functions:
Insert(x): Add object x to the data set.
Delete(x): Delete object x from the data set.
Search(x): Search object x in data set.
The Hash-Table is a useful data structure for implementing dictionary. The average time to search for an element in a Hash-Table is *O(1)*. A Hash Table generalizes the notion of an array.

HashSet implementation of JAVA Collections

HashSet <> is a class which implements Set<> interface which means that it store only unique elements. HashSet <> is implemented using a hash table. Since HashSet<> is implemented using a hash table its elements are not stored in sequential order.

Example 4.10

```java
import java.util.HashSet;

public class HashSetDemo {
    public static void main(String[] args) {

        // Create a hash set.
        HashSet<String> hs = new HashSet<String>();

        // Add elements to the hash set.
        hs.add("India");
        hs.add("USA");
        hs.add("Brazil");
        hs.add("Canada");
        hs.add("UK");
        hs.add("China");
        hs.add("France");
        hs.add("Spain");
        hs.add("Italy");
        System.out.println(hs);

        System.out.println("Hash Table contains USA : " + hs.contains("USA"));
        System.out.println("Hash Table contains Russia : " +
                                        hs.contains("Russia"));
        hs.remove("USA");
        System.out.println(hs);
        System.out.println("Hash Table contains USA : " + hs.contains("USA"));
    }
}
```

Output

```
[Canada, USA, UK, China, Italy, Brazil, France, India, Spain]
Hash Table contains USA : true
Hash Table contains Russia : false
[Canada, UK, China, Italy, Brazil, France, India, Spain]
Hash Table contains USA : false
```

LinkedHashSet implementation of JAVA Collections

LinkedHashSet <> is a class which implements Set<> interface which means that it store only unique elements. LinkedHashSet <> is implemented using a hash table and a linked list. Linked list is used to preserver order of elements based on insertion. Traversing the elements of the hash table is done in order of insertion.

Example 4.11

```java
import java.util.HashSet;
import java.util.LinkedHashSet;

public class LinkedHashSetDemo {
    public static void main(String[] args) {
        // Create a hash set.
        HashSet<String> hs = new HashSet<String>();
        // Add elements to the hash set.
        hs.add("India");
        hs.add("USA");
        hs.add("Brazil");
        hs.add("Canada");
        hs.add("UK");
        hs.add("China");
        hs.add("France");
        hs.add("Spain");
        hs.add("Italy");
        System.out.println("HashSet value:: " + hs);

        // Create a linked hash set.
        LinkedHashSet<String> lhs = new LinkedHashSet<String>();
        // Add elements to the linked hash set.
        lhs.add("India");
        lhs.add("USA");
        lhs.add("Brazil");
        lhs.add("Canada");
        lhs.add("UK");
        lhs.add("China");
        lhs.add("France");
        lhs.add("Spain");
        lhs.add("Italy");
        System.out.println("LinkedHashSet value:: " + lhs);
    }
}
```

Output

```
HashSet value:: [Canada, USA, UK, China, Italy, Brazil, France, India, Spain]
LinkedHashSet value:: [India, USA, Brazil, Canada, UK, China, France, Spain, Italy]
```

Comparison of various Set classes.

	TreeSet	HashSet	LinkedHashSet
Storage	Red-Black Tree	Hash Table	Hash Table with Linked List.
Performance	Slower than HashSet, O(log(N))	Fastest, constant time	More expensive then HashSet, have to maintain linked list.
Order of Iteration	Increasing Order	No order guarantee	Order of insertion

HashMap implementation in JAVA Collection

A Map<> is a data structure that maps keys to values. Also called a dictionary. A HashMap<> is an implementation of Map<> and is implemented using a hash table so the key value pairs are not stored in sorted order. Map<> does not allow duplicate keys buts values can be duplicate.

Example 4.12

```java
import java.util.HashMap;

public class HashMapDemo {

    public static void main(String[] args) {

        // Create a hash map.
        HashMap<String, Integer> hm = new HashMap<String, Integer>();

        // Put elements into the map
        hm.put("Mason", new Integer(55));
        hm.put("Jacob", new Integer(77));
        hm.put("William", new Integer(99));
        hm.put("Alexander", new Integer(80));
        hm.put("Michael", new Integer(50));
        hm.put("Emma", new Integer(65));
        hm.put("Olivia", new Integer(77));
        hm.put("Sophia", new Integer(88));
        hm.put("Emily", new Integer(99));
        hm.put("Isabella", new Integer(100));

        System.out.println("Total number of students in class :: " + hm.size());
        for(String key : hm.keySet()){
            System.out.println(key + " score marks :" + hm.get(key));
        }
        System.out.println("Emma score present::" + hm.containsKey("Emma"));
        System.out.println("John score  present:: " + hm.containsKey("John"));
    }
}
```

Output

```
Total number of students in class :: 10
Olivia score marks :77
Emily score marks :99
Alexander score marks :80
Mason score marks :55
```

```
Michael score marks :50
Isabella score marks :100
William score marks :99
Emma score marks :65
Sophia score marks :88
Jacob score marks :77
Emma present in class :: true
John present in class :: false
```

Dictionary / Symbol Table

A symbol table is a mapping between a string(key) and a value, which can be of any data type. A value can be an integer such as occurrence count, dictionary meaning of a word and so on.

Binary Search Tree (BST) for Strings

Binary Search Tree (BST) is the simplest way to implement symbol table. Simple string compare function can be used to compare two strings. If all the keys are random, and the tree is balanced. Then on an average key lookup can be done in logarithmic time.

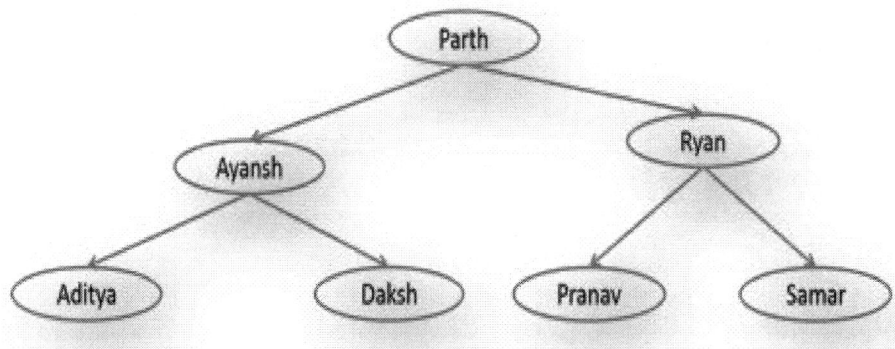

BINARY SEARCH TREE AS DICTIONARY

Hash-Table

The Hash-Table is another data structure, which can be used for symbol table implementation. Below Hash-Table diagram, we can see the name of that person is taken as the key, and their meaning is the value of the search. The first key is converted into a hash code by passing it to appropriate hash function. Inside hash function the size of Hash-Table is also passed, which is used to find the actual index where values will be stored. Finally, the value that is meaning of name is stored in the Hash-Table, or you can store a reference to the string which store meaning can be stored into the Hash-Table.

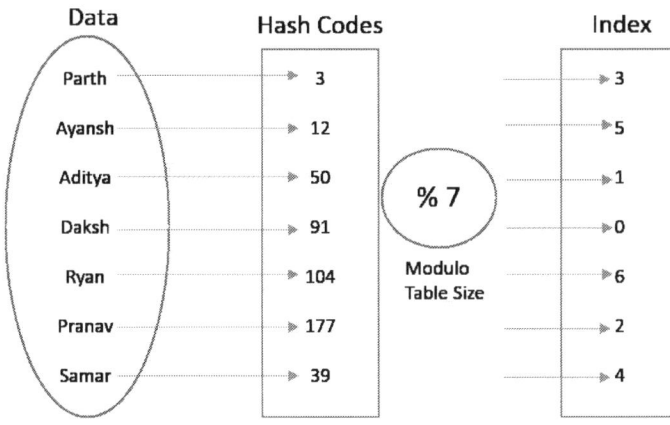

Hash-Table has an excellent lookup of constant time.

Let us suppose we want to implement autocomplete the box feature of Google search. When you type some string to search in google search, it proposes some complete string even before you have done typing. BST cannot solve this problem as related strings can be in both right and left subtree.

The Hash-Table is also not suited for this job. One cannot perform a partial match or range query on a Hash-Table. Hash function transforms string to a number. Moreover, a good hash function will give a distributed hash bode even for partial string and there is no way to relate two strings in a Hash-Table.

Trie and Ternary Search tree are a special kind of tree, which solves partial match, and range query problem well.

Trie

Trie is a tree, in which we store only one character at each node. This final key value pair is stored in the leaves. Each node has K children, one for each possible character. For simplicity purpose, let us consider that the character set is 26, corresponds to different characters of English alphabets.

Trie is an efficient data structure. Using Trie, we can search the key in O(M) time. Where M is the maximum string length. Trie is also suitable for solving partial match and range query problems.

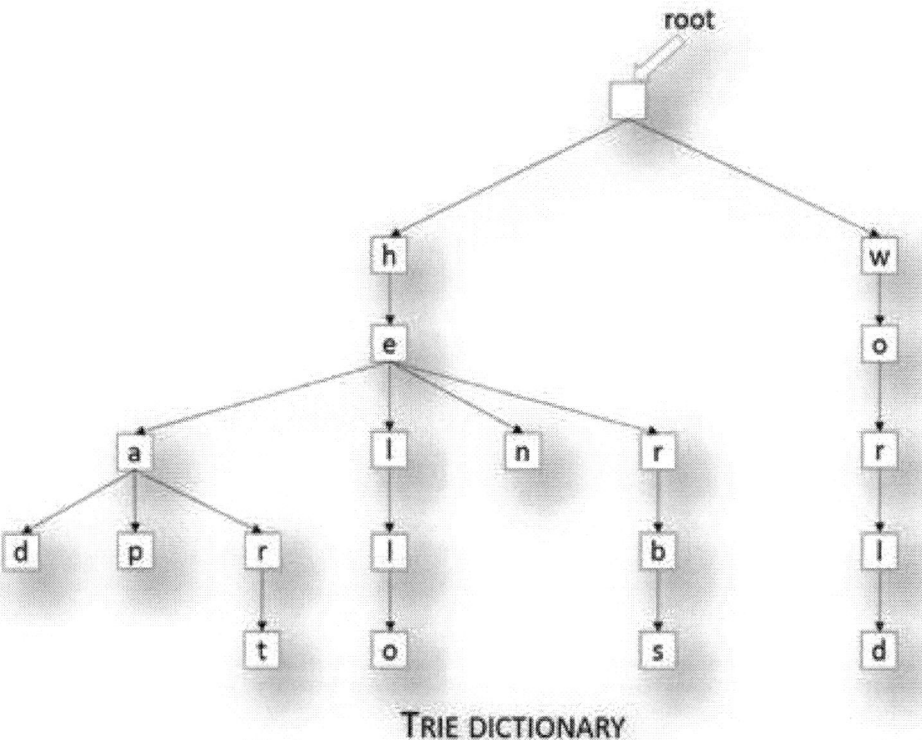

TRIE DICTIONARY

Ternary Search Trie/ Ternary Search Tree

Tries having a very good search performance of O(M) where M is the maximum size of the search string. However, tries having very high space requirement. Every node Trie contain references to multiple nodes, each reference corresponds to possible characters of the key. To avoid this high space requirement Ternary Search Trie (TST) is used.

A TST avoid the heavy space requirement of the traditional Trie while still keeping many of its advantages. In a TST, each node contains a character, an end of key indicator, and three references. The three references are corresponding to current char hold by the node(equal), characters less than and character greater than.

The *Time Complexity* of ternary search tree operation is proportional to the height of the ternary search tree. In the worst case, we need to traverse up to 3 times that many links. However, this case is rare.

Therefore, TST is a very good solution for implementing Symbol Table, Partial match and range query.

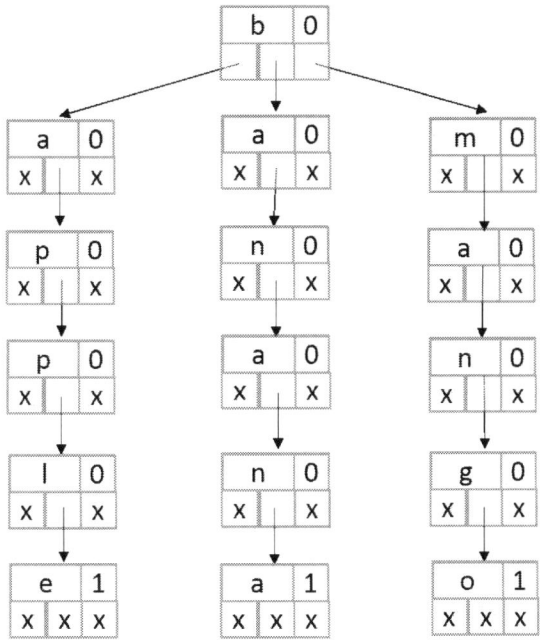

Ternary Search Tree

Graphs

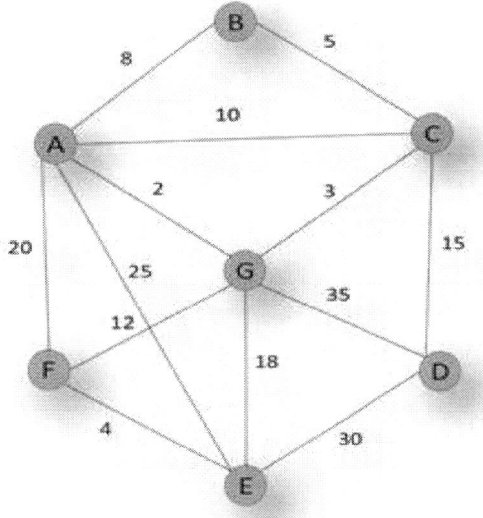

A graph is a data structure which represents a network, that connects a collection of nodes called vertices, and their connections, called edges. An edge can be seen as a path between two nodes. These edges can be either directed or undirected. If a path is directed then you can move only in one direction, while in an undirected path you can move in both the directions.

Graph Algorithms

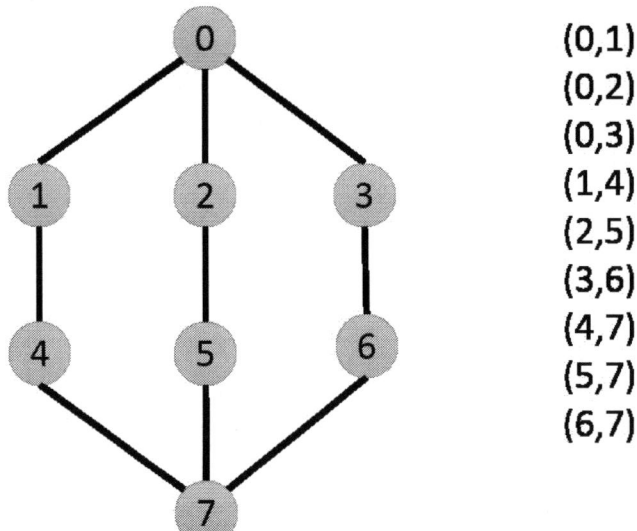

(0,1)
(0,2)
(0,3)
(1,4)
(2,5)
(3,6)
(4,7)
(5,7)
(6,7)

Depth-First Search (DFS)

The DFS algorithm we start from starting point and go into depth of graph until we reach a dead end and then move up to parent node (Backtrack). In DFS, we use stack to get the next vertex to start a search. Alternatively, we can use recursion (system stack) to do the same.

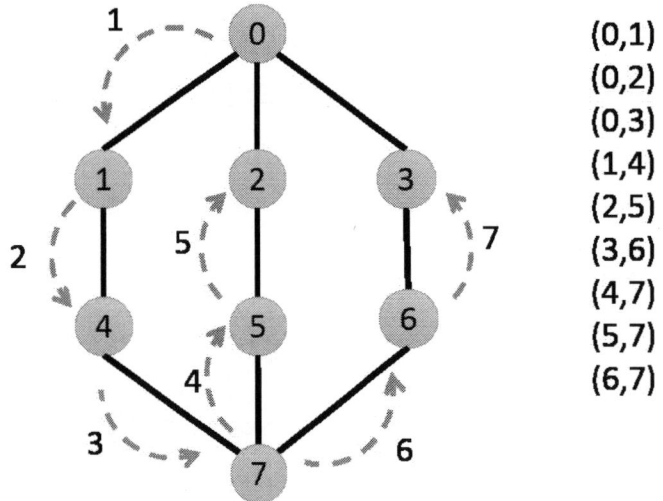

(0,1)
(0,2)
(0,3)
(1,4)
(2,5)
(3,6)
(4,7)
(5,7)
(6,7)

Depth First Traversal
0, 1, 4, 7 , 5, 2, 6, 3

Breadth-First Search (BFS)

In BFS algorithm, a graph is traversed in layer-by-layer fashion. The graph is traversed closer to the starting point. The queue is used to implement BFS.

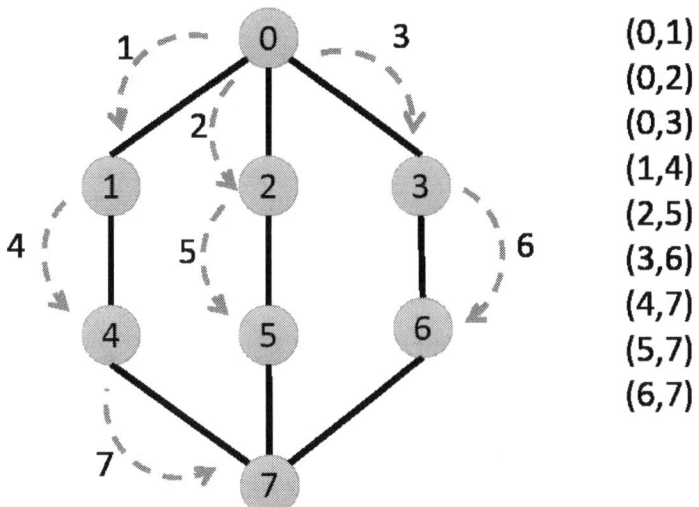

(0,1)
(0,2)
(0,3)
(1,4)
(2,5)
(3,6)
(4,7)
(5,7)
(6,7)

Breadth First Traversal
0, 1, 2, 3, 4, 5, 6, 7

Sorting Algorithms

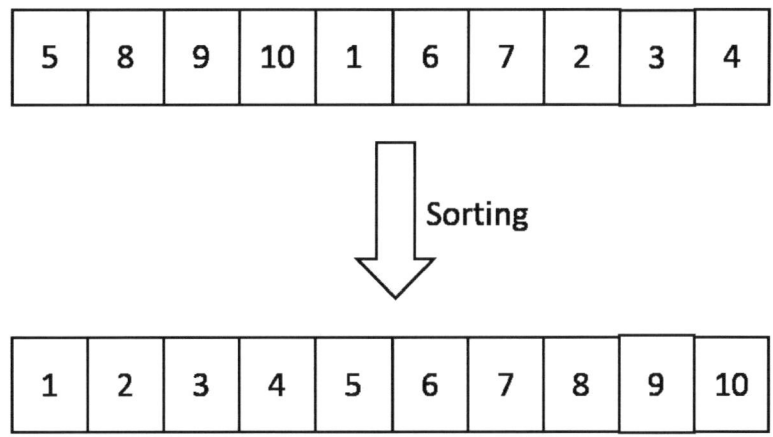

Sorting is the process of placing elements from a collection into ascending or descending order. Sorting arranges data elements in order so that searching become easier.
There are good sorting functions available which does sorting in *O(nlogn)* time, so in this book when we need sorting we will use sort() function and will assume that the sorting is done in *O(nlogn)* time.

Counting Sort

Counting sort is the simplest and most efficient type of sorting. Counting sort has a strict requirement of a predefined range of data.

Like, sort how many people are in which age group. We know that the age of people can vary between 1 and 130.

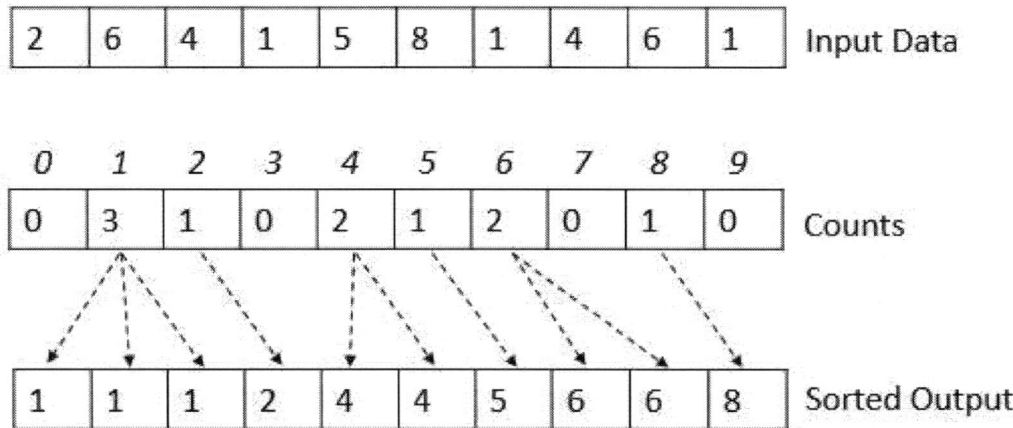

If we know the range of input, then sorting can be done using counting in O(n+k).

End note

This chapter has provided a brief introduction of the various data structures, algorithms and their complexities. In the coming chapters we will look into all these data structure in details. If you know the interface of the various data structures, then you can use them while solving other problems without knowing the internal details how they are implemented.

Chapter 5: Searching

Introduction

In Computer Science, Searching is the algorithmic process of finding a particular item in a collection of items. The item may be a keyword in a file, a record in a database, a node in a tree or a value in an array etc.

Why Searching?

Imagine you are in a library with millions of books. You want to get a specific book with specific title. How will you find? You will just start searching by initial letter of the book title. Then you continue matching with a whole book title until you find your desired book. (By doing this small heuristic you have reduced the search space by a factor of 26, consider we have an equal number of books whose title begin with particular char.)

Similarly, computer stores lots of information and to retrieve this information efficiently, we need very efficient searching algorithms. To make searching efficient, we keep the data in some proper order. There are certain ways of organizing the data. If you keep the data in proper order, it is easy to search required element. For example, Sorting is one of the process for making data organized.

Different Searching Algorithms

- Linear Search – Unsorted Input
- Linear Search – Sorted Input
- Binary Search (Sorted Input)
- String Search: Tries, Suffix Trees, Ternary Search.
- Hashing and Symbol Tables

Linear Search – Unsorted Input

When elements of an array are not ordered or sorted and we want to search for a particular value, we need to scan the full array unless we find the desired value. This kind of algorithm known as unordered linear search. The major problem with this algorithm is less performance or high Time Complexity in worst case.

Example 5.1

```
boolean linearSearchUnsorted(int[] arr, int size, int value) {
    int i = 0;
    for(i = 0 ; i < size ; i++) {
        if(value == arr[i] )
            return true;
    }
    return false;
}
```

Time Complexity: *O(n)*. As we need to traverse the complete array in worst case. Worst case is when your desired element is at the last position of the array. Here, 'n' is the size of the array.
Space Complexity: *O(1)*. No extra memory is used to allocate the array.

Linear Search – Sorted

If elements of the array are sorted either in increasing order or in decreasing order, searching for a desired element will be much more efficient than unordered linear search. In many cases, we do not need to traverse the complete array. Following example explains when you encounter a greater element from the increasing sorted array, you stop searching further. This is how this algorithm saves the time and improves the performance.

Example 5.2

```
boolean linearSearchSorted(int[] arr, int size, int value)
{
    int i = 0;
    for(i = 0 ; i < size ; i++)
    {
        if(value == arr[i] )
            return true;
        else if( value < arr[i] )
            return false;
    }
    return false;
}
```

Time Complexity: *O(n)*. As we need to traverse the complete array in worst case. Worst case is when your desired element is at the last position of the sorted array. However, in the average case this algorithm is more efficient even though the growth rate is same as unsorted.
Space Complexity: *O(1)*. No extra memory is used to allocate the array.

Binary Search

How do we search a word in a dictionary? In general, we go to some approximate page (mostly middle) and start searching from that point. If we see the word that we are searching is same then we are done with the search. Else, if we see the page is before the selected pages, then apply the same procedure for the first half otherwise to the second half. Binary Search also works in the same way. We get the middle point from the sorted array and start comparing with the desired value.

Note: Binary search requires the array to be sorted otherwise binary search cannot be applied.

Example 5.3

```
// Binary Search Algorithm – Iterative Way
boolean Binarysearch(int[] arr, int size, int value)
{
    int low = 0;
    int high = size-1;
    int mid;
```

```
        while (low <= high)
        {
                mid = low + (high-low)/2; // To avoid the overflow
                if (arr[mid] == value)
                        return true;
                else if (arr[mid] < value)
                        low = mid + 1;
                else
                        high = mid - 1;
        }
        return false;
}
```

Time Complexity: *O(logn)*. We always take half input and throwing out the other half. So the recurrence relation for binary search is T(n) = T(n/2) + c. Using master theorem (divide and conquer), we get T(n) = *O(logn)*
Space Complexity: O(1)

Example 5.4

```
// Binary Search Algorithm – Recursive Way
boolean BinarySearchRecursive(int[] arr, int low, int high, int value)
{
        if(low > high)
                return false;
        int mid = low + (high-low)/2; // To avoid the overflow

        if (arr[mid] == value)
                return true;
        else if (arr[mid] < value)
                return BinarySearchRecursive (arr, mid + 1, high, value);
        else
                return BinarySearchRecursive (arr, low, mid - 1 , value);
}
```

Time Complexity: O(logn). **Space Complexity**: **O(logn)** For system stack in recursion

String Searching Algorithms

Refer String chapter.

Hashing and Symbol Tables

Refer Hash-Table chapter.

How sorting is useful in Selection Algorithm?

Selection problems can be converted to sorting problems. Once the array is sorted, it is easy to find the minimum/maximum (or desired element) from the sorted array. The method 'Sorting and then Selecting' is inefficient for selecting a single element, but it is efficient when many selections need to be made from the array. It is because only one initial expensive sort is needed, followed by many cheap selection operations.

For example, if we want to get the maximum element from an array. After sorting the array, we can simply return the last element from the array. What if we want to get second maximum. Now, we do not have to sort the array again and we can return the second last element from the sorted array. Similarly, we can return the kth maximum element by just one scan of the sorted list.
So with the above discussion, sorting is used to improve the performance. In general this method requires *O(nlogn)* (for sorting) time. With the initial sorting, we can answer any query in one scan, *O(n)*.

Problems in Searching

Print Duplicates in Array

Given an array of n numbers, print the duplicate elements in the array.

First approach: Exhaustive search or Brute force, for each element in array find if there is some other element with the same value. This is done using two for loop, first loop to select the element and second loop to find its duplicate entry.

Example 5.5

```
void printRepeating(int[] arr, int size)
{
        int i, j;
        System.out.println(" Repeating elements are ");
        for(i = 0; i < size; i++)
                for(j = i+1; j < size; j++)
                        if(arr[i] == arr[j])
                                System.out.println(" " + arr[i]);
}
```

The Time Complexity is $O(n^2)$ and Space Complexity is $O(1)$

Second approach: Sorting, Sort all the elements in the array and after this in a single scan, we can find the duplicates.

Example 5.6

```
void printRepeating2(int[] arr, int size) {
        int i;
        Arrays.sort(arr); // Sort(arr,size);
        System.out.println(" Repeating elements are ");

        for(i = 1; i < size; i++)
        {
                if(arr[i] == arr[i-1])
                        System.out.println(" " + arr[i]);
        }
}
```

Sorting algorithms take $O(n.\log n)$ time and single scan take $O(n)$ time.
The *Time Complexity* of an algorithm is $O(n.\log n)$ and *Space Complexity* is $O(1)$

Third approach: Hash-Table, using Hash-Table, we can keep track of the elements we have already seen and we can find the duplicates in just one scan.

Example 5.7

```java
void printRepeating3(int[] arr, int size)
{
        HashSet<Integer> hs = new HashSet<Integer>();
        int i;
        System.out.println(" Repeating elements are ");
        for(i = 0; i < size; i++)
        {
                if(hs.contains(arr[i]))
                        System.out.println(" " + arr[i]);
                else
                        hs.add(arr[i]);
        }
}
```

Hash-Table insert and find take constant time $O(1)$ so the total *Time Complexity* of the algorithm is $O(n)$ time. *Space Complexity* is also $O(n)$.

Forth approach: Counting, this approach is only possible if we know the range of the input. If we know that, the elements in the array are in the range 0 to n-1. We can reserve and array of length n and when we see an element we can increase its count. In just one single scan, we know the duplicates. If we know the range of the elements, then this is the fastest way to find the duplicates.

Example 5.8

```java
void printRepeating4(int[] arr, int size)
{
        int[] count = new int[size];
        int i;
        for(i = 0; i < size; i++)
                count[i]=0;
        System.out.println(" Repeating elements are ");
        for(i = 0; i < size; i++)
        {
                if(count[arr[i]] == 1)
                        System.out.println(" " + arr[i]);
                else
                        count[arr[i]]++;
        }
}
```

Counting approach just uses an array so insert and find take constant time $O(1)$ so the total *Time Complexity* of the algorithm is $O(n)$ time. *Space Complexity* for creating count array is also $O(n)$.

Find max, appearing element in an array

Given an array of n numbers, find the element, which appears maximum number of times.

First approach: Exhaustive search or Brute force, for each element in array find how many times this particular value appears in array. Keep track of the maxCount and when some element count is greater than it then update the maxCount. This is done using two for loop, first loop to select the element and second loop to count the occurrence of that element.
The Time Complexity is $O(n^2)$ and Space Complexity is $O(1)$

Example 5.9

```
int getMax(int[] arr, int size)
{
    int i, j;
    int max = arr[0], count = 1, maxCount = 1;
    for (i = 0; i < size; i++)
    {
        count = 1;
        for (j = i + 1; j < size; j++)
            if (arr[i] == arr[j])
                count++;
        if (count > maxCount)
        {
            max = arr[i];
            maxCount = count;
        }
    }
    return max;
}
```

Second approach: Sorting, Sort all the elements in the array and after this in a single scan, we can find the counts. Sorting algorithms take $O(n.\log n)$ time and single scan take $O(n)$ time. The *Time Complexity* of an algorithm is $O(n.\log n)$ and *Space Complexity* is $O(1)$

Example 5.10

```
int getMax2(int[] arr, int size)
{
    int max = arr[0], maxCount = 1;
    int curr = arr[0], currCount = 1;
    int i;
    Arrays.sort(arr); // Sort(arr,size);
    for (i = 1; i < size; i++)
    {
        if (arr[i] == arr[i - 1])
            currCount++;
        else
        {
            currCount = 1;
            curr = arr[i];
        }
```

```
                if (currCount > maxCount)
                {
                        maxCount = currCount;
                        max = curr;
                }
        }
        return max;
}
```

Third approach: Counting, This approach is only possible if we know the range of the input. If we know that, the elements in the array are in the range 0 to n-1. We can reserve and array of length n and when we see an element we can increase its count. In just one single scan, we know the duplicates. If we know the range of the elements, then this is the fastest way to find the max count.

Counting approach just uses array so to increase count take constant time $O(1)$ so the total *Time Complexity* of the algorithm is $O(n)$ time. *Space Complexity* for creating count array is also $O(n)$.

Example 5.11

```
int getMax(int[] arr, int size, int range)
{
        int max = arr[0], maxCount = 1;
        int[] count = new int[range];
        int i;
        for (i = 0; i < size; i++)
        {
                count[arr[i]]++;
                if (count[arr[i]] > maxCount)
                {
                        maxCount = count[arr[i]];
                        max = arr[i];
                }
        }
        return max;
}
```

Majority element in an Array

Given an array of n elements. Find the majority element, which appears more than n/2 times. Return 0 in case there is no majority element.

First approach: Exhaustive search or Brute force, for each element in array find how many times this particular value appears in array. Keep track of the maxCount and when some element count is greater than it then update the maxCount. This is done using two for loop, first loop to select the element and second loop to count the occurrence of that element.

Once we have the final, maxCount we can see if it is greater than n/2, if it is greater than we have a majority if not we do not have any majority.

The Time Complexity is $O(n^2) + O(1) = O(n^2)$ and Space Complexity is $O(1)$

Example 5.12

```
int getMajority(int[] arr, int size)
{
        int i, j;
        int max=0, count=0 , maxCount=0;

        for(i = 0; i < size; i++)
        {
                for(j = i+1; j < size; j++)
                        if(arr[i] == arr[j])
                                count++;
                if(count > maxCount)
                {
                        max = arr[i];
                        maxCount = count;
                }
        }

        if (maxCount > size/2)
                return max;
        else
                return Integer.MIN_VALUE;
}
```

Second approach: Sorting, Sort all the elements in the array. If there is a majority than the middle element at the index n/2 must be the majority number. So just single scan can be used to find its count and see if the majority is there or not.

Sorting algorithms take $O(n.\log n)$ time and single scan take $O(n)$ time.
The *Time Complexity* of an algorithm is $O(n.\log n)$ and *Space Complexity* is $O(1)$

Example 5.13

```
int getMajority2(int[] arr, int size)
{
        int majIndex = size/2, count = 1;
        int i;
        int candidate;
        Arrays.sort(arr); // Sort(arr,size);
        candidate = arr[majIndex];
        count = 0;
        for (i = 0; i < size; i++)
                if(arr[i] == candidate)
                        count++;
        if (count > size/2)
                return arr[majIndex];
        else
                return Integer.MIN_VALUE;
}
```

Third approach: This is a cancelation approach (Moore's Voting Algorithm), if all the elements stand against the majority and each element is cancelled with one element of majority if there is majority then majority prevails.
- Set the first element of the array as majority candidate and initialize the count to be 1.
- Start scanning the array.
    - If we get some element whose value same as a majority candidate, then we increase the count.
    - If we get an element whose value is different from the majority candidate, then we decrement the count.
    - If count become 0, that means we have a new majority candidate. Make the current candidate as majority candidate and reset count to 1.
    - At the end, we will have the only probable majority candidate.
- Now scan through the array once again to see if that candidate we found above have appeared more than n/2 times.

Counting approach just scans throw array two times. The *Time Complexity* of the algorithm is $O(n)$ time. *Space Complexity* for creating count array is also $O(1)$.

Example 5.14

```
int getMajority3(int[] arr, int size)
{
        int majIndex = 0, count = 1;
        int i;
        int candidate;
        for(i = 1; i < size; i++)
        {
                if(arr[majIndex] == arr[i])
                        count++;
                else
                        count--;
                if(count == 0)
                {
                        majIndex = i;
                        count = 1;
                }
        }
        candidate = arr[majIndex];
        count = 0;
        for (i = 0; i < size; i++)
                if(arr[i] == candidate)
                        count++;
        if (count > size/2)
                return arr[majIndex];
        else
                return Integer.MIN_VALUE;
}
```

Find the missing number in an Array

Given an array of n-1 elements, which are in the range of 1 to n. There are no duplicates in the array. One of the *integer* is missing. Find the missing element.

First approach: Exhaustive search or Brute force, for each value in the range 1 to n, find if there is some element in array which have the same value. This is done using two for loop, first loop to select value in the range 1 to n and the second loop to find if this element is in the array or not.

The Time Complexity is $O(n^2)$ and Space Complexity is $O(1)$

Example 5.15

```java
int findMissingNumber(int[] arr, int size)
{
        int i, j, found = 0;
        for (i = 1; i <= size; i++)
        {
                found = 0;
                for (j = 0; j < size; j++)
                {
                        if (arr[j] == i)
                        {
                                found = 1;
                                break;
                        }
                }
                if (found == 0)
                        return i;
        }
        return Integer.MAX_VALUE;
}
```

Second approach: Sorting, Sort all the elements in the array and after this in a single scan, we can find the duplicates.

Sorting algorithms take $O(n.\log n)$ time and single scan take $O(n)$ time.
The *Time Complexity* of an algorithm is $O(n.\log n)$ and *Space Complexity* is $O(1)$

Third approach: Hash-Table, using Hash-Table, we can keep track of the elements we have already seen and we can find the missing element in just one scan.

Hash-Table insert and find take constant time $O(1)$ so the total *Time Complexity* of the algorithm is $O(n)$ time. *Space Complexity* is also $O(n)$.

Forth approach: Counting, we know the range of the input so counting will work. As we know that, the elements in the array are in the range 0 to n-1. We can reserve and array of length n and when we see an element we can increase its count. In just one single scan, we know the missing element.

Counting approach just uses an array so insert and find take constant time $O(1)$ so the total *Time Complexity* of the algorithm is $O(n)$ time. *Space Complexity* for creating count array is also $O(n)$.

Fifth approach: You are allowed to modify the given input array. Modify the given input array in such a way that in the next scan you can find the missing element.

When you scan through the array. When at index "index", the value stored in the array will be arr[index] so add the number "n + 1" to arr[arr[index]]. Always read the value from the array using a reminder operator "%". When you scan the array for the first time and modified all the values, then one single scan you can see if there is some value in the array which is smaller than "n+1" that index is the missing number.
In this approach, the array is scanned two times and the *Time Complexity* of this algorithm is $O(n)$. *Space Complexity* is $O(1)$.

Sixth approach: Summation formula to find the sum of n numbers from 1 to n. Subtract the values stored in the array and you will have your missing number.
The *Time Complexity* of this algorithm is $O(n)$. *Space Complexity* is $O(1)$.

Seventh approach: XOR approach to find the sum of n numbers from 1 to n. XOR the values stored in the array and you will have your missing number.
The *Time Complexity* of this algorithm is $O(n)$. *Space Complexity* is $O(1)$.

Example 5.16

```
int findMissingNumber(int[] arr, int size, int range)
{
    int i;
    int xorSum = 0;
    //get the XOR of all the numbers from 1 to range
    for (i = 1; i <= range; i++)
        xorSum ^= i;
    //loop through the array and get the XOR of elements
    for (i = 0; i<size; i++)
        xorSum ^= arr[i];
    return xorSum;
}
```

Note: Same problem can be asked in many forms (sometimes you have to do the xor of the range sometime you do not):
1. There are numbers in the range of 1-n out of which all appears single time but one that appear two times.
2. All the elements in the range 1-n are appearing 16 times and one element appear 17 times. Find the element that appears 17 times.
3.

Find Pair in an Array

Given an array of n numbers, find two elements such that their sum is equal to "value"

First approach: Exhaustive search or Brute force, for each element in array find if there is some other element, which sum up to the desired value. This is done using two for loop, first loop to select the element and second loop to find another element.

The Time Complexity is $O(n^2)$ and Space Complexity is $O(1)$

Example 5.17

```
int FindPair(int[] arr, int size, int value)
{
        int i, j;
        for (i = 0; i < size; i++)
                for (j = i + 1; j < size; j++)
                        if ((arr[i] + arr[j] ) == value)
                        {
                                System.out.println("The pair is : "+arr[i]+","+arr[j]);
                                return 1;
                        }
        return 0;
}
```

Second approach: Sorting, Steps are as follows:
1. Sort all the elements in the array.
2. Take two variable first and second. Variable first= 0 and second = size -1
3. Compute sum = arr[first]+arr[second]
4. If the sum is equal to the desired value, then we have the solution
5. If the sum is less than the desired value, then we will increase first
6. If the sum is greater than the desired value, then we will decrease the second
7. We repeat the above process till we get the desired pair or we get first >= second (don't have a pair)

Sorting algorithms take $O(n.\log n)$ time and single scan take $O(n)$ time.

The *Time Complexity* of an algorithm is $O(n.\log n)$ and *Space Complexity* is $O(1)$

Example 5.18

```
int FindPair2(int[] arr, int size, int value) {
        int first = 0, second = size - 1;
        int curr;
        Arrays.sort(arr);//Sort(arr, size);
        while (first < second) {
                curr = arr[first] + arr[second];
                if (curr == value)
                {
                        System.out.println("The pair is " + arr[first]+ "," + arr[second]);
                        return 1;
                }
                else if (curr < value)
                        first++;
                else
                        second--;
        }
        return 0;
}
```

Third approach: Hash-Table, using Hash-Table, we can keep track of the elements we have already seen and we can find the pair in just one scan.
1. For each element, insert the value in Hashtable. Let say current value is arr[index]
2. And see if the value - arr[index] is already in a Hashtable.
3. If value - arr[index] is in the Hashtable then we have the desired pair.
4. Else, proceed to the next entry in the array.

Hash-Table insert and find take constant time $O(1)$ so the total *Time Complexity* of the algorithm is $O(n)$ time. *Space Complexity* is also $O(n)$.

Example 5.19
```
int FindPair3(int[] arr, int size, int value)
{
    HashSet<Integer> hs = new HashSet<Integer>();
    int i;
    for (i = 0; i < size; i++)
    {
        if ( hs.contains(value - arr[i]))
        {
            System.out.println("The pair is : "+arr[i]+" , "+(value - arr[i]));
            return 1;
        }
        hs.add(arr[i]);
    }
    return 0;
}
```

Forth approach: Counting, This approach is only possible if we know the range of the input. If we know that, the elements in the array are in the range 0 to n-1. We can reserve and array of length n and when we see an element we can increase its count. In place of the Hashtable in the above approach, we will use this array and will find out the pair.

Counting approach just uses an array so insert and find take constant time $O(1)$ so the total *Time Complexity* of the algorithm is $O(n)$ time. *Space Complexity* for creating count array is also $O(n)$.

Find the Pair in two Arrays

Given two array X and Y. Find a pair of elements (x_i, y_i) such that $x_i \in X$ and $y_i \in Y$ where $x_i + y_i$ = value.

First approach: Exhaustive search or Brute force, loop through element xi of X and see if you can find (value – xi) in Y. Two for loop.
The Time Complexity is $O(n^2)$ and Space Complexity is $O(1)$

Second approach: Sorting, Sort all the elements in the second array Y. For each element if X you can see if that element is there in Y by using binary search.

Sorting algorithms take $O(m.\log m)$ and searching will take $O(n.\log m)$ time.
The Time Complexity of an algorithm is $O(n.\log m)$ or $O(m.\log m)$ and Space Complexity is $O(1)$

Third approach: Sorting, Steps are as follows:
1. Sort the elements of both X and Y in increasing order.
2. Take the sum of the smallest element of X and the largest element of Y.
3. If the sum is equal to value, we got our pair.
4. If the sum is smaller than value, take next element of X
5. If the sum is greater than value, take the previous element of Y

Sorting algorithms take $O(n.\log n) + O(m.\log m)$ for sorting and searching will take $O(n + m)$ time.
The Time Complexity of an algorithm is $O(n.\log n)$ Space Complexity is $O(1)$

Forth approach: Hash-Table, Steps are as follows:
1. Scan through all the elements in the array Y and insert them into Hashtable.
2. Now scan through all the elements of array X, let us suppose the current element is xi see if you can find (value - xi) in the Hashtable.
3. If you find the value, you got your pair.
4. If not, then go to the next value in the array X.

Hash-Table insert and find take constant time $O(1)$ so the total *Time Complexity* of the algorithm is $O(n)$ time. *Space Complexity* is also $O(n)$.

Fifth approach: Counting, This approach is only possible if we know the range of the input. Same as Hashtable implementation just use a simple array in place of Hashtable and you are done.

Counting approach just uses an array so insert and find take constant time $O(1)$ so the total *Time Complexity* of the algorithm is $O(n)$ time. *Space Complexity* for creating count array is also $O(n)$.

Two elements whose sum is closest to zero

Given an Array of *integer*s, both +ve and -ve. You need to find the two elements such that their sum is closest to zero.

First approach: Exhaustive search or Brute force, for each element in array find the other element whose value when added will give minimum absolute value. This is done using two for loop, first loop to select the element and second loop to find the element that should be added to it so that the absolute of the sum will be minimum or close to zero.

The Time Complexity is $O(n^2)$ and Space Complexity is $O(1)$
Example 5.20
```
void minabsSumPair(int[] arr, int size) {
    int l, r, minSum, sum, minFirst, minSecond;
    // Array should have at least two elements
    if(size < 2) {
        System.out.println("Invalid Input");
        return;
    }
```

```
        // Initialization of values
        minFirst = 0;
        minSecond = 1;
        minSum = Math.abs(arr[0] + arr[1]);
        for(l = 0; l < size - 1; l++)
        {
                for(r = l+1; r < size; r++)
                {
                        sum = Math.abs(arr[l] + arr[r]);
                        if(sum < minSum)
                        {
                                minSum = sum;
                                minFirst = l;
                                minSecond = r;
                        }
                }
        }
        System.out.println(" The two elements with minimum sum are : "
                                + arr[minFirst] + " , "+ arr[minSecond]);
}
```

Second approach: Sorting

Steps are as follows:
1. Sort all the elements in the array.
2. Take two variable firstIndex = 0 and secondIndex = size -1
3. Compute sum = arr[firstIndex]+arr[secondIndex]
4. If the sum is equal to the 0 then we have the solution
5. If the sum is less than the 0 then we will increase first
6. If the sum is greater than the 0 then we will decrease the second
7. We repeat the above process 3 to 6, till we get the desired pair or we get first >= second

Example 5.21

```
void minabsSumPair2(int[] arr, int size)
{
        int l, r, minSum, sum, minFirst, minSecond;
        // Array should have at least two elements
        if (size < 2)
        {
                System.out.println("Invalid Input");
                return;
        }
        Arrays.sort(arr);//Sort(arr, size);

        // Initialization of values
        minFirst = 0;
        minSecond = size - 1;
        minSum = Math.abs(arr[minFirst] + arr[minSecond]);
```

```
        for (l = 0, r = size - 1; l < r;)
        {
                sum = (arr[l] + arr[r]);
                if (Math.abs(sum) < minSum)
                {
                        minSum = Math.abs(sum);
                        minFirst = l;
                        minSecond = r;
                }
                if (sum < 0)
                        l++;
                else if (sum > 0)
                        r++;
                else
                        break;
        }
        System.out.println(" The two elements with minimum sum are : "
                        + arr[minFirst] + " , "+ arr[minSecond]);
}
```

Find maxima in a bitonic array

A bitonic array comprises of an increasing sequence of *integers* immediately followed by a decreasing sequence of *integer* s. Since the elements are sorted in some order, we should go for algorithm similar to binary search. The steps are as follows:
1. Take two variable for storing start and end index. Variable start=0 and end=size-1
2. Find the middle element of the array.
3. See if the middle element is the maxima. If yes, return the middle element.
4. Alternatively, If the middle element in increasing part, then we need to look for in mid+1 and end.
5. Alternatively, if the middle element is in the decreasing part, then we need to look in the start and mid-1.
6. Repeat step 2 to 5 until we get the maxima.

Example 5.22

```
int SearchBotinicArrayMax(int[] arr, int size) {
        int start = 0, end = size - 1;
        int mid = (start + end) / 2;
        int maximaFound = 0;
        if (size < 3)
        {
                System.out.println("error");
                return 0;
        }
        while (start <= end)
        {
                mid = (start + end) / 2;
                if (arr[mid - 1] < arr[mid] && arr[mid + 1] < arr[mid])//maxima
                {
                        maximaFound = 1;
                        break;
                }
```

```
            else if (arr[mid - 1] < arr[mid] && arr[mid] < arr[mid + 1])//increasing
            {
                    start = mid + 1;
            }
            else if (arr[mid - 1] > arr[mid] && arr[mid] > arr[mid + 1])//decreasing
            {
                    end = mid - 1;
            }
            else
            {
                    break;
            }
    }
    if (maximaFound == 0)
    {
            System.out.println("error");
            return 0;
    }
    return arr[mid];
}
```

Search element in a bitonic array

A bitonic array comprises of an increasing sequence of *integers* immediately followed by a decreasing sequence of *integer* s. To search an element in a bitonic array:
1. Find the index or maximum element in the array. By finding the end of increasing part of the array, using modified binary search.
2. Once we have the maximum element, search the given value in increasing part of the array using binary search.
3. If the value is not found in increasing part, search the same value in decreasing part of the array using binary search.

Example 5.23

```
int SearchBitonicArray(int[] arr, int size, int key) {
    int max = FindMaxBitonicArray(arr, size);
    int k = BinarySearch(arr, 0, max, key, true);
    if (k != -1)
            return k;
    else
            return BinarySearch(arr, max + 1, size - 1, key, false);
}
```

```
int FindMaxBitonicArray(int[] arr, int size) {
    int start = 0, end = size - 1, mid;
    if (size < 3)
    {
            System.out.println("error");
            return 0;
    }
```

```java
        while (start <= end)
        {
                mid = (start + end) / 2;
                if (arr[mid - 1] < arr[mid] && arr[mid + 1] < arr[mid])//maxima
                {
                        return mid;
                }
                else if (arr[mid - 1] < arr[mid] && arr[mid] < arr[mid + 1])//increasing
                {
                        start = mid + 1;
                }
                else if (arr[mid - 1] > arr[mid] && arr[mid] > arr[mid + 1])//increasing
                {
                        end = mid - 1;
                }
                else
                {
                        break;
                }
        }
        System.out.println("error");
        return 0;
}
```

```java
int BinarySearch(int[] arr, int start, int end, int key, boolean isInc)
{
        int mid;
        if (end < start)
                return -1;
        mid = (start + end) / 2;
        if (key == arr[mid])
                return mid;
        if (isInc != false && key < arr[mid] ||
                        isInc == false && key > arr[mid])
        {
                return BinarySearch(arr, start, mid - 1, key, isInc);
        }
        else
        {
                return BinarySearch(arr, mid + 1, end, key, isInc);
        }
}
```

Occurrence counts in sorted Array

Given a sorted array arr[] find the number of occurrences of a number.

First approach: Brute force, Traverse the array and in linear time we will get the occurrence count of the number. This is done using one loop.
The Time Complexity is $O(n)$ and Space Complexity is $O(1)$

Example 5.24

```
int findKeyCount(int[] arr, int size, int key)
{
        int i, count = 0;
        for (i = 0; i < size ; i++)
        {
                if (arr[i] == key)
                        count++;
        }
        return count;
}
```

Second approach: Since we have sorted array, we should think about some binary search.
1. First, we should find the first occurrence of the key.
2. Then we should find the last occurrence of the key.
3. Take the difference of these two values and you will have the solution.

Example 5.25

```
int findKeyCount2(int[] arr, int size, int key)
{
        int firstIndex, lastIndex;
        firstIndex = findFirstIndex(arr, 0, size -1, key);
        lastIndex = findLastIndex(arr, 0, size - 1, key);
        return (lastIndex - firstIndex + 1);
}
```

```
int findFirstIndex(int[] arr, int start, int end, int key)
{
        int mid;
        if (end < start)
                return -1;
        mid = (start + end) / 2;
        if (key == arr[mid] && (mid == start || arr[mid - 1] != key))
                return mid;
        if (key <= arr[mid])// <= is us the number.t in sorted array.
        {
                return findFirstIndex(arr, start, mid - 1, key);
        }
        else
        {
                return findFirstIndex(arr, mid + 1, end, key);
        }
}
```

```
int findLastIndex(int[] arr, int start, int end, int key)
{
        int mid;

        if (end < start)
                return -1;
```

```
        mid = (start + end) / 2;

        if (key == arr[mid] && (mid == end || arr[mid + 1] != key))
            return mid;

        if (key < arr[mid])// <
        {
            return findLastIndex(arr, start, mid - 1, key);
        }
        else
        {
            return findLastIndex(arr, mid + 1, end, key);
        }
}
```

Separate even and odd numbers in Array

Given an array of even and odd numbers, write a program to separate even numbers from the odd numbers.

First approach: allocate a separate array, then scan through the given array, and fill even numbers from the start and odd numbers from the end.

Second approach: Algorithm is as follows.
1. Initialize the two variable left and right. Variable left=0 and right= size-1.
2. Keep increasing the left index until the element at that index is even.
3. Keep decreasing the right index until the element at that index is odd.
4. Swap the number at left and right index.
5. Repeat steps 2 to 4 until left is less than right.

Example 5.26

```
void swap (int[] arr, int first, int second)
{
    int temp = arr[first];
    arr[first] = arr[second];
    arr[second] = temp;
}
```

```
void seperateEvenAndOdd(int[] arr, int size)
{
    int left = 0, right = size - 1;
    while (left < right)
    {
        if (arr[left] % 2 == 0 )
        {
            left++;
        }
```

```
                    else if (arr[right] % 2 == 1)
                    {
                            right--;
                    }
                    else
                    {
                            swap(arr, left, right);
                            left++;
                            right--;
                    }
            }
    }
```

Stock purchase-sell problem

Given an array, whose nth element is the price of the stock on nth day. You are asked to buy once and sell once, on what date you will be buying and at what date you will be selling to get maximum profit.
Or
Given an array of numbers, you need to maximize the difference between two numbers, such that you can subtract the number, which appear before form the number that appear after it.

First approach: Brute force, for each element in array find if there is some other element whose difference is maximum. This is done using two for loop, first loop to select, buy date index and the second loop to find its selling date entry.
The Time Complexity is $O(n^2)$ and Space Complexity is $O(1)$

Second approach: Another clever solution is to keep track of the smallest value seen so far from the start. At each point, we can find the difference and keep track of the maximum profit. This is a linear solution.
The *Time Complexity* of the algorithm is $O(n)$ time. *Space Complexity* for creating count array is also $O(1)$.

Example 5.27

```
void maxProfit(int stocks[], int size) {
    int buy = 0, sell = 0;
    int curMin = 0;
    int currProfit=0;
    int maxProfit = 0;
    int i;
    for (i = 0; i < size; i++) {
        if (stocks[i] < stocks[curMin])
            curMin = i;
        currProfit = stocks[i] - stocks[curMin];
        if (currProfit > maxProfit) {
            buy = curMin;
            sell = i;
            maxProfit = currProfit;
        }
    }
}
```

```
            System.out.println("Purchase day is- "+ buy +" at price " + stocks[buy]);
            System.out.println("Sell day is- "+sell+" at price " + stocks[sell]);
}
```

Find a median of an array

Given an array of numbers of size n, if all the elements of the array are sorted then find the element, which lie at the index n/2.

First approach: Sort the array and return the element in the middle.
Sorting algorithms take $O(n.\log n)$.
The *Time Complexity* of an algorithm is $O(n.\log n)$ and *Space Complexity* is $O(1)$

Example 5.28

```
int getMedian(int[] arr, int size)
{
        Arrays.sort(arr); //Sort(arr, size);
        return arr[size / 2];
}
```

Second approach: Use QuickSelect algorithm. This algorithm we will look into the next chapter. In QuickSort algorithm just skip the recursive call that we do not need.

The average *Time Complexity* of this algorithm will be $O(1)$

Find median of two sorted arrays.

First approach: Keep track of the index of both the array, say the index are i and j. keep increasing the index of the array which ever have a smaller value. Use a counter to keep track of the elements that we have already traced.

The *Time Complexity* of an algorithm is $O(n)$ and *Space Complexity* is $O(1)$

Example 5.29

```
int findMedian(int[] arrFirst, int sizeFirst, int[] arrSecond, int sizeSecond)
{
        int medianIndex = ((sizeFirst + sizeSecond) +
                (sizeFirst + sizeSecond) % 2) / 2;//cealing function.
        int i = 0, j = 0;
        int count = 0;
        while (count < medianIndex - 1)
        {
                if (i < sizeFirst - 1 && arrFirst[i] < arrSecond[j])
                        i++;
                else
                        j++;
                count++;
        }
```

```
        if (arrFirst[i] < arrSecond[j])
                return arrFirst[i];
        else
                return arrSecond[j];
}
```

Search 01 Array

Given an array of 0's and 1's. All the 0's come before 1's. Write an algorithm to find the index of the first 1.
Or
You are given an array which contains either 0 or 1, and they are in sorted order Ex. a [] = { 1,1,1,1,0,0,0} How will you count no of 1`s and 0's?

First approach: Binary Search, since the array is sorted using binary search to find the desired index. The Time Complexity of an algorithm is $O(\log n)$ and Space Complexity is $O(1)$

Example 5.30

```
int BinarySearch01Wrapper(int[] arr, int size)
{
        if (size == 1 && arr[0] == 1)
                return 0;
        return BinarySearch01(arr, 0, size - 1);
}
```

```
int BinarySearch01(int[] arr, int start, int end)
{
        int mid;
        if (end < start)
                return -1;
        mid = (start + end) / 2;
        if (1 == arr[mid] && 0 == arr[mid - 1])
                return mid;
        if (0 == arr[mid])
        {
                return BinarySearch01(arr, mid + 1, end);
        }
        else
        {
                return BinarySearch01(arr, start, mid - 1);
        }
}
```

Search in sorted rotated Array

Given a sorted array s of n integer. s is rotated an unknown number of times. Find an element in the array.

First approach: Since the array is sorted, we can use modified binary search to find the element. The Time Complexity of an algorithm is $O(\log n)$ and Space Complexity is $O(1)$

Example 5.31

```
int BinarySearchRotateArray(int[] arr, int start, int end, int key)
{
        int mid;
        if (end < start)
                return -1;
        mid = (start + end) / 2;
        if (key == arr[mid])
                return mid;
        if (arr[mid] > arr[start])
        {
                if (arr[start] <= key && key < arr[mid])
                {
                        return BinarySearchRotateArray(arr, start, mid - 1, key);
                }
                else
                {
                        return BinarySearchRotateArray(arr, mid + 1, end, key);
                }
        }
        else
        {
                if (arr[mid] < key && key <= arr[end])
                {
                        return BinarySearchRotateArray(arr, mid + 1, end, key);
                }
                else
                {
                        return BinarySearchRotateArray(arr, start, mid - 1, key);
                }
        }
}

int BinarySearchRotateArrayWrapper(int[] arr, int size, int key)
{
        return BinarySearchRotateArray(arr, 0, size - 1, key);
}
```

First Repeated element in the array

Given an unsorted array of n elements, find the first element, which is repeated.

First approach: Exhaustive search or Brute force, for each element in array find if there is some other element with the same value. This is done using two for loop, first loop to select the element and second loop to find its duplicate entry.

The Time Complexity is $O(n^2)$ and Space Complexity is $O(1)$

Example 5.32

```
int FirstRepeated(int[] arr, int size)
{
    int i, j;
    for (i = 0; i < size; i++)
        for (j = i + 1; j < size; j++)
            if (arr[i] == arr[j])
                return arr[i];
    return 0;
}
```

Second approach: Hash-Table, using Hash-Table, we can keep track of the number of times a particular element came in the array. First scan just populate the Hashtable. In the second, scan just look the occurrence of the elements in the Hashtable. If occurrence is more for some element, then we have our solution and the first repeated element.

Hash-Table insert and find take constant time $O(1)$ so the total *Time Complexity* of the algorithm is $O(n)$ time. *Space Complexity* is also $O(n)$ for maintaining hash.

Transform Array

How would you swap elements of an array like [a1 a2 a3 a4 b1 b2 b3 b4] to convert it into [a1 b1 a2 b2 a3 b3 a4 b4]?

Approach:
- First swap elements in the middle pair
- Next swap elements in the middle two pairs
- Next swap elements in the middle three pairs
- Iterate n-1 steps.

Ex: with n = 4.
a1 a2 a3 a4 b1 b2 b3 b4
a1 a2 a3 b1 a4 b2 b3 b4
a1 a2 b1 a3 b2 a4 b3 b4
a1 b1 a2 b2 a3 b3 a4 b4

Example 5.33

```
void transformArrayAB1(int[] arr, int size) {
    int N = size/2, i, j;
    for (i = 1; i < N; i++) {
        for (j = 0; j < i; j++) {
            swap(arr, N-i+2*j, N-i+2*j+1);
        }
    }
}
```

Find 2nd largest number in an array with minimum comparisons

Suppose you are given an unsorted array of n distinct elements. How will you identify the second largest element with minimum number of comparisons?

First approach: Find the largest element in the array. Then replace the last element with the largest element. Then search the second largest element int the remaining n-1 elements.
The total number of comparisons is: (n-1) + (n-2)

Second approach: Sort the array and then give the (n-1) element. This approach is still more inefficient.

Third approach: Using priority queue / Heap. This approach we will look into heap chapter. Use buildHeap() function to build heap from the array. This is done in n comparisons. Arr[0] is the largest number , and the grater among arr[1] and arr[2] is the second largest.
The total number of comparisons are: (n-1) + 1 = n

Check if two arrays are permutation of each other

Given two integer arrays. You have to check whether they are permutation of each other.

First approach: Sorting, Sort all the elements of both the arrays and Compare each element of both the arrays from beginning to end. If there is no mismatch, return true. Otherwise, false.
Sorting algorithms take $O(n.\log n)$ time and comparison take $O(n)$ time.
The *Time Complexity* of an algorithm is $O(n.\log n)$ and *Space Complexity* is $O(1)$

Example 5.34

```
boolean checkPermutation(int[] array1, int size1, int[] array2, int size2){
    if (size1 != size2)
        return false;
    Arrays.sort(array1); // Sort(array1, size1);
    Arrays.sort(array2); // Sort(array2, size2);
    for (int i = 0; i < size1; i++) {
        if (array1[i] != array2[i])
            return false;
    }
    return true;
}
```

Second approach: Hash-Table (Assumption: No duplicates).
1. Create a Hash-Table for all the elements of the first array.
2. Traverse the other array from beginning to the end and search for each element in the Hash-Table.
3. If all the elements are found in, the Hash-Table return true, otherwise return false.

Hash-Table insert and find take constant time $O(1)$ so the total *Time Complexity* of the algorithm is $O(n)$ time. *Space Complexity* is also $O(n)$
Time Complexity = **O(n)** (For creation of Hash-Table and look-up), *Space Complexity* = **O(n)** (For creation of Hash-Table).

Example 5.35

```
boolean checkPermutation2(int[] array1, int size1, int[] array2, int size2) {
        int i;
        if (size1 != size2)
                return false;

        ArrayList<Integer> al = new ArrayList<Integer>();

        for (i = 0; i < size1; i++)
                al.add(array1[i]);

        for (i = 0; i < size2; i++)
        {
                if (al.contains(array2[i]) == false)
                        return false;
                al.remove(array2[i]);
        }
        return true;
}
```

Remove duplicates in an integer array

First approach: Sorting, Steps are as follows:
1. Sort the array.
2. Take two references. A subarray will be created with all unique elements starting from 0 to the first reference (The first reference points to the last index of the subarray). The second reference iterates through the array from 1 to the end. Unique numbers will be copied from the second reference location to first reference location and the same elements are ignored.

Time Complexity calculation:
Time to sort the array = ***O(nlogn)***.
Time to remove duplicates = ***O(n)***. Overall *Time Complexity* = ***O(nlogn)***.
No additional space is required so *Space Complexity* is ***O(1)***.

Example 5.36

```
int removeDuplicates(int array[], int size) {
        int j = 0;
        int i;
        if (size == 0)
                return 0;
        Arrays.sort(array); // Sort(array,size);
        for (i = 1; i < size; i++) {
                if (array[i] != array[j]) {
                        j++;
                        array[j] = array[i];
                }
        }
        return j + 1;
}
```

Searching for an element in a 2-d sorted array

Given a 2 dimensional array. Each row and column are sorted in ascending order. How would you find an element in it?

The algorithm works as:
1. Start with element at last column and first row
2. If the element is the value we are looking for, return true.
3. If the element is greater than the value we are looking for, go to the element at previous column but same row.
4. If the element is less than the value we are looking for, go to the element at next row but same column.
5. Return false, if the element is not found after reaching the element of the last row of the first column. Condition (row < r && column >= 0) is false.

Running time = *O(N)*.

Example 5.37

```
int FindElementIn2DArray(int[] arr[], int r, int c, int value)
{
    int row = 0;
    int column = c - 1;
    while (row < r && column >= 0){
        if (arr[row][column] == value)
            return 1;
        else if (arr[row][column] > value)
            column--;
        else
            row++;
    }
    return 0;
}
```

Exercise

1. Given an array of n elements, find the first repeated element. Which of the following methods will work for us. And which, if the method will not work for us. If a method work, then implements it.
   - Brute force exhaustive search.
   - Use Hash-Table to keep an index of the elements and use the second scan to find the element.
   - Sorting the elements.
   - If we know the range of the element then we can use counting technique.

Hint: When order in which elements appear in input is important, we cannot use sorting.

2. Given an array of n elements, write an algorithm to find three elements in an array whose sum is a given value.

Hint: Try to do this problem using a brute force approach. Then try to apply the sorting approach along with a brute force approach. The *Time Complexity* will be $O(n^2)$

3. Given an array of –ve and +ve numbers, write a program to separate –ve numbers from the +ve numbers.

4. Given an array of 1's and 0's, write a program to separate 0's from 1's.
   Hint: QuickSelect, counting

5. Given an array of 0's, 1's and 2's, write a program to separate 0's , 1's and 2's.

6. Given an array whose elements is monotonically increasing with both negative and positive numbers. Write an algorithm to find the point at which list becomes positive.

7. Given a sorted array, find a given number. If found return the index if not, find the index of that number if it is inserted into the array.

8. Find max in sorted rotated array.

9. Find min in the sorted rotated array.

10. Find kth Smallest Element in the Union of Two Sorted Arrays

Chapter 6: Sorting

Introduction

Sorting is the process of placing elements from a collection into ascending or descending order. For example, when we play cards, sort cards, according to their value so that we can find the required card easily.

When we go to some library, the books are arranged according to streams (Algorithm, Operating systems, Networking etc.). Sorting arranges data elements in order so that searching become easier. When books are arranged in proper indexing order, then it is easy to find a book we are looking for.

This chapter discusses algorithms for sorting a set of N items. Understanding sorting algorithms are the first step towards understanding algorithm analysis. Many sorting algorithms are developed and analysed.

A sorting algorithm like *Bubble-Sort*, *Insertion-Sort* and *Selection-Sort* are easy to implement and are suitable for the small input set. However, for large dataset they are slow.

A sorting algorithm like *Merge-Sort*, *Quick-Sort* and *Heap-Sort* are some of the algorithms that are suitable for sorting large dataset. However, they are overkill if we want to sort the small dataset.

Some algorithm, which is suitable when we have some range information on input data.

Some other algorithm is there to sort a huge data set that cannot be stored in memory completely, for which external sorting technique is developed.

Before we start a discussion of the various algorithms one by one. First, we should look at comparison function that is used to compare two values.

Less function will return 1 if value1 is less than value2 otherwise, it will return 0.

```
private boolean less(int value1, int value2)
{
    return value1 < value2;
}
```

More function will return 1 if value1 is more than value2 otherwise it will return 0.

```
private boolean more(int value1, int value2)
{
    return value1 > value2;
}
```

The value in various sorting algorithms is compared using one of the above functions and it will be swapped depending upon the return value of these functions. If more() comparison function is used, then sorted output will be increasing in order and if less() is used than resulting output will be in descending order.

Type of Sorting

Internal Sorting: All the elements can be read into memory at the same time and sorting is performed in memory.
1. Selection-Sort
2. Insertion-Sort
3. Bubble-Sort
4. Quick-Sort

External Sorting: In this, the dataset is so big that it is impossible to load the whole dataset into memory so sorting is done in chunks.
1. Merge-Sort

Three things to consider in choosing, sorting algorithms for application:
1. Number of elements in list
2. A number of different orders of list required
3. The amount of time required to move the data or not move the data

Bubble-Sort

Bubble-Sort is the slowest algorithm for sorting, but it is heavily used, as it is easy to implement.

In Bubble-Sort, we compare each pair of adjacent values. We want to sort values in increasing order so if the second value is less than the first value then we swap these two values. Otherwise, we will go to the next pair.
Thus, smaller values bubble to the start of the array.

We will have N number of passes to get the array completely sorted.
After the first pass, the largest value will be in the rightmost position.

First Pass

5	1	2	4	3	7	6	Swap
1	5	2	4	3	7	6	Swap
1	2	5	4	3	7	6	Swap
1	2	4	5	3	7	6	Swap
1	2	4	3	5	7	6	No Swap
1	2	4	3	5	7	6	Swap
1	2	4	3	5	6	7	

Example 6.1

```java
public class BubbleSort {

    private int[] arr;

    public BubbleSort(int[] array)
    {
        arr = array;
    }

    private boolean less(int value1, int value2)
    {
        return value1 < value2;
    }

    private boolean more(int value1, int value2)
    {
        return value1 > value2;
    }

    public void sort()
    {
        int size = arr.length;

        int i, j, temp;
        for (i = 0 ; i < ( size - 1 ); i++)
        {
            for (j = 0 ; j < size - i - 1; j++)
            {
                if (more(arr[j], arr[j+1]))
                {
                    /* Swapping */
                    temp= arr[j];
                    arr[j]= arr[j+1];
                    arr[j+1] = temp;
                }
            }
        }
    }
}
```

Analysis:
- The outer for loops represents the number of swaps that are done for comparison of data.
- The inner loop is actually used to do the comparison of data. At the end of each inner loop iteration, the largest value is moved to the end of the array. In the first iteration the largest value, in the second iteration the second largest and so on.
- *more()* function is used for comparison which means when the value of the first argument is greater than the value of the second argument then perform a swap. By this we are sorting in increasing order if we have, the *less()* function in place of more() than we will get decreasing order sorting.

Have a look into *more()* function in case you forgot

```
private boolean more(int value1, int value2)
{
    return value1 > value2;
}
```

Complexity Analysis:
Each time the inner loop execute for (n-1), (n-2), (n-3)…
(n-1) + (n-2) + (n-3) + ….. + 3 + 2 + 1 = n(n-1)/2

Worst case performance	$O(n^2)$
Average case performance	$O(n^2)$
Space Complexity	*O(1)* as we need only one temp variable
Stable Sorting	Yes

Modified (improved) Bubble-Sort

When there is no more swap in one pass of the outer loop. It indicates that all the elements are already in order so we should stop sorting. This sorting improvement in Bubble-Sort is extremely useful when we know that, except few elements rest of the array is already sorted.

Example 6.2

```
public void sort()
{
    int size = arr.length;
    int i, j, temp, swapped=1;
    for (i = 0; i < (size - 1) && swapped == 1; i++)
    {
        swapped = 0;
        for (j = 0; j < size - i - 1; j++)
        {
            if (more(arr[j], arr[j + 1]))
            {
                /* Swapping */
                temp = arr[j];
                arr[j] = arr[j + 1];
                arr[j + 1] = temp;
                swapped = 1;
            }
        }
    }
}
```

By applying this improvement, best case of this algorithm, when an array is nearly sorted, is improved. Best case is *O(n)*

Complexity Analysis:

Worst case performance	$O(n^2)$
Average case performance	$O(n^2)$
Space Complexity	$O(1)$
Adaptive: When array is nearly sorted	$O(n)$
Stable Sorting	Yes

Insertion-Sort

Insertion-Sort *Time Complexity* is $O(n^2)$ which is same as *Bubble-Sort* but perform a bit better than it. It is the way we arrange our playing cards. We keep a sorted subarray. Each value is inserted into its proper position in the sorted sub-array in the left of it.

5	6	2	4	7	3	1	Insert 5
5	6	2	4	7	3	1	Insert 6
2	5	6	4	7	3	1	Insert 2
2	4	5	6	7	3	1	Insert 4
2	4	5	6	7	3	1	Insert 7
2	3	4	5	6	7	1	Insert 3
1	2	3	4	5	6	7	Insert 1

Example 6.3

```java
public class InsertionSort {
    private int[] arr;
    public InsertionSort(int[] array) {
        arr = array;
    }
    private boolean more(int value1, int value2)
    {
        return value1 > value2;
    }
    public void sort()
    {
        int size = arr.length;

        int temp,j;
        for(int i=1; i<size; i++)
        {
            temp=arr[i];
            for(j=i; j>0 && more(arr[j-1], temp); j--)
            {
                arr[j]=arr[j-1];
            }
            arr[j]=temp;
        }
    }
}
```

```java
public class InsertionSortDemo
{
    public static void main(String[] args)
    {
        int[] array = {9,1,8,2,7,3,6,4,5};
        InsertionSort bs = new InsertionSort(array);
        bs.sort();
        for(int i=0;i<array.length ;i++)
        {
            System.out.print(array[i] + " ");
        }
    }
}
```

Analysis:

- The outer loop is used to pick the value we want to insert into the sorted left array.

- The value we want to insert we have picked and saved in a temp variable.

- The inner loop is doing the comparison using the more() function. The values are shifted to the right until we find the proper position of the temp value for which we are doing this iteration.

- Finally the value is placed into the proper position. In each iteration of the outer loop, the length of the sorted array increase by one. When we exit the outer loop the whole array is sorted.

Complexity Analysis:

Worst case Time Complexity	$O(n^2)$
Best case Time Complexity	O(n)
Average case Time Complexity	$O(n^2)$
Space Complexity	O(1)
Stable sorting	Yes

Selection-Sort

Selection-Sort searches the whole unsorted array and put the largest value at the end of it. This algorithm is having the same *Time Complexity*, but performs better than both bubble and Insertion-Sort as less number of comparisons required. The sorted array is created backward in Selection-Sort.

5	6	2	4	7	3	1	Swap
5	6	2	4	1	3	7	Swap
5	3	2	4	1	6	7	Swap
1	3	2	4	5	6	7	No Swap
1	3	2	4	5	6	7	Swap
1	2	3	4	5	6	7	No Swap
1	2	3	4	5	6	7	

Example 6.4:

```java
public class SelectionSort
{
    private int[] arr;
    public SelectionSort(int[] array) {
        arr = array;
    }
    //back array
    public void sort() {
        int size = arr.length;
        int i, j, max, temp;
        for (i = 0; i < size - 1; i++) {
            max = 0;
            for (j = 1; j < size -1 - i ; j++) {
                if (arr[j] > arr[max])         {
                    max = j;
                }
            }
```

```
                    temp = arr[size - 1 - i];
                    arr[size - 1 - i] = arr[max];
                    arr[max] = temp;
            }
        }
}
```

Analysis:
- The outer loop decide the number of times the inner loop will iterate. For a input of N elements the inner loop will iterate N number of times.
- In each iteration of the inner loop the largest value is calculated and is placed placed at the end of the array.
- This is the final replacement of the maximum value to the proper location. The sorted array is created backward.

Complexity Analysis:

Worst Case Time Complexity	$O(n^2)$
Best Case Time Complexity	$O(n^2)$
Average case Time Complexity	$O(n^2)$
Space Complexity	$O(1)$
Stable Sorting	No

The same algorithm can be implemented by creating the sorted array in the front of the array.

Example 6.5:

```
//front array varient
void sort2() {
    int size = arr.length;
    int i, j, min, temp;
    for (i = 0; i < size - 1; i++){
        min = i;
        for (j = i + 1; j < size; j++) {
            if (arr[j] < arr[min]) {
                min = j;
            }
        }
        temp = arr[i];
        arr[i] = arr[min];
        arr[min] = temp;
    }
}
```

Merge-Sort

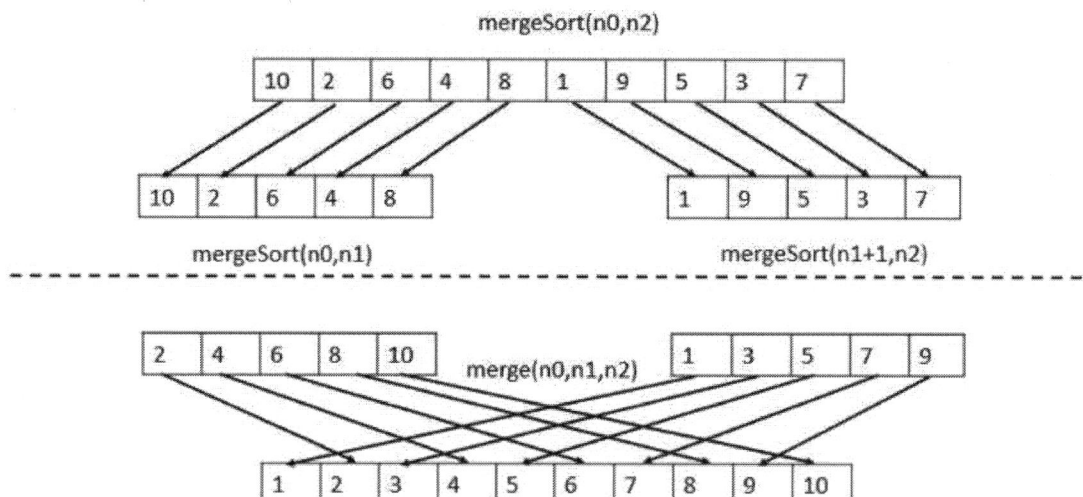

Merge sort divide the input into half recursive in each step. It sort the two parts separately recursively and finally combine the result into final sorted output.

Example 6.6:

```java
public class MergeSort {
    private int[] arr;

    private void merge(int[] arr,int[] tempArray, int lowerIndex, int middleIndex, int upperIndex)
    {
        int lowerStart=lowerIndex;
        int lowerStop=middleIndex;
        int upperStart=middleIndex+1;
        int upperStop=upperIndex;
        int count=lowerIndex;
        while(lowerStart<=lowerStop && upperStart<=upperStop)
        {
            if(arr[lowerStart]<arr[upperStart])
                tempArray[count++]=arr[lowerStart++];
            else
                tempArray[count++]=arr[upperStart++];
        }
        while(lowerStart<=lowerStop)
        {
            tempArray[count++]=arr[lowerStart++];
        }
        while( upperStart<=upperStop)
        {
            tempArray[count++]=arr[upperStart++];
        }
        for(int i=lowerIndex;i<=upperIndex;i++)
            arr[i]=tempArray[i];
    }
}
```

```java
        private void mergeSrt(int[] arr,int[] tempArray, int lowerIndex, int upperIndex)
        {
                if(lowerIndex >= upperIndex)
                        return;
                int middleIndex=(lowerIndex+upperIndex)/2;
                mergeSrt(arr,tempArray,lowerIndex,middleIndex);
                mergeSrt(arr,tempArray,middleIndex+1,upperIndex);
                merge(arr,tempArray,lowerIndex,middleIndex,upperIndex);
        }
```

```java
        public void sort() {
                int size = arr.length;
                int[] tempArray= new int[size];
                mergeSrt(arr,tempArray,0,size-1);
        }

        public MergeSort(int[] array) {
                        arr = array;
        }
}
```

```java
public class MergeSortDemo {

        public static void main(String[] args) {
                int[] array={3,4,2,1,6,5,7,8,1,1};
                MergeSort m = new MergeSort(array);
                m.sort();
                for(int i=0;i<array.length ;i++)
                {
                        System.out.print(array[i] + " ");
                }
        }
}
```

- The *Time Complexity* of Merge-Sort is ***O(nlogn)*** in all 3 cases (best, average and worst) as Merge-Sort always divides the array into two halves and take linear time to merge two halves.
- It requires the equal amount of additional space as the unsorted list. Hence, it is not at all recommended for searching large unsorted lists.
- It is the best Sorting technique for sorting Linked Lists.

Complexity Analysis:

Worst Case Time Complexity	O(nlogn)
Best Case Time Complexity	O(nlogn)
Average Time Complexity	O(nlogn)
Space Complexity	O(n)
Stable Sorting	Yes

Quick-Sort

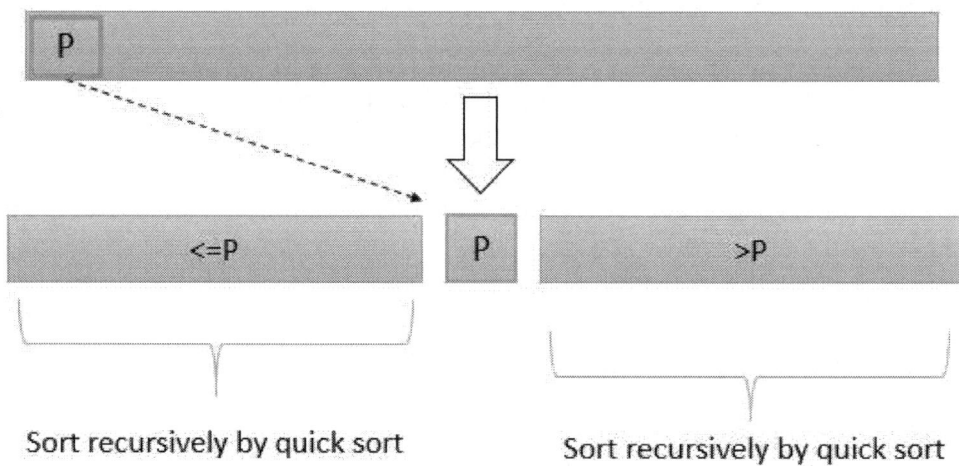

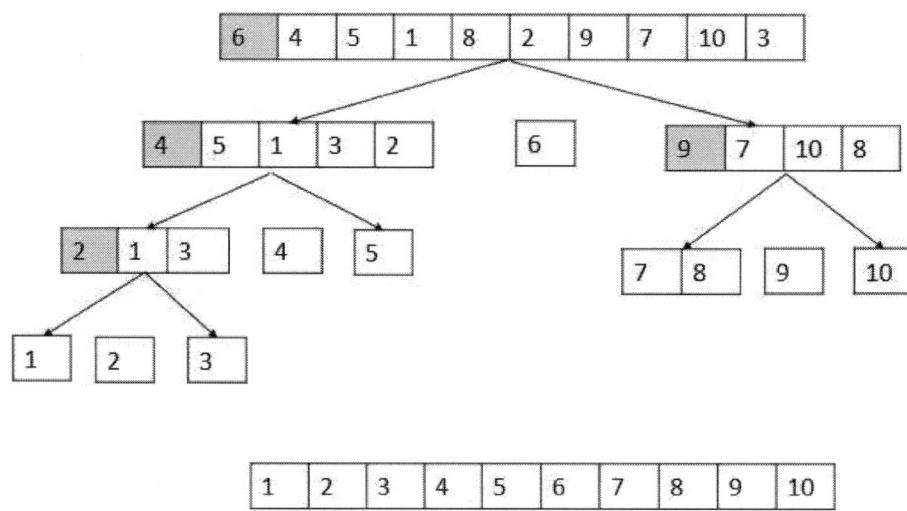

Quick sort is also a recursive algorithm.
- In each step we select a pivot (let us say first element of array).
- Then we traverse the rest of the array and copy all the elements of the array which are smaller than the pivot to the left side of array
- We copy all the elements of the array which are grater then pivot to the right side of the array. Obviously the pivot is at its sorted position.
- Then we sort the left and right subarray separately.
- When the algorithm returns the whole array is sorted.

Example 6.7:

```java
public class QuickSort {
    private int[] arr;
    public QuickSort(int[] array) {
        arr = array;
    }
    private void swap(int arr[], int first, int second){
        int temp = arr[first];
        arr[first] = arr[second];
        arr[second] = temp;
    }
```

```java
    private void quickSortUtil (int arr[], int lower, int upper)
    {
        if (upper<=lower)
            return;
        int pivot = arr[lower];
        int start = lower;
        int stop = upper;

        while ( lower < upper)
        {
            while (arr[lower] <= pivot && lower < upper)
            {
                lower++;
            }
            while (arr[upper] > pivot && lower <= upper)
            {
                upper--;
            }
            if (lower < upper)
            {
                swap(arr,upper,lower);
            }
        }
        swap(arr, upper, start); // upper is the pivot position
        quickSortUtil (arr, start, upper - 1); // pivot -1 is the upper for left sub array.
        quickSortUtil (arr, upper + 1, stop); // pivot + 1 is the lower for right sub array.
    }

    public void sort(){
    int size = arr.length;
    quickSortUtil (arr, 0, size - 1);
    }
}
```

```java
public class QuickSortDemo {
    public static void main(String[] args) {
        int[] array={3,4,2,1,6,5,7,8,1,1};
        QuickSort m = new QuickSort(array);
        m.sort();
        for(int i=0;i<array.length ;i++) {
            System.out.print(array[i] + " ");
        }
    }
}
```

- The space required by Quick-Sort is very less, only **O(nlogn)** additional space is required.
- Quicksort is not a stable sorting technique, so it might change the occurrence of two similar elements in the list while sorting.

Complexity Analysis:

Worst Case Time Complexity	$O(n^2)$
Best Case Time Complexity	O(nlogn)
Average Time Complexity	O(nlogn)
Space Complexity	O(nlogn)
Stable Sorting	No

Quick Select

Quick select is very similar to Quick-Sort in place of sorting the whole array we just ignore the one-half of the array at each step of Quick-Sort and just focus on the region of array on which we are interested.

Example 6.8:

```java
public class QuickSelect {
    public static void swap(int arr[], int first, int second){
        int temp = arr[first];
        arr[first] = arr[second];
        arr[second] = temp;
    }
    public static void quickSelect(int arr[], int lower, int upper,int k)
    {
        if (upper <= lower)
            return;
        int pivot = arr[lower];
        int start = lower;
        int stop = upper;

        while (lower < upper)
        {
            while (arr[lower] <= pivot && lower < upper)
                lower++;
```

```
                while (arr[upper] > pivot && lower <= upper)
                    upper--;

                if (lower < upper)
                    swap(arr, upper, lower);
            }
            swap(arr, upper, start); //upper is the pivot position
            if (k<upper)
                quickSelect(arr, start, upper - 1, k); //pivot -1 is the upper for left sub array.
            if (k>upper)
                quickSelect(arr, upper + 1, stop, k); // pivot + 1 is the lower for right sub array.
        }
```

```
        public static void quickSelect(int arr[], int k){
            quickSelect(arr, 0, arr.length - 1, k);
        }
}
```

```
public class QuickSelectDemo {
    public static void main(String[] args) {
        int[] array={3,4,2,1,6,5,7,8,10,9};
        QuickSelect.quickSelect(array, 5);
        System.out.print("value at index 5 is : "+ array[4]);
    }
}
```

Complexity Analysis:

Worst Case Time Complexity	$O(n^2)$
Best Case Time Complexity	$O(logn)$
Average Time Complexity	$O(logn)$
Space Complexity	$O(nlogn)$

Bucket Sort

Bucket sort is the simplest and most efficient type of sorting. Bucket sort has a strict requirement of a predefined range of data.

Like, sort how many people are in which age group. We know that the age of people can vary between 1 and 130.

| 2 | 6 | 4 | 1 | 5 | 8 | 1 | 4 | 6 | 1 | Input Data

0	1	2	3	4	5	6	7	8	9
0	3	1	0	2	1	2	0	1	0	Counts

| 1 | 1 | 1 | 2 | 4 | 4 | 5 | 6 | 6 | 8 | Sorted Output

Example 6.9:

```
public class BucketSort {
    int[] array ;
    int range;
    int lowerRange;

    public BucketSort (int[] arr,int lowerRange, int upperRange){
        array = arr;
        range = upperRange - lowerRange;
        this.lowerRange = lowerRange;
    }
}
```

```
public void sort()
{
    int i, j;
    int size = array.length;
    int[] count = new int[range];

    for (i = 0; i < range; i++)
    {
        count[i] = 0;
    }

    for (i = 0; i < size; i++)
    {
        count[array[i] - lowerRange]++;
    }

    j = 0;
```

```
            for (i = 0; i < range; i++)
            {
                    for (; count[i]>0; (count[i])--)
                    {
                            array[j++] = i + lowerRange;
                    }
            }
    }
}
```

```
public class BucketSortDemo {
        public static void main(String[] args) {
            int[] array={23,24,22,21,26,25,27,28,21,21};

            BucketSort m = new BucketSort(array,20,30);
            m.sort();
            for(int i=0;i<array.length ;i++)
            {
                    System.out.print(array[i] + " ");
            }
        }
}
```

Analysis:

- We have created a count array to store counts.
- count array elements are initialized to zero.
- Index corresponding to input array is incremented.
- Finally, the information stored in count array is saved in the array.

Complexity Analysis:

Data structure	Array
Worst case performance	O(n+k)
Average case performance	O(n+k)
Worst case Space Complexity	O(k)

Where k - is number of distinct elements.
n – is the total number of elements in array.

Generalized Bucket Sort

There are cases when the element falling into a bucket are not unique but are in the same range. When we want to sort an index of a name, we can use the reference bucket to store names.

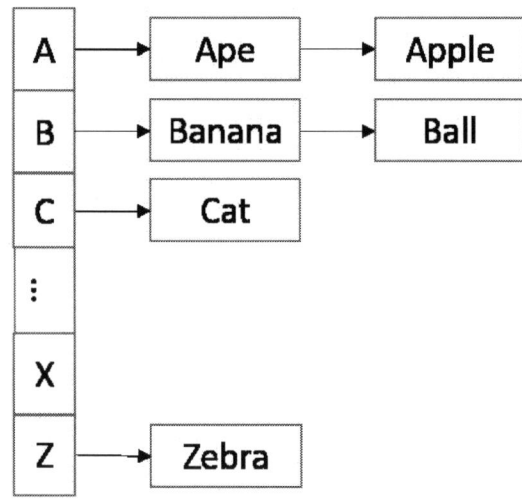

The buckets are already sorted and the elements inside each bucket can be kept sorted by using an Insertion-Sort algorithm. We are leaving this generalized bucket sort implementation to the reader of this book. The similar data structure will be defined in the coming chapter of Hash-Table using separate chaining.

Heap-Sort

Heap-Sort we have already studied in the Heap chapter.

Complexity Analysis:

Data structure	Array
Worst case performance	O(nlogn)
Average case performance	O(nlogn)
Worst case Space Complexity	O(1)

Tree Sorting

In-order traversal of the binary search tree can also be seen as a sorting algorithm. We will see this in binary search tree section of tree chapter.

Complexity Analysis:

Worst Case Time Complexity	$O(n^2)$
Best Case Time Complexity	O(nlogn)
Average Time Complexity	O(nlogn)
Space Complexity	O(n)
Stable Sorting	Yes

External Sort (External Merge-Sort)

When data need to be sorted is huge. Moreover, it is not possible to load it completely in memory (RAM) for such a dataset we use external sorting. Specific data is sorted using external Merge-Sort algorithm. First data are picked in chunks and it is sorted in memory. Then this sorted data is written back to disk. Whole data are sorted in chunks using *Merge-Sort*. Now we need to combine these sorted chunks into final sorted data.

Then we create queues for the data, which will read from the sorted chunks. Each chunk will have its own queue. We will pop from this queue and these queues are responsible for reading from the sorted chunks. Let us suppose we have K different chunks of sorted data each of length M.

The third step is using a Min-Heap, which will take input data from each of this queue. It will take one element from each queue. The minimum value is taken from the Heap and added to the final sorted element output. Then queue from which this min element is inserted in the heap will again popped and one more element from that queue is added to the Heap. Finally, when the data is exhausted from some queue that queue is removed from the input list. Finally, we will get a sorted data came out from the heap.

We can optimize this process further by adding an output buffer, which will store data coming out of Heap and will do a limited number of the write operation in the final Disk space.

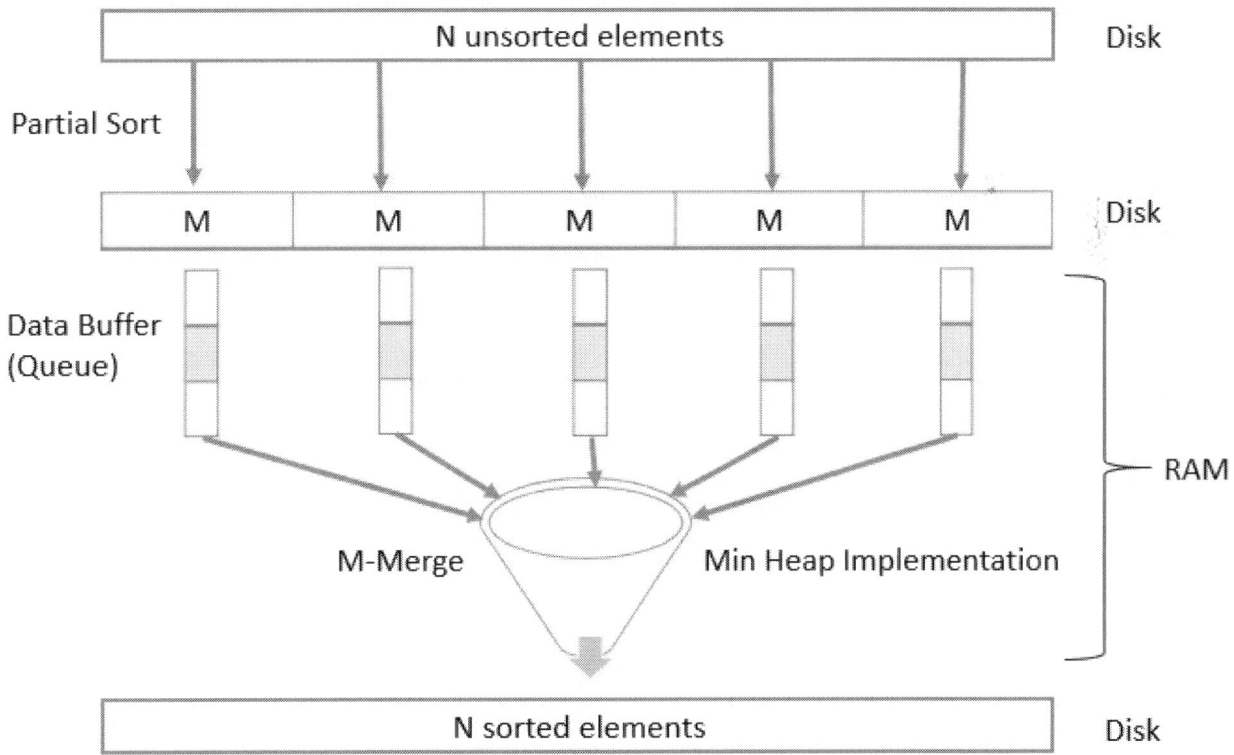

Note: No one will be asking to implement external sorting in an interview, but it is good to know about it.

Comparisons of the various sorting algorithms.

Sort	Average Time	Best Time	Worst Time	Space	Stability
Bubble-Sort	$O(n^2)$	$O(n^2)$	$O(n^2)$	$O(1)$	Stable
Modified Bubble-Sort	$O(n^2)$	$O(n)$	$O(n^2)$	$O(1)$	Stable
Selection-Sort	$O(n^2)$	$O(n^2)$	$O(n^2)$	$O(1)$	Unstable
Insertion-Sort	$O(n^2)$	$O(n)$	$O(n^2)$	$O(1)$	Stable
Heap-Sort	$O(n*\log(n))$	$O(n*\log(n))$	$O(n*\log(n))$	$O(1)$	UnStable
Merge-Sort	$O(n*\log(n))$	$O(n*\log(n))$	$O(n*\log(n))$	$O(n)$	Stable
Quick-Sort	$O(n*\log(n))$	$O(n*\log(n))$	$O(n^2)$	$O(n)$ worst case $O(\log(n))$ average case	Unstable
Bucket Sort	$O(n\ k)$	$O(n\ k)$	$O(n\ k)$	$O(n\ k)$	Stable

Selection of Best Sorting Algorithm

No sorting algorithm is perfect. Each of them has their own pros and cons. Let us read one by one:

Quick-Sort: When you do not need a stable sort and average case performance matters more than worst-case performance. When data is random, we prefer the Quick-Sort. Average case *Time Complexity* of Quick-Sort is ***O(nlogn)*** and worst-case *Time Complexity* is ***O(n²)***. *Space Complexity* of Quick-Sort is ***O(logn)*** auxiliary storage, which is stack space used in recursion.

Merge-Sort: When you need a stable sort and *Time Complexity* of ***O(nlogn)***, Merge-Sort is used. In general, Merge-Sort is slower than Quick-Sort because of lot of copy happening in the merge phase. There are two uses of Merge-Sort when we want to merge two sorted linked lists and Merge-Sort is used in external sorting.

Heap-Sort: When you do not need a stable sort and you care more about worst-case performance than average case performance. It has guaranteed to be ***O(nlogn)***, and uses ***O(1)*** auxiliary space, meaning that you will not unexpectedly run out of memory on very large inputs.

Insertion-Sort: When we need a stable sort, When N is guaranteed to be small, including as the base case of a Quick-Sort or Merge-Sort. Worst-case *Time Complexity* is ***O(n²)***, it has a very small constant, so for smaller input size it performs better than Merge-Sort or Quick-Sort. It is also useful when the data is already pre-sorted in this case its best case running time is ***O(N)***.

Bubble-Sort: Where we know the data is very nearly sorted. Say only two elements are out of place. Then in one pass, Bubble Sort will make the data sorted and in the second pass, it will see everything is sorted and then exit. Only takes 2 passes of the array.

Selection-Sort: Best Worst Average Case running time all $O(n^2)$. It is only useful when you want to do something quick. They can be used when you are just doing some prototyping.

Counting-Sort: When you are sorting data within a limited range.

Radix-Sort: When log(N) is significantly larger than K, where K is the number of radix digits.

Bucket-Sort: When your input is more or less uniformly distributed.

Note: A stable sort is one that has guaranteed not to reorder elements with identical keys.

Exercise

1. Given a text file, print the words with their frequency. Now print the kth word in term of frequency.

    1. *First approach* may be you can use the sorting and return the kth element.
    2. *Second approach*: You can use the kth element quick select algorithm.
    3. *Third approach* You can use Hashtable or Trie to keep track of the frequency. Use Heap to get the Kth element.

2. Given K input streams of number in sorted order. You need to make a single output stream, which contains all the elements of the K streams in sorted order. The input streams support ReadNumber() operation and output stream support WriteNumber() operation.

    1. Read the first number from all the K input streams and add them to a Priority Queue. (Nodes should keep track of the input stream)
    2. Dequeue one element at a time from PQ, Put this element value to the output stream, Read the input stream number and from the same input stream add another element to PQ.
    3. If the stream is empty, just continue
    4. Repeat till PQ is empty.

3. Given K sorted arrays of fixed length M. Also, given a final output array of length M*K. Give an efficient algorithm to merge all the arrays into the final array, without using any extra space.
    Hint: you can use the end of the final array to make PQ.

4. How will you sort 1 PB numbers? 1 PB = 1000 TB.

5. What will be the complexity of the above solution?

6. Any other improvement on question 3 solution if the number of cores is eight.

7. Given an *integer* array that support three function findMin, findMax, findMedian. Sort the array.

8. Given a pile of patient files of High, mid and low priority. Sort these files such that higher priority comes first, then mid and last low priority.
   Hint: Bucket sort.

9. Write pros and cons of Heap-Sort, Merge-Sort and Quick-Sort.

10. Given a rotated - sorted array of N *integer* s. (The array was sorted then it was rotated some arbitrary number of times.) If all the elements in the array were unique the find the index of some value.
    Hint: Modified binary search

11. In the problem 9, what if there are repetitions allowed and you need to find the index of the first occurrence of the element in the rotated-sorted array.

12. Merge two sorted arrays into a single sorted array.
    Hint: Use merge method of Merge-Sort.

13. Given an array contain 0's and 1's, sort the array such that all the 0's come before 1's.

14. Given an array of English characters, sort the array in linear time.

15. Write a method to sort an array of strings so that all the anagrams are next to each other.

> 1. Loop through the array.
> 2. For each word, sort the characters and add it to the hash map with keys as sorted word and value as the original word. At the end of the loop, you will get all anagrams as the value to a key (which is sorted by its constituent chars).
> 3. Iterate over the hashmap, print all values of a key together and then move to the next key.
>   *Space Complexity:* O(n), *Time Complexity:* O(n)

Chapter 7: Linked List

Introduction

Let us suppose we have an array that contains following five elements 1, 2, 4, 5, 6. We want to insert a new element with value "3" in between "2" and "4". In the array, we cannot do so easily. We need to create another array that is long enough to store the current values and one more space for "3". Then we need to copy these elements in the new space. This copy operation is inefficient. To remove this fixed length constraint linked list is used.

Linked List

The linked list is a list of items, called nodes. Nodes have two parts, value part and link part. Value part is used to stores the data. The value part of the node can be either a basic data-type like an integer or it can be some other data-type like an object of some class.
The link part is a reference, which is used to store addresses of the next element in the list.

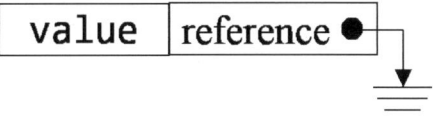

Types of Linked list

There are different types of linked lists. The main difference among them is how their nodes refer to each other.

Singly Linked List

Each node (Except the last node) has a reference to the next node in the linked list. The link portion of node contains the address of the next node. The link portion of the last node contains the value null.

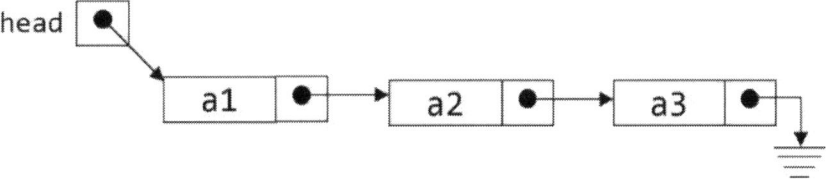

Doubly Linked list

The node in this type of linked list has reference to both previous and the next node in the list.

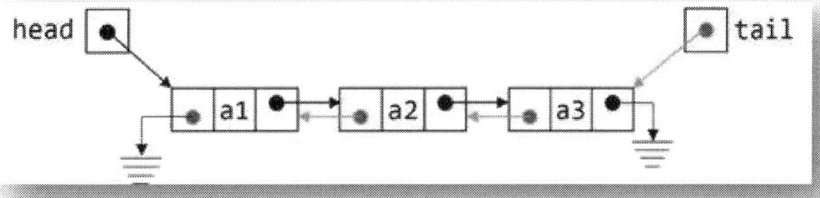

Circular Linked List

This type is similar to the singly linked list except that the last element have reference to the first node of the list. The link portion of the last node contains the address of the first node.

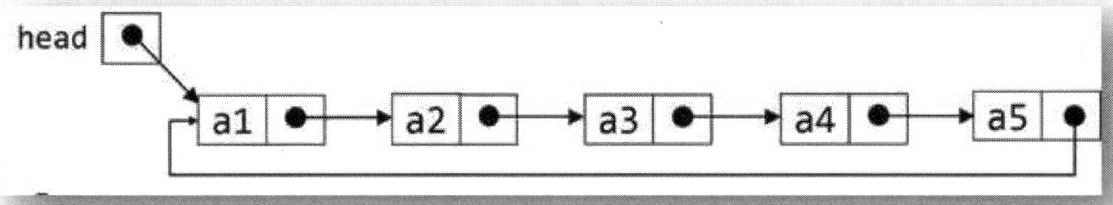

The various parts of linked list

1. Head: Head is a reference that holds the address of the first node in the linked list.
2. Nodes: Items in the linked list are called nodes.
3. Value: The data that is stored in each node of the linked list.
4. Link: Link part of the node is used to store the reference of other node.
    a. We will use "next" and "prev" to store address of next or previous node.

Singly Linked List

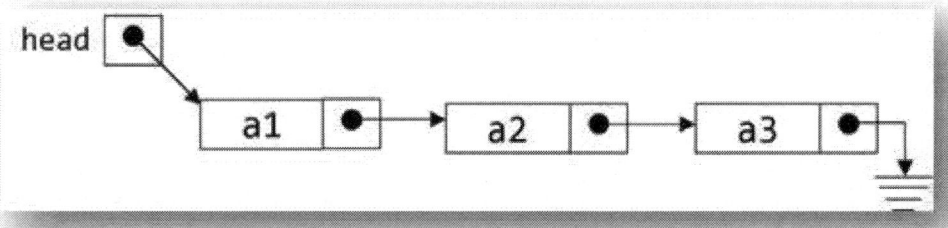

Let us look an example of Node, in this example, the value is of type int, but it can be of some other data-type. The link is named as next in the below class definition.

```
private static class Node {
    private int value;
    private Node next;
    public Node( int v, Node n){
        value = v;
        next = n;
    }
    public Node( int v)  {
        value = v;
        next = null;
    }
}
```

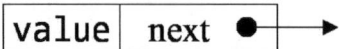

Note: For a singly linked, we should always test these three test cases before saying that the code is good to go. This one node and zero node case is used to catch boundary cases. It is always to take care of these cases before submitting code to the reviewer.
- Zero element / Empty linked list.
- One element / Just single node case.
- General case.

The various basic operations that we can perform on linked lists, many of these operations require list traversal:
- Insert an element in the list, this operation is used to create a linked list.
- Print various elements of the list.
- Search an element in the list.
- Delete an element from the list.
- Reverse a linked list.

You cannot use Head to traverse a linked list because if we use the head, then we lose the nodes of the list. We have to use another reference variable of same data-type as the head.

```java
public class LinkedList {
    private static class Node {
        private int value;
        private Node next;
        public Node( int v, Node n){
            value = v;
            next = n;
        }
    }
    private Node head;
    private int size = 0;
    //Other Methods.
}
```

Size of List

```java
public int size(){
    return size;
}
```

IsEmpty function

```java
public boolean isEmpty(){
    return size == 0;
}
```

Insert element in linked list

An element can be inserted into a linked list in various orders. Some of the example cases are mentioned below:

1. Insertion of an element at the start of linked list
2. Insertion of an element at the end of linked list
3. Insertion of an element at the Nth position in linked list
4. Insert element in sorted order in linked list

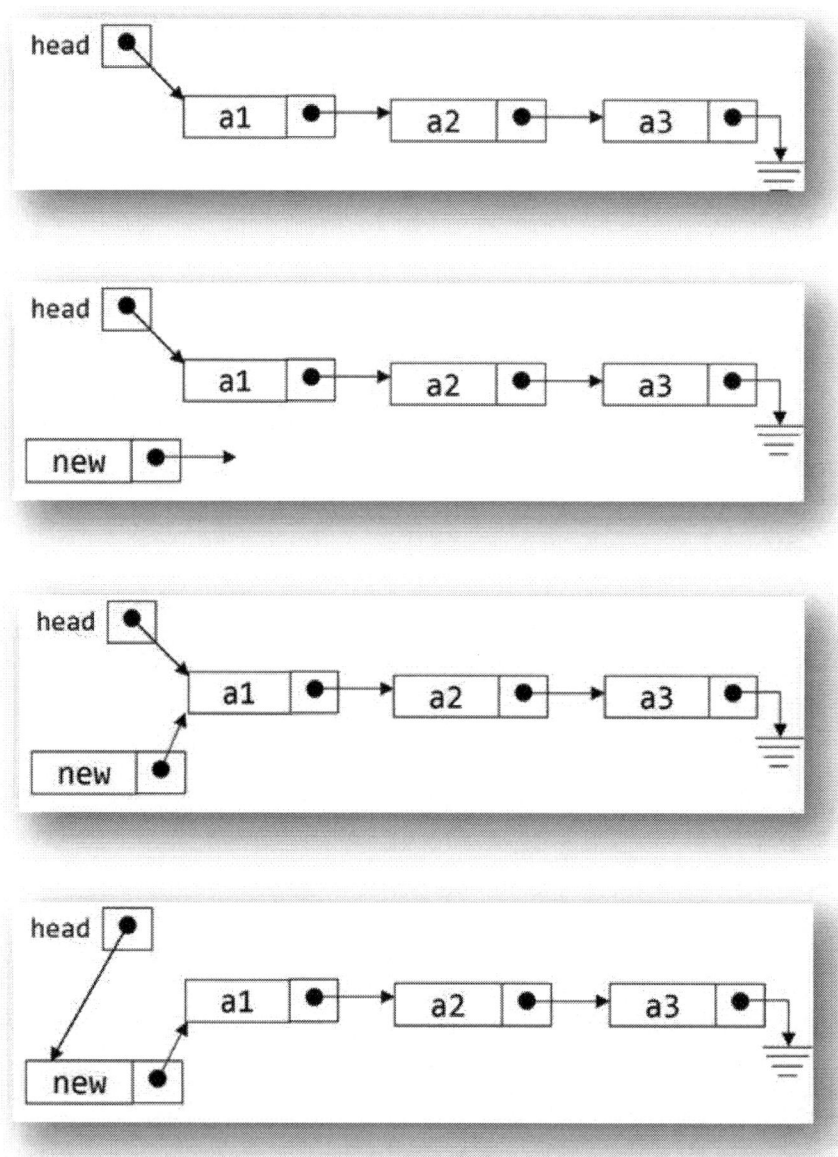

Insert element at the Head

```
public void addHead(int value)
{
    head = new Node(value, head);
    size++;
}
```

Analysis:
- We need to create a new node with the value passed to the function as argument.

- While creating the new node the reference stored in head is passed as argument to Node() constructor so that the next reference will start pointing to the node or null which is referenced by the head node.
- The newly created node will become head of the linked list.
- Size of the list is increased by one.

Insertion of an element at the end

Example 7.5: Insertion of an element at the end of linked list

```
public void addTail(int value) {
        Node newNode = new Node(value, null);
        Node curr=head;

        if(head==null)
                head=newNode;

        while(curr.next != null)
        {
                curr=curr.next;
        }
        curr.next = newNode;
}
```

Analysis:
- New node is created and the value is stored inside it. If the list is empty. Next of new node is null.
- If list is empty then head will store the reference to the newly created node.
- If list is not empty then we will traverse until the end of the list.
- Finally, new node is added to the end of the list.

Note: This operation is un-efficient as each time you want to insert an element you have to traverse to the end of the list. Therefore, the complexity of creation of the list is n^2. So how to make it efficient we have to keep track of the last element by keeping a tail reference. Therefore, if it is required to always insert element at the end of linked list, then we will keep track of the tail reference also.

Traversing Linked List

Example 7.2: Print various elements of a linked list

```
public void print(){
        Node temp = head;
        while(temp != null){
                System.out.print(temp.value+" ");
                temp = temp.next;
        }
}
```

Analysis:
We will store the reference of head in a temporary variable temp.
We will traverse the list by printing the content of list and always incrementing the temp by pointing to its next node.

Complete code for list creation and printing the list.

Example 7.3:

```
public class LinkedListDemo {
    public static void main(String[] args) {
        LinkedList ll = new LinkedList();
        ll.addHead(1);
        ll.addHead(2);
        ll.addHead(3);
        ll.addHead(1);
        ll.addHead(2);
        ll.addHead(3);
        ll.print();
    }
}
```

Analysis:
New instance of linked list is created.
Various elements are added to list by calling addHead() method.
Finally all the content of list is printed to screen by calling print() method.

Sorted Insert

Insert an element in sorted order in linked list given Head reference

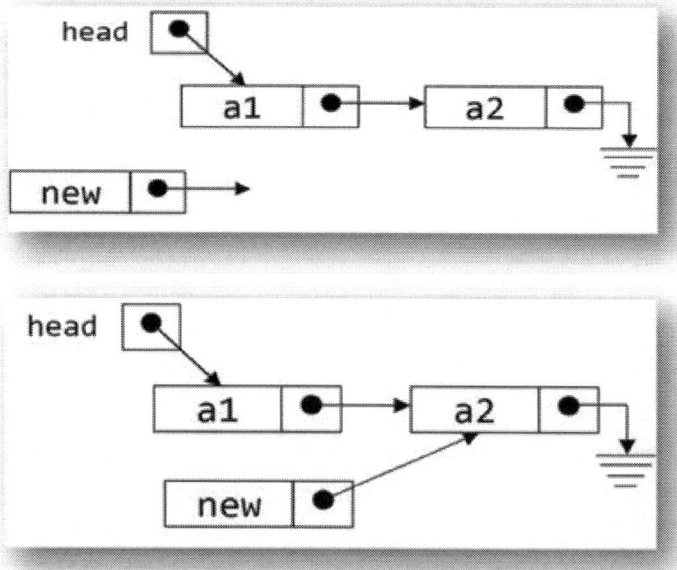

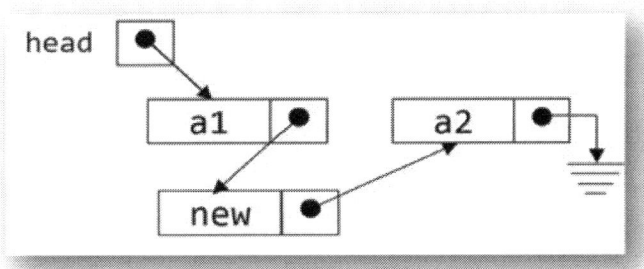

Example 7.4:

```
public void sortedInsert(int value){
    Node newNode = new Node(value, null);
    Node curr = head;

    if(curr == null || curr.value > value){
        newNode.next = head;
        head = newNode;
        return;
    }
    while(curr.next != null && curr.next.value < value){
        curr = curr.next;
    }

    newNode.next = curr.next;
    curr.next =newNode;
}
```

Analysis:
- Head of the list is stored in curr.
- A new empty node of the linked list is created. And initialized by storing an argument value into its value. Next of the node will point to null.
- It checks if the list was empty or if the value stored in the first node is greater than the current value. Then this new created node will be added to the start of the list. And head need to be modified.
- We iterate through the list to find the proper position to insert the node.
- Finally, the node will be added to the list.

Search Element in a Linked-List

Search element in linked list. Given a head reference and value. Returns true if value found in list else returns false.

Note: Search in a single linked list can be only done in one direction. Since all elements in the list have reference to the next item in the list. Therefore, traversal of linked list is linear in nature.

Example 7.5:

```
public boolean isPresent(int data){
    Node temp = head;
    while(temp != null){
        if(temp.value == data)
            return true;
        temp = temp.next;
    }
    return false;
}
```

Analysis:
- We create a temp variable which will point to head of the list.
- Using a while loop we will iterate through the list.
- Value of each element of list is compared with the given value. If value is found, then the function will return true.
- If the value is not found, then false will be returned from the function in the end.

Delete element from the linked list

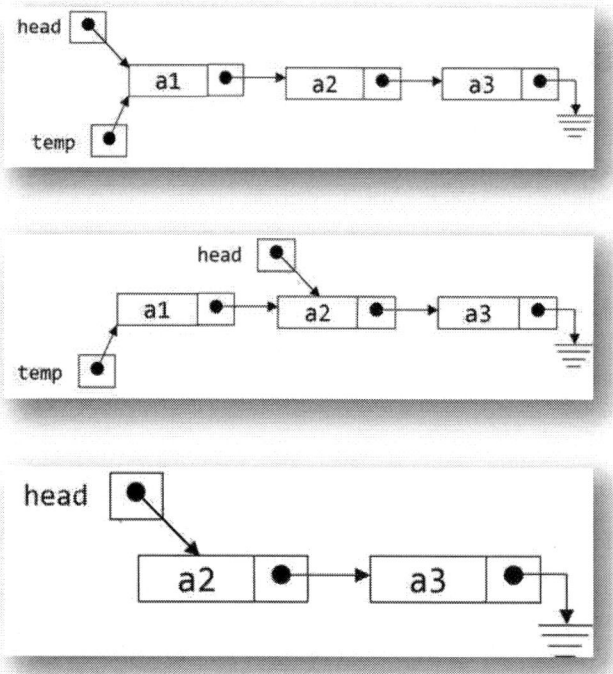

Delete First element in a linked list.

Example 7.6:

```
public int removeHead() throws IllegalStateException {
    if(isEmpty())
        throw new IllegalStateException("EmptyListException");
    int value = head.value;
    head = head.next;
    size--;
    return value;
}
```

Analysis:
- First, we need to check if the list is already empty. If list is already empty then throw EmptyListException.
- If list is not empty then store the value of head node in a temporary variable value.
- We need to find the second element of the list and assign it as head of the linked list.
- Since the first node is no longer referenced so it will be automatically deleted.
- Decrease the size of list. And return the value stored in temporary variable value.

Delete node from the linked list given its value.

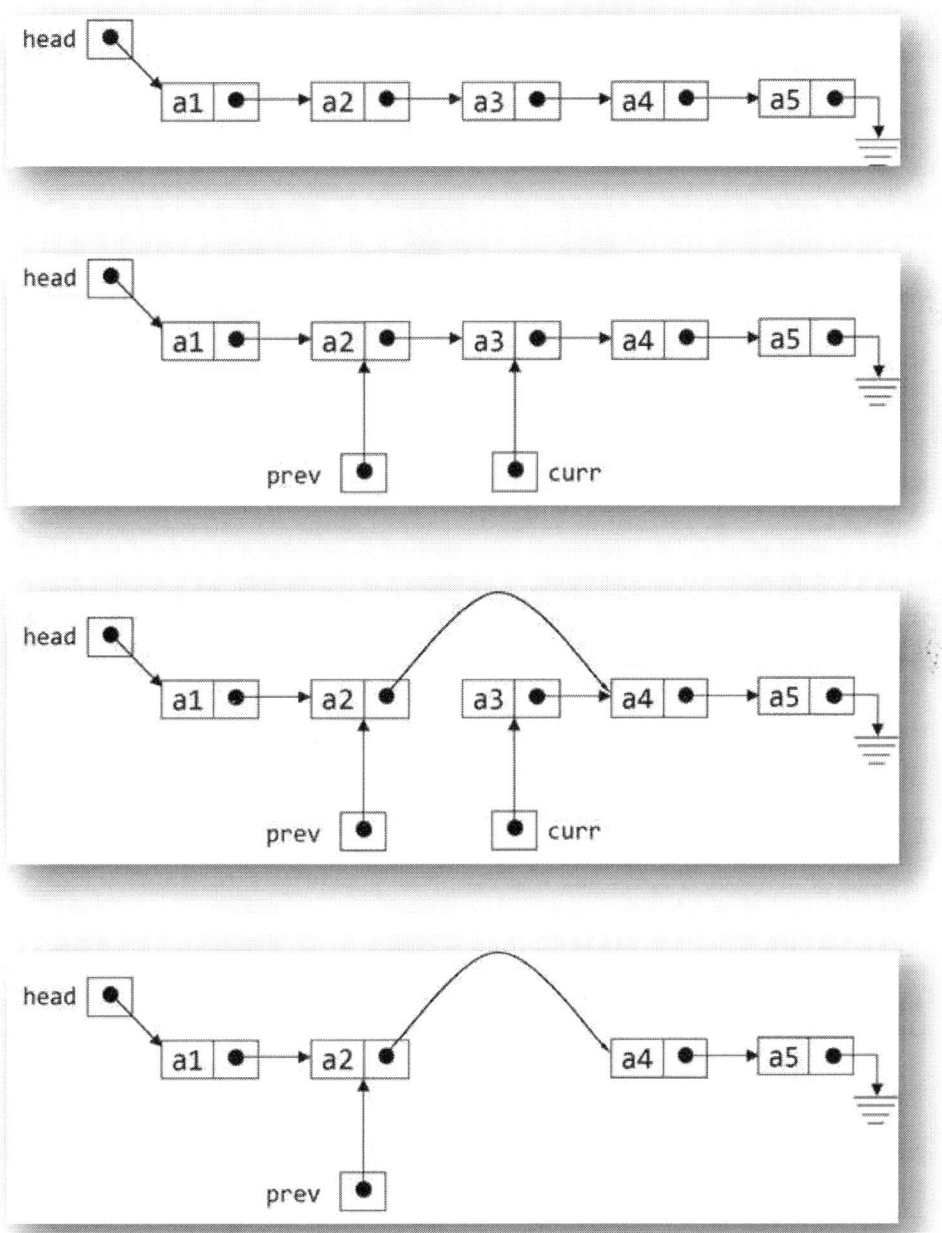

Example 7.7:
```
public boolean deleteNode(int delValue){
        Node temp = head;

        if(isEmpty())
              return false;

        if(delValue == head.value){
              head=head.next;
              size--;
              return true;
        }

        while(temp.next != null){
              if(temp.next.value == delValue){
                    temp.next = temp.next.next;
                    size--;
                    return true;
              }
              temp = temp.next;
        }
        return false;
}
```

Analysis:
- If the list is empty then we will return false from the function which indicate that the deleteNode() method executed with error.
- If the node that need to be deleted is head node than head reference need to be modified and point to the next node.
- In a while loop we will traverse the link list and try to find the node that need to be deleted. If the node is found that we will point its reference to the node next to it. And return true.
- If the node is not found then we will return false.

Delete all the occurrence of particular value in linked list.

Example 7.8:
```
public void deleteNodes(int delValue)
{
        Node currNode = head;
        Node nextNode;

        while(currNode != null && currNode.value == delValue)/*first node */
        {
              head = currNode.next;
              currNode = head;
        }
```

```
        while(currNode != null)
        {
                nextNode = currNode.next;
                if(nextNode != null && nextNode.value == delValue)
                {
                        currNode.next = nextNode.next;
                }
                else
                {
                        currNode = nextNode;
                }
        }
}
```

Analysis:
- In the first while loop will delete all the nodes that are at the front of the list, which have valued equal to delValue. In this, we need to update head of the list.
- In the second while loop, we will be deleting all the nodes that are having value equal to the delValue. Remember that we are not returning even though we have the node that we are looking for.

Delete a single linked list

Delete all the elements of a linked list, given a reference to head of linked list.

Example 7.9:
```
public void freeList(){
      head = null;
      size = 0;
}
```

Analysis:
We just need to point head to null. The reference to the list is lost so it will automatically deleted.

Reverse a linked list.

Reverse a singly linked List iteratively using three Pointers

Example 7.10:
```
public void reverse(){
      Node curr=head;
      Node prev=null;
      Node next=null;
      while(curr!=null)
      {
              next = curr.next;
              curr.next = prev;
              prev=curr;
              curr=next;
      }
      head = prev;
}
```

Analysis:
The list is iterated. Make next equal to the next node of the curr node. Make curr node's next point to prev node. Then iterate the list by making prev point to curr and curr point to next.

Recursively Reverse a singly linked List

Example 7.11: Recursively Reverse a singly linked List Arguments are current node and its next value.

```
public Node reverseRecurseUtil (Node currentNode, Node nextNode){
        Node ret;
        if(currentNode == null)
                return null;
        if(currentNode.next == null)
        {
                currentNode.next = nextNode;
                return currentNode;
        }

        ret= reverseRecurseUtil (currentNode.next, currentNode);
        currentNode.next = nextNode;
        return ret;
}
public void reverseRecurse() {
        head=reverseRecurseUtil(head,null);
}
```

Analysis:
- ReverseRecurse function will call a reverseRecurseUtil function to reverse the list and the reference returned by the reverseRecurseUtil will be the head of the reversed list.
- The current node will point to the nextNode that is previous node of the old list.

Note: A linked list can be reversed using two approaches the one approach is by using three references. The *Second approach* is using recursion both are linear solution, but three-reference solution is more efficient.

Remove duplicates from the linked list

Remove duplicate values from the linked list. The linked list is sorted and it contains some duplicate values, you need to remove those duplicate values. (You can create the required linked list using SortedInsert() function)

Example 7.12:

```
public void removeDuplicate() {
        curr = head;
        while(curr!= null)   {
                if(curr.next!=null && curr.value == curr.next.value)
                        curr.next = curr.next.next;
                else
                        curr = curr.next;
        }
}
```

Analysis: While loop is used to traverse the list. Whenever there is a node whose value is equal to the next node's value, that current node next will point to the next of next node. Which will remove the next node from the list.

Copy List Reversed

Copy the content of linked list in another linked list in reverse order. If the original linked list contains elements in order 1,2,3,4, the new list should contain the elements in order 4,3,2,1.

Example 7.13:

```
public LinkedList CopyListReversed() {
        Node tempNode = null;
        Node tempNode2 = null;
        Node curr = head;
        while (curr != null) {
                tempNode2 = new Node(curr.value, tempNode);
                curr = curr.next;
                tempNode = tempNode2;
        }
        LinkedList ll2 = new LinkedList();
        ll2.head = tempNode;
        return ll2;
}
```

Analysis: Traverse the list and add the node's value to the new list. Since the list is traversed in the forward direction and each node's value is added to another list so the formed list is reverse of the given list.

Copy the content of given linked list into another linked list

Copy the content of given linked list into another linked list. If the original linked list contains elements in order 1,2,3,4, the new list should contain the elements in order 1,2,3,4.

Example 7.14:

```
public LinkedList copyList() {
        Node headNode = null;
        Node tailNode = null;
        Node tempNode = null;
        Node curr = head;

        if (curr == null)
                return null;

        headNode = new Node(curr.value, null);
        tailNode = headNode;
        curr = curr.next;
```

```
        while (curr != null) {
                tempNode = new Node(curr.value, null);
                tailNode.next = tempNode;
                tailNode = tempNode;
                curr = curr.next;
        }
        LinkedList ll2 = new LinkedList();
        ll2.head = headNode;
        return ll2;
}
```

Analysis: Traverse the list and add the node's value to new list, but this time always at the end of the list. Since the list is traversed in the forward direction and each node's value is added to the end of another list. Therefore, the formed list is same as the given list.

Compare List

Example 7.15: Compare two list given

```
public boolean compareList(LinkedList ll) {
        return compareList(head, ll.head);
}
```

```
public boolean compareList(Node head1, Node head2) {
        if( head1==null && head2==null )
                return true;
        else if( (head1==null) || (head2==null) || (head1.value!=head2.value))
                return false;
        else
                return compareList(head1.next,head2.next);
}
```

Analysis:
- List is compared recursively. Moreover, if we reach the end of the list and both the lists are null. Then both the lists are equal and so return true.
- List is compared recursively. If either one of the list is empty or the value of corresponding nodes is unequal, then this function will return false.
- Recursively calls compare list function for the next node of the current nodes.

Find Length

Example 7.16: Find the length of given linked list.

```
public int findLength() {
        Node curr=head;
        int count = 0;
        while (curr!=null)   {
                count++;
                curr = curr.next;
        }
        return count;
}
```

Analysis: Length of linked list is found by traversing the list till we reach the end of list

Nth Node from Beginning

Example 7.17: : Find Nth node from beginning

```java
public int nthNodeFromBegining(int index) {
    if (index > size() || index < 1)
        return Integer.MAX_VALUE;
    int count = 0;
    Node curr = head;
    while (curr != null && count < index - 1) {
        count++;
        curr = curr.next;
    }
    return curr.value;
}
```

Analysis: Nth node can be found by traversing the list N-1 number of time and then return the node. If list does not have N elements method return null.

Nth Node from End

Example 7.18: Find Nth node from end

```java
public int nthNodeFromEnd(int index) {
    int size = findLength();
    int startIndex;
    if (size != 0 && size < index) {
        return Integer.MAX_VALUE;
    }
    startIndex = size - index + 1;
    return nthNodeFromBegining(startIndex);
}
```

Analysis: First find the length of list, then nth node from end will be (length – nth +1) node from the beginning.

Example 7.19:

```java
public int nthNodeFromEnd2(int index) {
    int count = 1;
    Node forward = head;
    Node curr = head;
    while (forward != null && count <= index ) {
        count++;
        forward = forward.next;
    }

    if (forward == null)
        return Integer.MAX_VALUE;

    while (forward != null) {
        forward = forward.next;
        curr = curr.next;
    }
    return curr.value;
}
```

Analysis: Second approach is to use two references one is N steps / nodes ahead of the other when forward reference reach the end of the list then the backward reference will point to the desired node.

Loop Detect

1. Traverse through the list.
2. If the current node is, not there in the Hash-Table then insert it into the Hash-Table.
3. If the current node is already in the Hashtable then we have a loop.

Loop Detect

We have to find if there is a loop in the linked list. There are two ways to find if there is a loop in a linked list. One way is called "Slow reference and fast reference approach (SPFP)" the other is called "Reverse list approach". Both approaches are linear in nature, but still in SPFP approach, we do not require to modify the linked list so it is preferred.

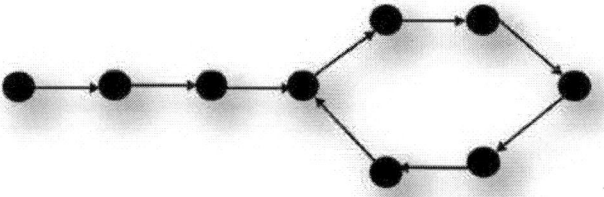

Find if there is a loop in a linked list. If there is a loop, then return 1 if not, then return 0. Use slow reference fast reference approach.

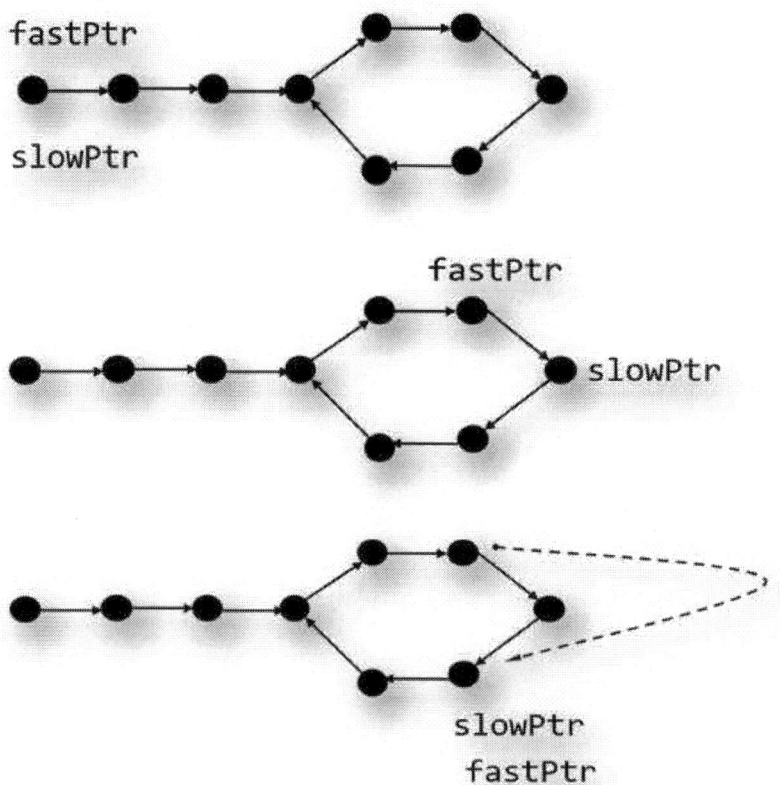

Example 7.20: Find if there is a loop in a linked list. If there is a loop, then return true if not, then return false.

```java
public boolean loopDetect()
{
    Node slowPtr;
    Node fastPtr;
    slowPtr = fastPtr = head;

    while(fastPtr.next != null && fastPtr.next.next != null)
    {
        slowPtr = slowPtr.next;
        fastPtr = fastPtr.next.next;
        if(slowPtr == fastPtr)
        {
            System.out.println("loop found" );
            return true;
        }
    }
    System.out.println("loop not found" );
    return false;
}
```

Analysis:
- The list is traversed with two references, one is slow reference and another is fast reference. Slow reference always moves one-step. Fast reference always moves two steps. If there is no loop, then control will come out of while loop. So return false.
- If there is a loop, then there came a point in a loop where the fast reference will come and try to pass slow reference and they will meet at a point. When this point arrives, we come to know that there is a loop in the list. So return true.

Reverse List Loop Detect

Example 7.21: Find if there is a loop in a linked list. Use reverse list approach.

```java
public boolean  reverseListLoopDetect()
{
    Node  tempHead=head;
    reverse();
    if(tempHead==head)
    {
        reverse();
        System.out.println("loop found" );
        return true;
    }
    else
    {
        reverse();
        System.out.println("loop not found" );
        return false;
    }
}
```

Analysis:
- Store reference of the head of list in a temp variable.
- Reverse the list
- Compare the reversed list head reference to the current list head reference.
- If the head of reversed list and the original list are same then reverse the list back and return true.
- If the head of the reversed list and the original list are not same, then reverse the list back and return false. Which means there is no loop.

Loop Type Detect

Find if there is a loop in a linked list. If there is no loop, then return 0, if there is loop return 1, if the list is circular then 2. Use slow reference fast reference approach.

Example 7.22:

```java
public int loopTypeDetect()
{
        Node slowPtr;
        Node fastPtr;
        slowPtr = fastPtr = head;

        while(fastPtr.next != null && fastPtr.next.next != null)
        {
                if(head == fastPtr.next || head==fastPtr.next.next)
                {
                        System.out.println("circular list loop found" );
                        return 2;
                }
                slowPtr = slowPtr.next;
                fastPtr = fastPtr.next.next;
                if(slowPtr == fastPtr)
                {
                        System.out.println("loop found" );

                        return 1;
                }
        }
        System.out.println("loop not found" );
        return 0;
}
```

Analysis: This program is same as the loop detect program only if it is a circular list than the fast reference reach the slow reference at the head of the list this means that there is a loop at the beginning of the list.

Remove Loop

Example 7.23: Given there is a loop in linked list remove the loop.

```java
public Node loopPointDetect()
{
    Node slowPtr;
    Node fastPtr;
    slowPtr = fastPtr = head;

    while(fastPtr.next != null &&
                fastPtr.next.next != null)
    {
        slowPtr=slowPtr.next;
        fastPtr=fastPtr.next.next;
        if(slowPtr==fastPtr)
        {
            return slowPtr;
        }
    }
    return null;
}
```

```java
public void removeLoop()
{
    Node loopPoint = loopPointDetect();
    if(loopPoint == null)
        return ;

    Node firstPtr = head;
    if(loopPoint == head)  // circular list case.
    {
        while(firstPtr.next != head)
            firstPtr = firstPtr.next;
        firstPtr.next = null;
        return;
    }

    Node secondPtr = loopPoint;
    while(firstPtr.next != secondPtr.next) // general loop case.
    {
        firstPtr = firstPtr.next;
        secondPtr = secondPtr.next;
    }
    secondPtr.next = null;
}
```

Analysis:
- Loop through the list by two reference, one fast reference and one slow reference. Fast reference jumps two nodes at a time and slow reference jump one node at a time. The point where these two reference intersect is a point in the loop.
- If that intersection point is head of the list, this is a circular list case and you need to again traverse through the list and make the node before head point to null.
- In the other case you need to use two reference variable one start from head and another start form loop point. They both will meet at the point of loop. (You can mathematically prove it ;))

Find Intersection

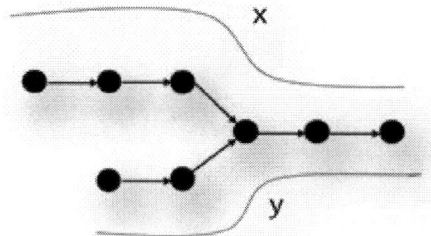

Example 7.24:

```
public Node findIntersection(Node head, Node head2)
{
    int l1 = 0;
    int l2 = 0;
    Node tempHead = head;
    Node tempHead2 = head2;
    while(tempHead != null)
    {
        l1++;
        tempHead = tempHead.next;
    }
    while(tempHead2 != null)
    {
        l2++;
        tempHead2 = tempHead2.next;
    }

    int diff;
    if(l1<l2)
    {
        Node temp = head;
        head = head2;
        head2 = temp;
        diff = l2-l1;
    }
    else
        diff = l1-l2;

    for(;diff>0;diff--)
        head = head.next;

    while(head!=head2)
    {
        head = head.next;
        head2 = head2.next;
    }

    return head;
}
```

Analysis: Find length of both the lists. Find the difference of length of both the lists. Increment the longer list by diff steps, and then increment both the lists and get the intersection point.

Doubly Linked List

In a Doubly Linked list, there are two references in each node. These references are called prev and next. The prev reference of the node will point to the node before it and the next reference will point to the node next to the given node.

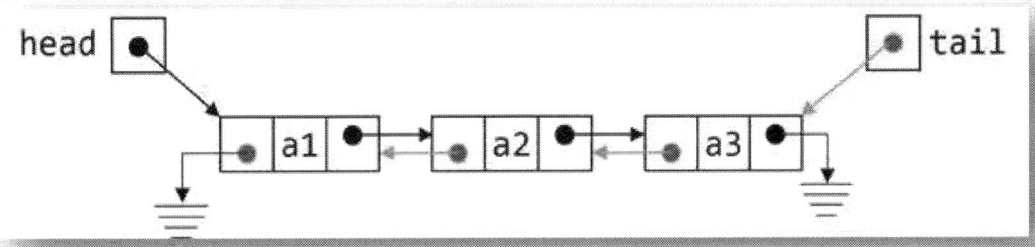

Let us look an example of Node, in this example, the value is of type int, but it can be of some other data-type. The two link references are prev and next.

Example 7.25:

```
private static class Node {
        private int value;
        private Node next;
        private Node prev;

        public Node( int v, Node nxt, Node prv)  {
                value = v;
                next = nxt;
                prev = prv;
        }

        public Node( int v) {
                value = null;
                next = null;
                prev = null;
        }
}
```

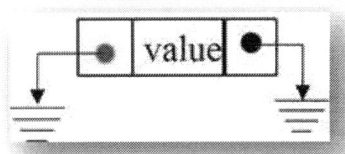

Search in a single linked list can be only done in one direction. Since all elements in the list has reference to the next item in the list. Therefore, traversal of linked list is linear in nature.

In a doubly linked list, we keep track of both head of the linked list and tail of linked list.

For a doubly linked list linked list, there are few cases that we need to keep in mind while coding:
- Zero element case (head and tail both can be modified)
- Only element case (head and tail both can be modified)
- First element (head can be modified)
- General case
- The last element (tail can be modified)

Note: Any program which is likely to change head reference or tail reference is to be passed as a double reference, which is pointing to head or tail reference.

Basic operations of Linked List

Basic operation of a linked list requires traversing a linked list. The various operations that we can perform on linked lists, many of these operations require list traversal:
- Insert an element in the list, this operation is used to create a linked list.
- Print various elements of the list.
- Search an element in the list.
- Delete an element from the list.
- Reverse a linked list.

For any linked list there are only three cases zero element, one element, and generally
For doubly linked list, we have a few more things
1. null values (head and tail both can be modified)
2. Only element (head and tail both can be modified)
3. First element (head can be modified)
4. General case
5. Last element (tail can be modified)

Example 7.26:
```
public class DoublyLinkedList {

    private static class Node {
        ..................
    }
    private Node head;
    private Node tail;
    private int size = 0;
}
```

Example 7.27:
```
public int size(){
    return size;
}
```

Example 7.28:
```
public boolean isEmpty(){
    return size == 0;
}
```

Example 7.29:
```
public int peek(){
    if(isEmpty())
        throw new IllegalStateException("EmptyListException");
    return head.value;
}
```

Insert at Head

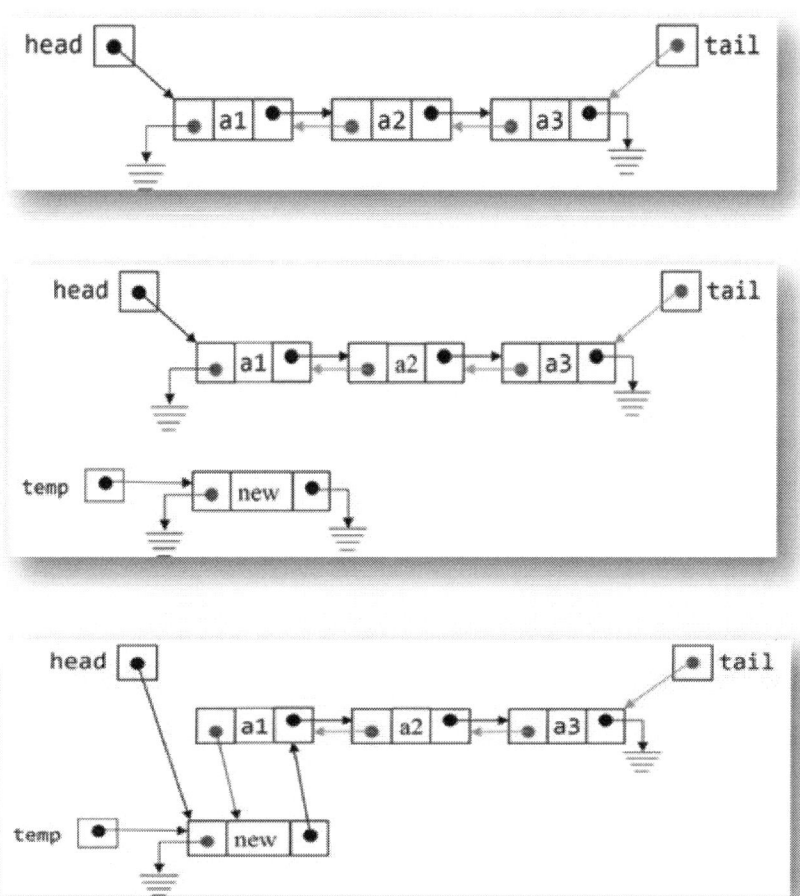

Example 7.30:
```
public void addHead(int value){
    Node newNode = new Node(value, null, null);
    if(size == 0)
        tail = head = newNode;
    else{
        head.prev = newNode;
        newNode.next = head;
        head = newNode;
    }
    size++;
}
```

Analysis: Insert in double linked list is same as insert in a singly linked list.
- Create a node assign null to prev reference of the node.

- If the list is empty then tail and head will point to the new node.
- If the list is not empty then prev of head will point to newNode and next of newNode will point to head. Then head will be modified to point to newNode.

Insert at Tail

Example 7.31: Insert an element at the end of the list.

```java
public void addTail(int value){
    Node newNode = new Node(value, null, null);
    if(size == 0)
        head = tail = newNode;
    else {
        newNode.prev = tail;
        tail.next = newNode;
        tail = newNode;
    }
    size++;
}
```

Analysis: Find the proper location of the node and add it to the list. Manage next and prev reference of the node so that list always remain double linked list.

Remove Head of doubly linked list

Example 7.32:

```java
public int removeHead(){
    if(isEmpty())         // empty list case.
        throw new IllegalStateException("EmptyListException");

    int value = head.value;
    head = head.next;

    if(head == null)      // single node case.
        tail = null;
    else
        head.prev = null;

    size--;
    return value;
}
```

Analysis:
- If the list is empty then EmptyListException will be raised.
- Now head will point to its next.
- If head is null then this was single node list case tail also need to be made null.
- In all the general case head. Prev will be set to null.
- Size of list will be reduced by one and value of node is returned.

Delete a node given its value

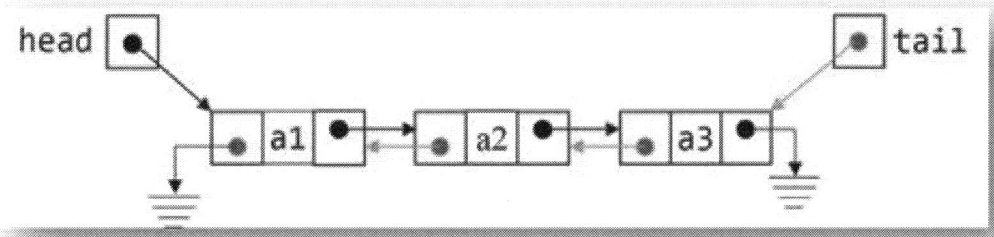

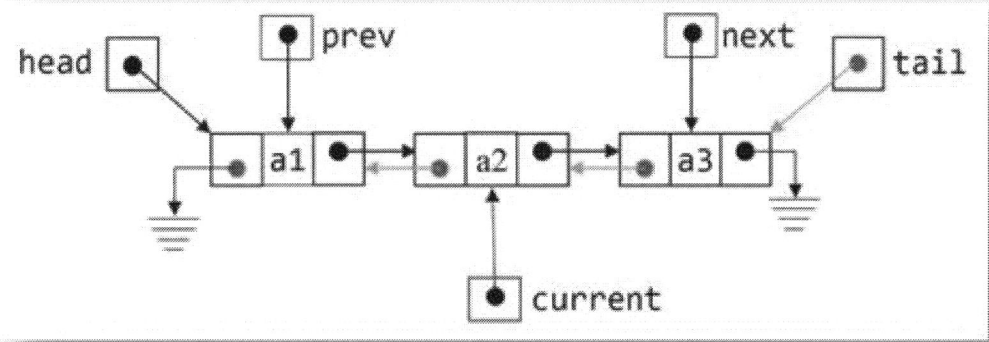

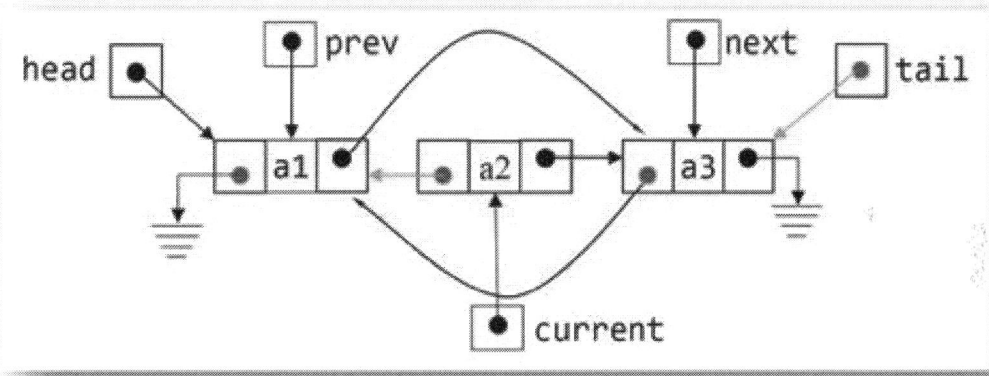

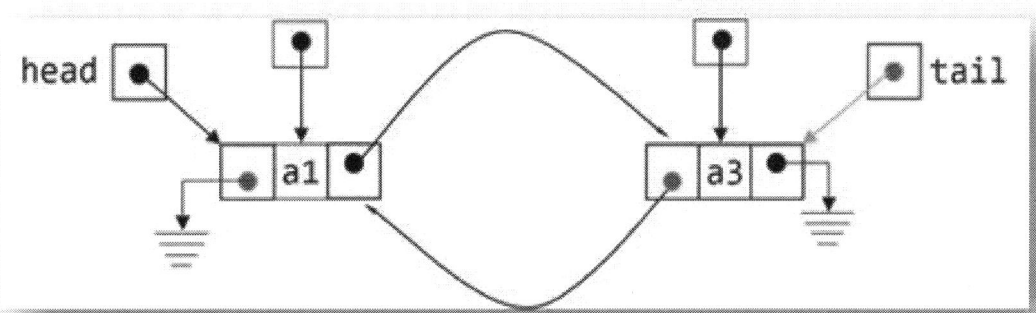

Example 7.33: Delete node in linked list

```java
public boolean removeNode(int key){
    Node curr = head;

    if(curr == null) //empty list
        return false;

    if(curr.value == key)//head is the node with value key.
    {
        head = head.next;
        size--;
        if(head == null)
            head.prev = null;
        else
            tail = null; // only one element in list.

        return true;
    }

    while(curr.next != null){
        if(curr.next.value == key)
        {
            curr.next = curr.next.next;
            if(curr.next == null)//last element case.
                tail = curr;
            else
                curr.next = curr;
            size--;
            return true;
        }
        curr = curr.next;
    }
    return false;
}
```

Analysis: Traverse the list find the node which need to be deleted. Then remove it and adjust next reference of the node previous to it and prev reference of the node next to it.

Search list

Example 7.34:

```java
public boolean isPresent(int key){
    Node temp = head;
    while(temp != null){
        if(temp.value == key)
            return true;
        temp = temp.next;
    }
    return false;
}
```

Analysis: Traverse the list and find if some value is resent or not.

Free List

Example 7.35:
```java
public void freeList(){
    head = null;
    tail = null;
    size = 0;
}
```
Analysis: Just head and tail references need to point to null. The rest of the list will automatically deleted by garbage collection.

Print list

Example 7.36:
```java
public void print(){
    Node temp = head;
    while(temp != null){
        System.out.print(temp.value + " ");
        temp = temp.next;
    }
}
```
Analysis: Traverse the list and print the value of each node.

Reverse a doubly linked List iteratively

Example 7.37:
```java
public void reverseList()
{
    Node curr=head;
    Node tempNode;
    while(curr!=null)
    {
        tempNode=curr.next;
        curr.next=curr.prev;
        curr.prev = tempNode;

        if(curr.prev == null)
        {
            tail=head;
            head=curr;
            return;
        }

        curr=curr.prev;
    }
    return;
}
```
Analysis: Traverse the list. Swap the next and prev. then traverse to the direction curr.prev which was next before swap. If you reach the end of the list then set head and tail.

Copy List Reversed

Example 7.38: Copy the content of the list into another list in reverse order.

```
public DoublyLinkedList copyListReversed()
{
        DoublyLinkedList dll = new DoublyLinkedList();
        Node curr = head;

        while(curr != null)
        {
                dll.addHead(curr.value);
                curr=curr.next;
        }
        return dll;
}
```

Analysis:
- Create a DoublyLinkedList class object dll.
- Traverse through the list and copy the value of the nodes into another list by calling addHead() method.
- Since the new nodes are added to the head of the list, the new list formed have nodes order reverse there by making reverse list.

Copy List

Example 7.39:

```
public DoublyLinkedList copyList()
{
        DoublyLinkedList dll = new DoublyLinkedList();
        Node curr = head;

        while(curr != null)
        {
                dll.addTail(curr.value);
                curr=curr.next;
        }
        return dll;
}
```

Analysis:
- Create a DoublyLinkedList class object dll.
- Traverse through the list and copy the value of the nodes into another list by calling addTail() method.
- Since the new nodes are added to the tail of the list, the new list formed have nodes order same as the original list.

Sorted Insert

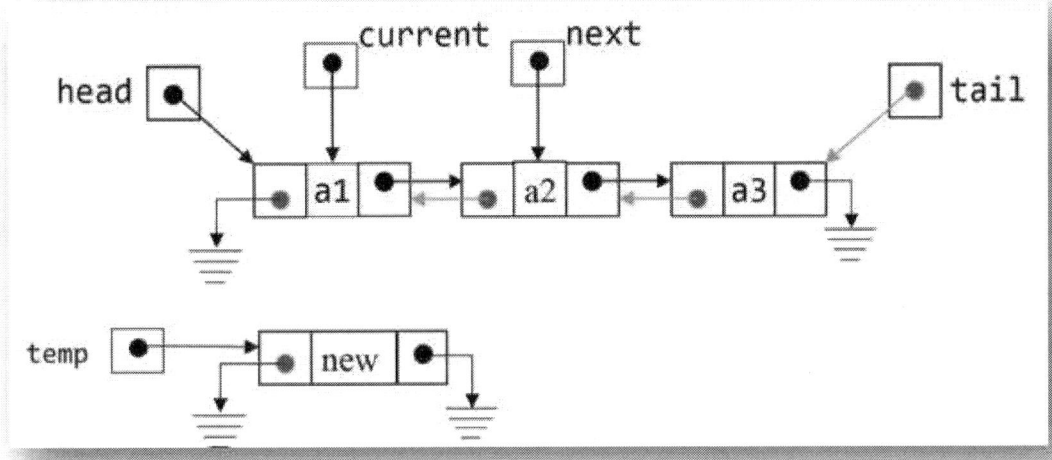

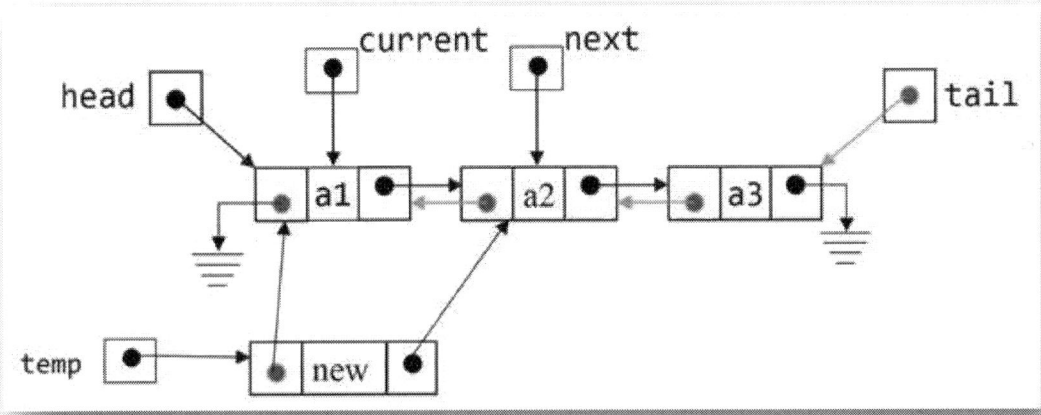

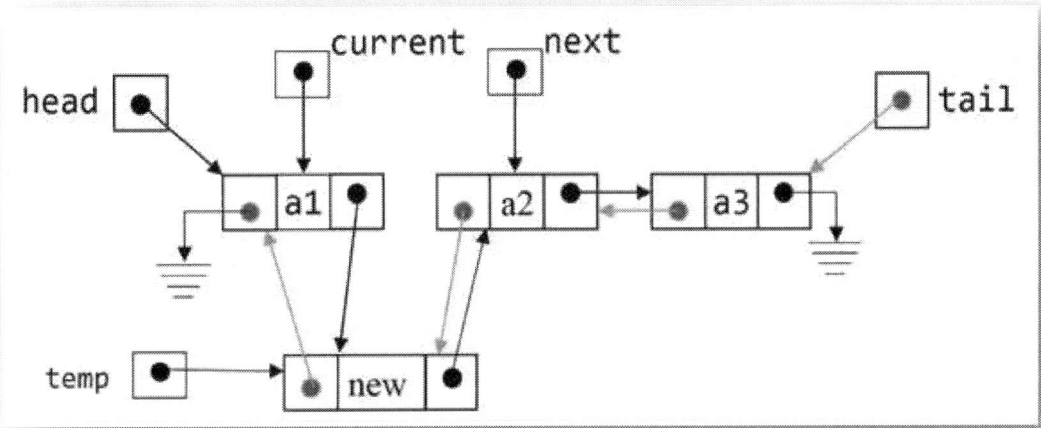

Example 7.40:

```java
//SORTED INSERT DECREASING
public void sortedInsert(int value)
{
        Node temp= new Node(value);

        Node curr=head;
        if(curr == null)  //only element
        {
                head=temp;
                tail=temp;
        }

        if(head.value <= value)  //at the begining
        {
                temp.next=head;
                head.prev=temp;
                head=temp;
        }

        while(curr.next != null && curr.next.value > value)  //treversal
        {
                curr=curr.next;
        }

        if(curr.next == null)  //at the end
        {
                tail=temp;
                temp.prev=curr;
                curr.next=temp;
        }
        else  //all other
        {
                temp.next=curr.next;
                temp.prev=curr;
                curr.next=temp;
                temp.next.prev=temp;
        }
}
```

Analysis:
- We need to consider only element case first. In this case, both head and tail will modify.
- Then we need to consider the case when head will be modified when new node is added to the beginning of the list.
- Then we need to consider general cases
- Finally, we need to consider the case when tail will be modified.

Remove Duplicate

Example 7.41: Consider the list as sorted remove the repeated value nodes of the list.

```
public void removeDuplicate() {
    Node curr = head;
    Node deleteMe;
    while(curr!=null)   {
        if((curr.next != null) && curr.value==curr.next.value)
        {
            deleteMe=curr.next;
            curr.next=deleteMe.next;
            curr.next.prev=curr;
            if(deleteMe == tail)
                tail = curr;
        }
        else
            curr=curr.next;
    }
}
```

Analysis:
- Remove duplicate is same as single linked list case.
- Head can never modify only the tail can modify when the last node is removed.

Circular Linked List

This type is similar to the singly linked list except that the last element points to the first node of the list. The link portion of the last node contains the address of the first node.

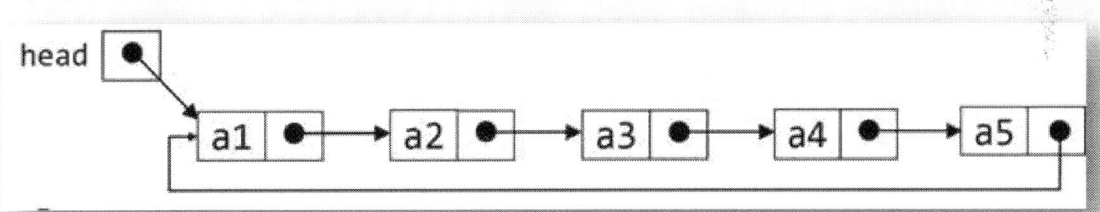

Example 7.42:

```
private static class Node {
    private int value;
    private Node next;

    public Node( int v, Node n) {
        value = v;
        next = n;
    }
    public Node( int v) {
        value = v;
        next = null;
    }
}
```

Example 7.43:

```
public class CircularLinkedList {

    private static class Node {
        // Node methods and fields.
    }
    private Node tail;
    private int size = 0;

    public int size(){
        return size;
    }

    public boolean isEmpty(){
        return size == 0;
    }

    public int peek(){
        if(isEmpty())
            throw new IllegalStateException("EmptyListException");
        return tail.next.value;
    }
    // Other methods
}
```

Analysis: In the circular linked list, we just need the pointer to the tail node. As head node can be easily reached from tail node. Size(), isEmpty() and peek() functions remains the same.

Insert element in front

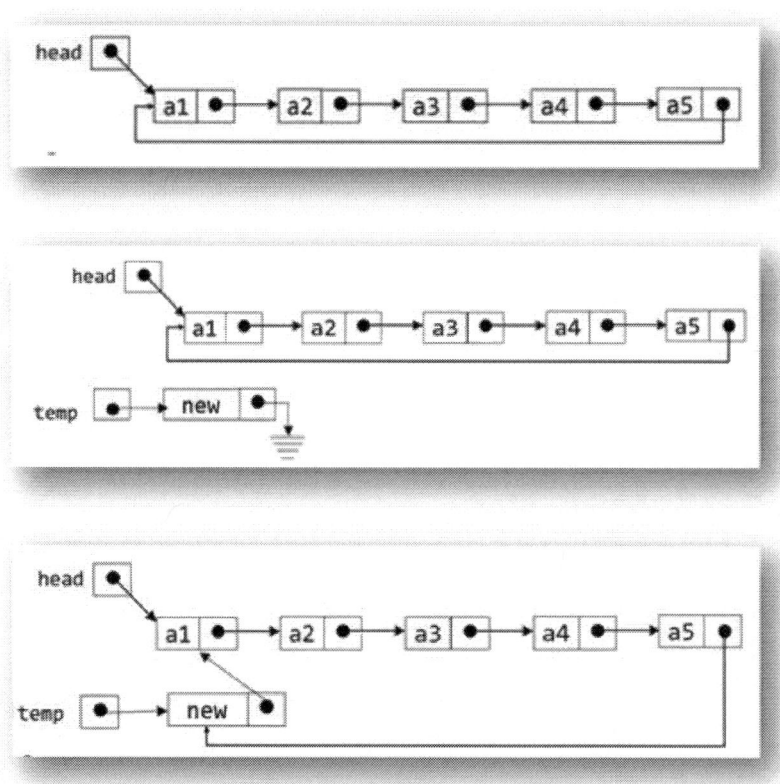

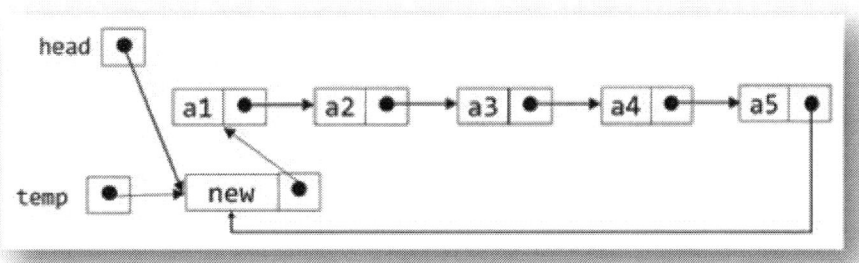

Example 7.44:

```
public void addHead(int value){
        Node temp = new Node(value, null);
        if(isEmpty())
        {
                tail=temp;
                temp.next = temp;
        }
        else{
                temp.next = tail.next;
                tail.next = temp;
        }
        size++;
}
```

```
public class CircularLinkedListDemo {
        public static void main(String[] args) {
                CircularLinkedList ll = new CircularLinkedList();
                ll.addHead(1);
                ll.addHead(2);
                ll.addHead(3);
                ll.addHead(1);
                ll.addHead(2);
                ll.addHead(3);
                ll.print();
        }
}
```

Analysis:
- First, we create node with given value and its next pointing to null.
- If the list is empty then tail of the list will point to it. And the next of node will point to itself.
- If the list is not empty then the next of the new node will be next of the tail. And tail next will start pointing to the new node.
- Thus, the new node is added to the head of the list.
- The demo program create an instance of CircularLinkedList class. Then add some value to it and finally print the content of the list.

Insert element at the end

Example 7.45:
```java
public void addTail(int value){
    Node temp = new Node(value, null);
    if(isEmpty()) {
        tail=temp;
        temp.next = temp;
    }
    else{
        temp.next = tail.next;
        tail.next = temp;
        tail=temp;
    }
    size++;
}
```

Analysis: Adding node at the end is same as adding at the beginning. Just need to modify tail reference in place of the head reference.

Remove element in the front

Example 7.46:
```java
public int removeHead() throws IllegalStateException {
    if(isEmpty()){
        throw new IllegalStateException("EmptyListException");
    }
    int value = tail.next.value;
    if(tail == tail.next) // single node case
        tail=null;
    else
        tail.next = tail.next.next;

    size--;
    return value;
}
```

Analysis:
- If the list is empty then exception will be thrown. Then the value stored in head is stored in local variable value.
- If tail is equal to its next node that means there is only one node in the list so the tail will become null.
- In all the other cases, the next of tail will point to next element of the head.
- Finally, the value is returned.

Search element in the list

Example 7.47:

```
public boolean isPresent(int data){
    Node temp = tail;
    for(int i=0;i<size;i++){
        if(temp.value == data)
            return true;
        temp = temp.next;
    }
    return false;
}
```

Analysis: Iterate through the list to find if particular value is there or not.

Print the content of list

Example 7.48:

```
public void print(){
    if(isEmpty()){
        return;
    }
    Node temp = tail.next;
    while(temp != tail){
        System.out.print(temp.value+" ");
        temp = temp.next;
    }
    System.out.print(temp.value);
}
```

Analysis: In circular list, end of list is not there so we cannot check with null. In place of null, tail is used to check end of the list.

Delete List

Example 7.49:

```
public void freeList(){
    tail = null;
    size = 0;
}
```

Analysis: The reference to the list is tail. By making tail null, the whole list is deleted.

Delete a node given its value

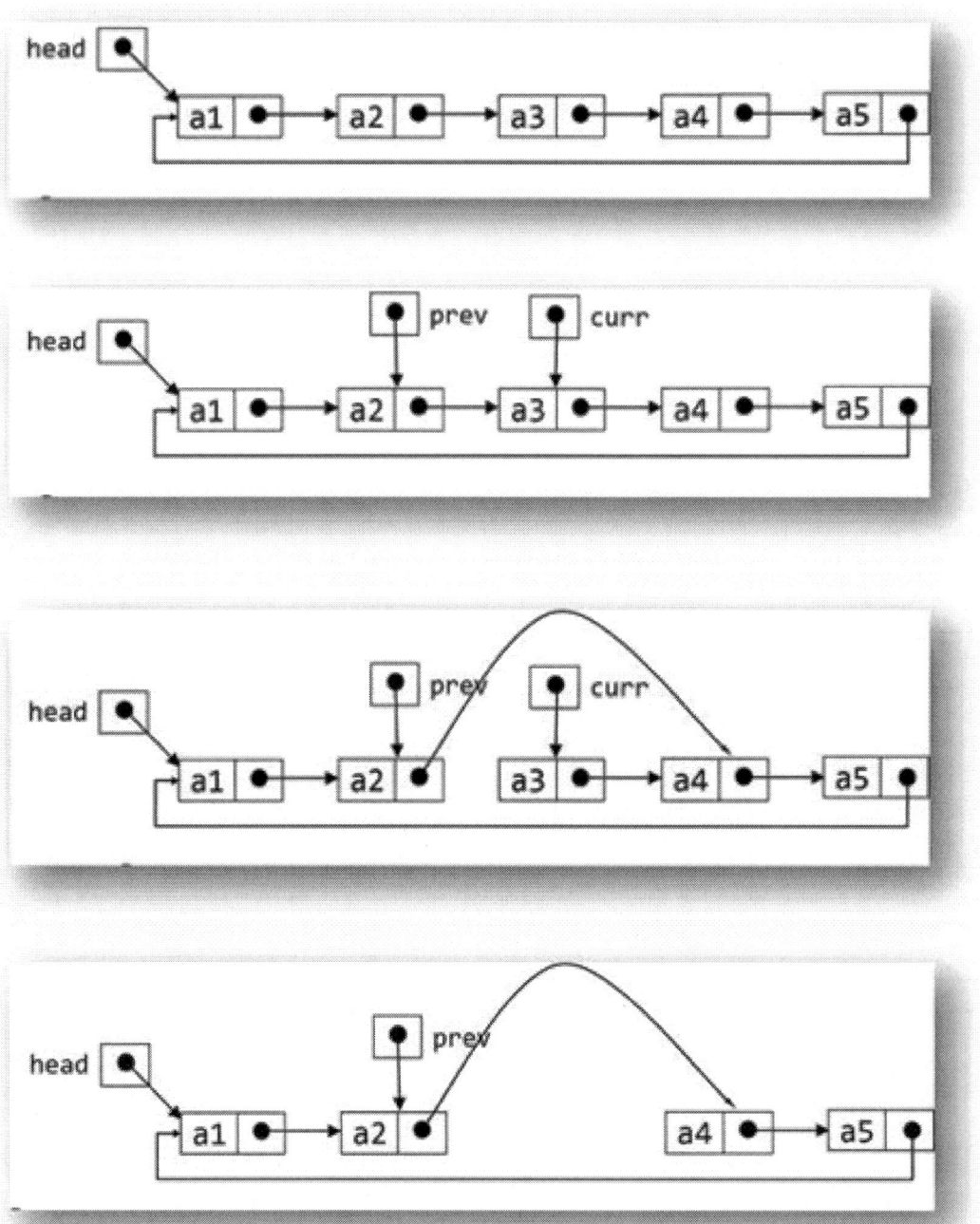

Example 7.50:
```
public boolean removeNode(int key){
    if(isEmpty()){
        return false;
    }
    Node prev = tail;
    Node curr = tail.next;
    Node head = tail.next;
```

```
            if(curr.value == key)//head and single node case.
            {
                    if(curr == curr.next)//single node case
                            tail=null;
                    else // head case
                            tail.next = tail.next.next;
                    return true;
            }

            prev = curr;
            curr = curr.next;

            while(curr != head)
            {
                    if(curr.value == key)
                    {
                            if(curr == tail)
                                    tail = prev;
                            prev.next = curr.next;
                            return true;
                    }
                    prev = curr;
                    curr = curr.next;
            }

            return false;
    }
```

Analysis: Find the node that need to free. Only difference is that while traversing the list end of list is tracked by the head reference in place of null.

Copy List Reversed

Example 7.51:

```
public CircularLinkedList copyListReversed()
{
        CircularLinkedList cl = new CircularLinkedList();
        Node curr = tail.next;
        Node head = curr;
        if (curr != null)
        {
                cl.addHead(curr.value);
                curr = curr.next;
        }
        while (curr != head)
        {
                cl.addHead(curr.value);
                curr = curr.next;
        }
        return cl;
}
```

Analysis: The list is traversed and nodes are added to new list at the beginning. There by making the new list reverse of the given list.

Copy List

Example 7.52:

```
public CircularLinkedList copyList()
{
        CircularLinkedList cl = new CircularLinkedList();
        Node curr = tail.next;
        Node head = curr;

        if (curr != null)
        {
                cl.addTail(curr.value);
                curr = curr.next;
        }
        while (curr != head)
        {
                cl.addTail(curr.value);
                curr = curr.next;
        }
        return cl;
}
```

Analysis:
List is traversed and nodes are added to the new list at the end. There by making the list whose value are same as the input list.

Doubly Circular list

1. For any linked list there are only three cases zero element, one element, general case
2. To doubly linked list we have a few more things
    a) null values
    b) Only element (it generally introduces an if statement with null)
    c) Always an "if" before "while". Which will check from this head.
    d) General case (check with the initial head kept)
    e) Avoid using recursion solutions it makes life harder

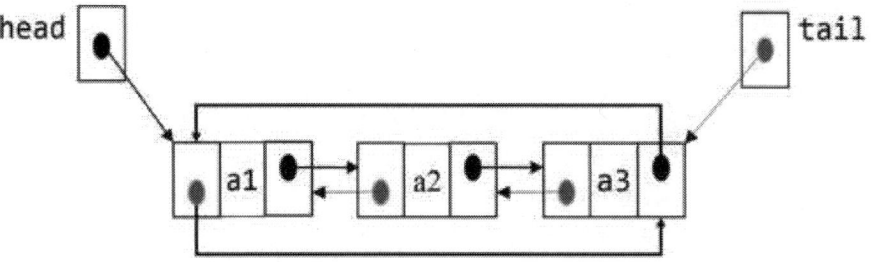

Example 7.53:

```java
private static class Node {
    private int value;
    private Node next;
    private Node prev;

    public Node(int v, Node nxt, Node prv) {
        value = v;
        next = nxt;
        prev = prv;
    }

    public Node(int v) {
        value = v;
        next = this;
        prev = this;
    }
}
```

```java
public class DoublyCircularLinkedList {

    private static class Node {
        ...........
    }

    private Node head=null;
    private Node tail=null;
    private int size = 0;

    public int size(){
        return size;
    }

    public boolean isEmpty(){
        return size == 0;
    }

    public int peekHead(){
        if(isEmpty())
            throw new IllegalStateException("EmptyListException");
        return head.value;
    }
}
```

Search value

Example 7.54:

```
public boolean isPresent(int key){
        Node temp = head;
        if(head == null)
                return false;

        do{
                if(temp.value == key)
                        return true;
                temp = temp.next;
        }while(temp != head);

        return false;
}
```

Analysis: Traverse through the list and see if given key is present or not. We use do..while loop as initial state is our termination state too.

Delete list

Example 7.55:

```
public void freeList(){
        head = null;
        tail = null;
        size = 0;
}
```

Analysis: Remove the reference and list will be freed.

Print List

Example 7.56:

```
public void print() {
        if (isEmpty()) {
                return;
        }
        Node temp = head;
        while (temp != tail) {
                System.out.print(temp.value + " ");
                temp = temp.next;
        }
        System.out.print(temp.value);
}
```

Analysis: Traverse the list and print its content. Do..while is used as we want to terminate when temp is head. And want to process head node once.

Insert Node at head

Example 7.57: Insert value at the front of the list.

```
public void addHead(int value) {
        Node newNode = new Node(value, null, null);
        if(size == 0) {
                tail = head = newNode;
                newNode.next = newNode;
                newNode.prev = newNode;
        }
        Else {
                newNode.next = head;
                newNode.prev = head.prev;
                head.prev = newNode;
                newNode.prev.next = newNode;
                head = newNode;
        }
        size++;
}
```

Analysis:
- A new node is created and if the list is empty then head and tail will point to it. The newly created newNode's next and prev also point to newNode.
- If the list is not empty then the pointers are adjested and a new node is added to the front of the list. Only head need to be changed in this case.
- Size of the list is increased by one.

Insert Node at tail

Example 7.58:

```
public void addTail(int value) {
        Node newNode = new Node(value, null, null);
        if(size == 0) {
                head = tail = newNode;
                newNode.next = newNode;
                newNode.prev = newNode;
        }
        else {
                newNode.next = tail.next;
                newNode.prev = tail;
                tail.next = newNode;
                newNode.next.prev = newNode;
                tail = newNode;
        }
        size++;
}
```

Analysis:
- A new node is created and if the list is empty then head and tail will point to it. The newly created newNode's next and prev also point to newNode.
- If the list is not empty then the pointers are adjested and a new node is added to the end of the list. Only tail need to be changed in this case.
- Size of the list is increased by one.

Delete head node

Example 7.59:

```java
public int removeHead(){
    if(size == 0)
        throw new IllegalStateException("EmptyListException");

    int value = head.value;
    size--;

    if(size == 0)
    {
        head = null;
        tail = null;
        return value;
    }

    Node next = head.next;
    next.prev = tail;
    tail.next = next;
    head = next;
    return value;
}
```

Analysis: Delete node in a doubly circular linked list is just same as delete node in a circular linked list. Just few extra next reference need to be adjusted.

Delete tail node

Example 7.60:

```java
public int removeTail(){
    if(size == 0)
        throw new IllegalStateException("EmptyListException");

    int value = tail.value;
    size--;

    if(size == 0)
    {
        head = null;
        tail = null;
        return value;
    }

    Node prev = tail.prev;
    prev.next = head;
    head.prev = prev;
    tail=prev;
    return value;
}
```

Analysis: Delete node in a doubly circular linked list is just same as delete node in a circular linked list. Just few extra prev reference need to be adjusted.

Exercise

1) Insert an element k^{th} position from the start of linked list. Return 1 if success and if list is not long enough, then return -1.
   Hint: Take a reference advance it K steps forward, then inserts the node.

2) Insert an element k^{th} position from the end of linked list. Return 1 if success and if list is not long enough, then return -1.
   Hint: Take a reference advance it K steps forward, then take another reference and advance both of them simultaneously, so that when the first reference reach the end of a linked list that is the point where you need to insert the node.

3) Consider there is a loop in a linked list, Write a program to remove loop if there is a loop in this linked list.

4) In the above SearchList program return, the count of how many instances of same value found else if value not found then return 0. For example, if the value passed is "4". The elements in the list are 1,2,4,3 & 4. The program should return 2.
   Hint: In place of return 1 in the above program increment a counter and then return counter at the end.

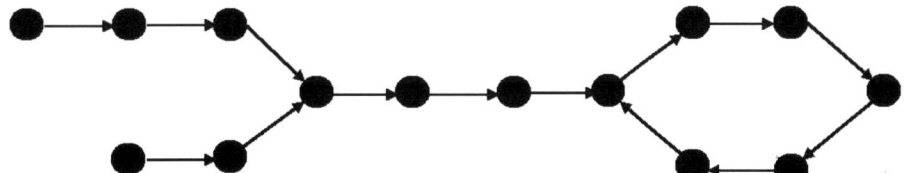

5) If linked list having a loop is given. Count the number of nodes in the linked list

6) We were supposed to write the complete code for the addition of polynomials using Linked Lists. This takes time if you do not have it by heart, so revise it well.

7) Given two linked lists. We have to find that whether the data in one is reverse that of data in another. No extra space should be used and traverse the linked lists only once.

8) Find the middle element in a singly linked list. Tell the complexity of your solution.
   Hint:-
   - *Approach 1:* find the length of linked list. Then find the middle element and return it.
   - *Approach 2:* use two reference one will move fast and one will move slow make sure you handle border case properly. (Even length and odd length linked list cases.)

9) Print list in reverse order.
   Hint: Use recursion.

Chapter 8: Stack

Introduction

A stack is a basic data structure that organized items in last-in-first-out (LIFO) manner. Last element inserted in a stack will be the first to be removed from it.

The real-life analogy of the stack is "chapattis in hotpot", "stack of plates". Imagine a stack of plates in a dining area everybody takes a plate at the top of the stack, thereby uncovering the next plate for the next person.

Stack allow to only access the top element. The elements that are at the bottom of the stack are the one that is going to stay in the stack for the longest time.

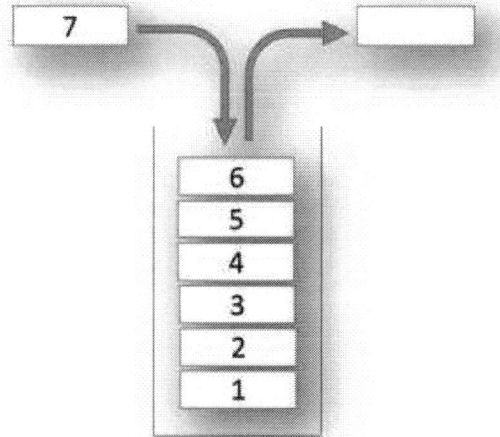

Computer science also has the common example of a stack. Function call stack is a good example of a stack. Function main() calls function foo() and then foo() calls bar(). These function calls are implemented using stack first bar() exists, then go() and then finally main().

As we navigate from web page to web page, the URL of web pages are kept in a stack, with the current page URL at the top. If we click back button, then each URL entry is popped one by one.

The Stack Abstract Data Type

Stack abstract data type is defined as a class, which follows LIFO or last-in-first-out for the elements, added to it.
The stack should support the following operation:
  1. Push(): which add a single element at the top of the stack
  2. Pop(): which remove a single element from the top of a stack.
  3. Top(): Reads the value of the top element of the stack (does not remove it)
  4. isEmpty(): Returns 1 if stack is empty
  5. Size(): returns the number of elements in a stack.

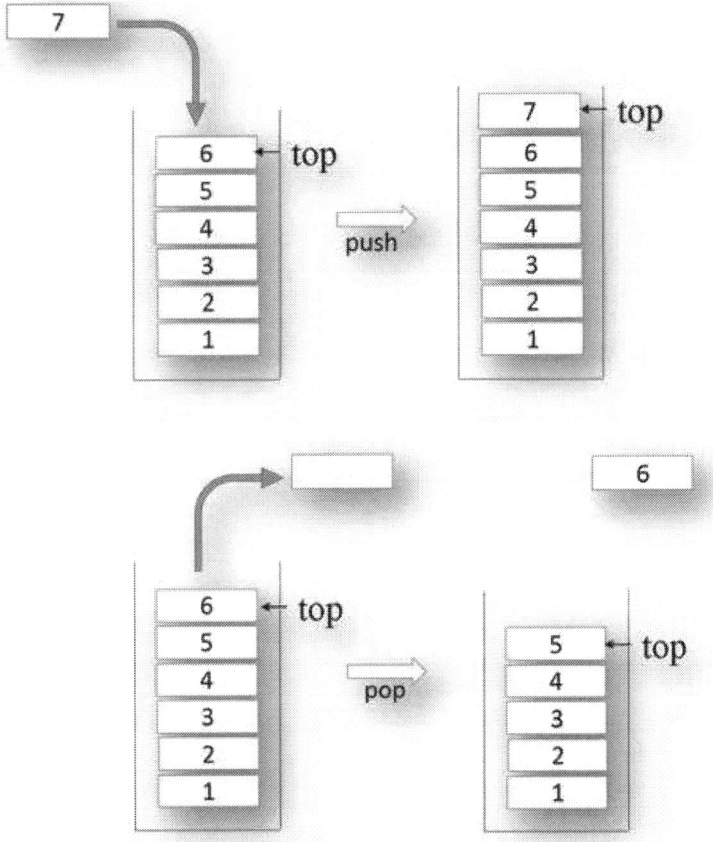

Add n to the top of a stack
void push(int value);.
Remove the top element of the stack and return it to the caller function.
int pop();

The stack can be implemented using an array or a linked list.
In array case, there are two types of implementations
- One in which array size is fixed, so it the capacity of the stack.
- Another approach is variable size array in which memory of the array is allocated using malloc and when the array is filled the size if doubled using realloc (when the stack size decreases below half the capacity is again reduced).

In case of a linked list, there is no such limit on the number of elements it can contain.

When a stack is implemented, using an array top of the stack is managed using an index variable called top.

When a stack is implemented using a linked list, push() and pop() is implemented using insert at the head of the linked list and remove from the head of the linked list.

Stack using Array

Implement a stack using a fixed length array.

Example 8.1:

```java
public class Stack {

    private static final int CAPACITY = 1000;
    private int[ ] data;
    private int top = -1;

    public Stack() {
        this(CAPACITY);
    }

    public Stack(int capacity) {
        data = new int[capacity];
    }
}
```

If user does not provide the max capacity of the array. Then an array of 1000 elements is created. The top is the index to the top of the stack.

Number of elements in the stack is governed by the "top" index and top is initialized to -1 when a stack is initialized. Top index value of -1 indicates that the stack is empty in the beginning.

```java
public boolean isEmpty(){
    return (top == -1);
}
```

isEmpty() function returns 1 if stack is empty or 0 in all other cases. By comparing the top index value with -1.

```java
public int size() {
    return (top + 1);
}
```

size() function returns the number of elements in the stack. It just returns "top+1". As the top is referring the array index of the stack top variable so we need to add one to it.

```java
public void print(){
    for(int i=top;i>-1;i--)
        System.out.print(" "+ data[i]);
}
```

The print function will print the elements of the array.

```java
public void push(int value) throws IllegalStateException {
    if (size( ) == data.length)
        throw new IllegalStateException("StackOvarflowException");
    top++;
    data[top] = value;
}
```

push() function checks whether the stack has enough space to store one more element, then it increases the "top" by one. Finally sort the data in the stack "data" array. In case, stack is full then "stack overflow" message is printed and that value will not be added to the stack and will be ignored.

```java
public int pop( ) {
    if (isEmpty( ))
        throw new IllegalStateException("StackEmptyException");
    int topVal = data[top];
    top--;
    return topVal;
}
```

The pop() function is implemented, first it will check that there are some elements in the stack by checking its top index. If some element is there in the stack, then it will store the top most element value in a variable "value". The top index is reduced by one. Finally, that value is returned.

```java
public int top( ) throws IllegalStateException {
    if (isEmpty( ))
        throw new IllegalStateException("StackEmptyException");
    return data[top];
}
```

top()function returns the value of stored in the top element of stack (does not remove it)

```java
public static void main(String[] args) {
    Stack s = new Stack(1000);
    for(int i=1;i<=100;i++)
    {
        s.push(i);
    }
    for(int i=1;i<=50;i++)
    {
        s.pop();
    }
    s.print();
}
```

Analysis:
- The user of the stack will create a stack local variable.
- Use push() and pop() functions to add / remove variables to the stack.
- Read the top element using the top() function call.
- Query regarding size of the stack using size() function call
- Query if stack is empty using isEmpty() function call

Stack using Array (Growing-Reducing capacity implementation)

In the above dynamic array implementation of a stack. Make the capacity of stack variable so that when it is nearly filled, then double the capacity of the stack.

Example 8.2:

```
public class Stack {
    private static final int MIN_CAPACITY = 1000;
    private int[ ] data;
    private int top = -1;
    private int minCapacity;
    private int maxCapacity;

    public Stack() {
        this(MIN_CAPACITY);
        maxCapacity = minCapacity = MIN_CAPACITY;
    }

    public Stack(int capacity) {
        data = new int[capacity];
        maxCapacity = minCapacity = capacity;
    }
}
```

```
public void push(int value) throws IllegalStateException {
    if (size( ) == maxCapacity)
    {
        System.out.println("size dubbelled");
        int[] newData = new int[maxCapacity*2];
        System.arraycopy(data,0,newData,0,maxCapacity);
        data=newData;
        maxCapacity=maxCapacity*2;
    }
    top++;
    data[top] = value;
}
```

```java
public int pop( ) {
    if (isEmpty( ))
        throw new IllegalStateException("StackEmptyException");

    int topVal = data[top];
    top--;

    if(size() == maxCapacity/2 && maxCapacity > minCapacity)
    {
        System.out.println("size halfed");
        maxCapacity = maxCapacity/2;
        int[] newData = new int[maxCapacity];
        System.arraycopy(data,0,newData,0,maxCapacity);
        data=newData;
    }
    return topVal;
}
```

Analysis:
- In the **push()** function we double the size of the stack when stack is full. First, an array of double capacity is created. All the data of the old array is copied to the new array. Finally the reference to the array of the stack is changed to the newly array.
- In the **pop()** function when size of the stack is half the max capacity of the stack and is grater then the minimum threshold then the an array of half size is created and the value of the old bigger array is copied to the newly created array.
- Apart from **push()** and **pop()** all the other functions are same as the before fixed size implementation of stack.

Stack using linked list

Example 8.3: Implement stack using a linked list.

```java
public class Stack {

    private static class Node {
        private int value;
        private Node next;

        public Node( int v, Node n){
            value = v;
            next = n;
        }
    }

    private Node head = null;
    private int size = 0;

    //Other Methods
}
```

```java
public int size() {
    return size;
}
```

```java
public boolean isEmpty() {
    return size == 0;
}
```

```java
public int peek() throws IllegalStateException {
    if(isEmpty())
        throw new IllegalStateException("StackEmptyException");
    return head.value;
}
```

```java
public void push(int value) {
    head = new Node(value, head);
    size++;
}
```

```java
public int pop() throws IllegalStateException {
    if(isEmpty())
        throw new IllegalStateException("StackEmptyException");
    int value = head.value;
    head = head.next;
    size--;
    return value;
}
```

```java
public void insertAtBottom(int value) {
    if (isEmpty())
        push(value);
    else
    {
        int temp = pop();
        insertAtBottom(value);
        push(temp);
    }
}
```

```java
public void print(){
    Node temp = head;
    while(temp != null){
        System.out.print(temp.value +" ");
        temp = temp.next;
    }
}
```

```java
public static void main(String[] args) {
    Stack s = new Stack();
    for(int i=1;i<=100;i++)
        s.push(i);

    for(int i=1;i<=50;i++)
        s.pop();
    s.print();
}
```

Analysis:
- Stack implemented using a linked list is simply insertion and deletion at the head of a singly linked list.
- In push() function, memory is created for one node. Then the value is stored into that node. Finally, the node is inserted at the beginning of the list.
- In pop() function, the head of the linked list starts pointing to the second node there by releasing the memory allocated to the first node (Garbage collection.).

Problems in Stack

Balanced Parenthesis

Example 8.4: Stacks can be used to check a program for balanced symbols (such as {}, (), []). The closing symbol should be matched with the most recently seen opening symbol.
Example: {()} is legal, {() ({})} is legal, but {((} and {()} are not legal

```java
public static boolean isBalancedParenthesis (String expn) {
    ArrayDeque<Character> stk = new ArrayDeque<Character>();
    for (char ch : expn.toCharArray( )) {
        switch (ch) {
        case '{':
        case '[':
        case '(':
            stk.push(ch);
            break;
        case '}':
            if (stk.pop() != '{')
                return false;
            break;
        case ']':
            if (stk.pop() != '[')
                return false;
            break;
        case ')':
            if (stk.pop() != '(')
                return false;
            break;
        }
    }
    return stk.isEmpty();
}
```

```java
public static void main(String[] args) {
    String expn = "{()}[";
    boolean value = isBalancedParenthesis (expn);
    System.out.println("Given Expn:"+expn);
    System.out.println("Result after isParenthesisMatched:"+value);
}
```

Analysis:
- Traverse the input string when we get an opening parenthesis we push it into stack. And when we get a closing parenthesis then we pop a parenthesis from the stack and compare if it is the corresponding to the one on the closing parenthesis.
- We return false if there is a mismatch of parenthesis.
- If at the end of the whole staring traversal, we reached to the end of the string and the stack is empty then we have balanced parenthesis.

Infix, Prefix and Postfix Expressions

When we have an algebraic expression like A + B then we know that the variable is being added to variable B. This type of expression is called **infix** expression because the operator "+" is between operands A and operand B.

Now consider another infix expression A + B * C. In the expression there is a problem that in which order + and * works. Does A and B are added first and then the result is multiplied. Or B and C are multiplied first and then the result is added to A. This makes the expression ambiguous. To deal with this ambiguity we define the precedence rule or use parentheses to remove ambiguity.

So if we want to multiply B and C first and then add the result to A. Then the same expression can be written unambiguously using parentheses as A + (B * C). On the other hand, if we want to add A and B first and then the sum will be multiplied by C we will write it as (A + B) * C. So in the infix expression to make the expression unambiguous, we need parenthesis.

Infix expression: In this notation, we place operator in the middle of the operands.
< Operand > < operator > < operand >

Prefix expressions: In this notation, we place operator at the beginning of the operands.
< Operator > < operand > < operand >

Postfix expression: In this notation, we place operator at the end of the operands.
< Operand > < operand > < operator >

Infix Expression	Prefix Expression	Postfix Expression
A + B	+ A B	A B +
A + (B * C)	+ A * B C	A B C * +
(A + B) * C	* + A B C	A B + C *

Now comes the most obvious question why we need so unnatural Prefix or Postfix expressions when we already have infix expressions which words just fine for us.

The answer to this is that infix expressions are ambiguous and they need parenthesis to make them unambiguous. While postfix and prefix notations do not need any parenthesis.

Infix-to-Postfix Conversion

Example 8.5:

```java
public static char[] infixToPostfix(char[] expn)
{
        ArrayDeque<Character> stk = new ArrayDeque<Character>();

        String output="";
        char out;

        for (char ch : expn) {
            if( ch <= '9' && ch >= '0')
            {
                    output = output + ch ;
            }
            else
            {
                    switch (ch)
                    {
                    case '+':
                    case '-':
                    case '*':
                    case '/':
                    case '%':
                    case '^':
                            while (stk.isEmpty() == false &&
                            precedence(ch) <= precedence(stk.peek()))
                            {
                                    out = stk.pop();
                                    output = output + " " + out ;
                            }
                            stk.push(ch);
                            output = output + " ";
                            break;
                    case '(':
                            stk.push(ch);
                            break;
                    case ')':
                            while (stk.isEmpty() == false &&
                                (out = stk.pop()) != '(')
                            {
                                    output = output + " " + out + " ";
                            }
                            break;
                    }
            }
        }
}
```

```java
            while(stk.isEmpty() == false)
            {
                    out = stk.pop();
                    output = output + out + " ";
            }
            return output.toCharArray();
    }
```

```java
    public static String infixToPostfix(String expn) {
            String output="";
            char[] out = infixToPostfix(expn.toCharArray( ));
            for(char ch : out)
                    output = output + ch ;
            return output;
    }
```

```java
    public static void main(String[] args) {
            String expn = "10+((3))*5/(16-4)";
            String value = infixToPostfix (expn);
            System.out.println("Infix Expn: "+expn);
            System.out.println("Postfix Expn: "+value);
    }
```

Analysis:
- Print operands in the same order as they arrive.
- If the stack is empty or contains a left parenthesis "(" on top, we should push the incoming operator in the stack.
- If the incoming symbol is a left parenthesis "(", push left parenthesis in the stack.
- If the incoming symbol is a right parenthesis ")", pop from the stack and print the operators till you see a left parenthesis ")". Discard the pair of parentheses.
- If the precedence of incoming symbol is higher than the operator at the top of the stack, then push it to the stack.
- If the incoming symbol has, an equal precedence compared to the top of the stack, use association. If the association is left to right, then pop and print the symbol at the top of the stack and then push the incoming operator. If the association is right to left, then push the incoming operator.
- If the precedence of incoming symbol is lower than the operator on the top of the stack, then pop and print the top operator. Then compare the incoming operator against the new operator at the top of the stack.
- At the end of the expression, pop and print all operators on the stack.

Infix-to-Prefix Conversion

Example 8.6:

```java
public static String infixToPrefix(String expn)
{
        char[] arr = expn.toCharArray();
        reverseString(arr);
        replaceParanthesis(arr);
        arr=infixToPostfix(arr);
        reverseString(arr);
        expn = new String(arr);
        return expn;
}
```

```java
public static void replaceParanthesis(char[] a)
{
        int lower=0;
        int upper=a.length-1;
        while(lower<=upper)
        {
                if(a[lower] == '(')
                        a[lower] = ')';
                else if(a[lower] == ')')
                        a[lower] = '(';
                lower++;
        }
}
```

```java
public static void reverseString(char[] expn)
{
        int lower=0;
        int upper=expn.length - 1;
        char tempChar;
        while(lower<upper)
        {
                tempChar=expn[lower];
                expn[lower]=expn[upper];
                expn[upper]=tempChar;
                lower++;
                upper--;
        }
}
```

```java
public static void main(String[] args) {
        String expn = "10+((3))*5/(16-4)";
        String value = infixToPrefix (expn);
        System.out.println("Infix Expn: "+expn);
        System.out.println("Prefix Expn: "+value);
}
```

Analysis:
1. Reverse the given infix expression.
2. Replace '(' with ')' and ')' with '(' in the reversed expression.
3. Now apply infix to postfix subroutine already discussed.
4. Reverse the generated postfix expression and this will give required prefix expression.

Postfix Evaluate

Write a postfixEvaluate() function to evaluate a postfix expression. Such as: 1 2 + 3 4 + *

Example 8.7:

```java
public static int postfixEvaluate(String expn)
{
    ArrayDeque<Integer> stk = new ArrayDeque<Integer>();

    Scanner tokens = new Scanner(expn);

    while(tokens.hasNext()){
        if(tokens.hasNextInt()){
            stk.push(tokens.nextInt());
        }
        else{
            int num1 = stk.pop();
            int num2 = stk.pop();
            char op=tokens.next().charAt(0);
            switch (op)
            {
            case '+':
                stk.push(num1 + num2);
                break;
            case '-':
                stk.push(num1 - num2);
                break;
            case '*':
                stk.push(num1 * num2);
                break;
            case '/':
                stk.push(num1 / num2);
                break;
            }
        }
    }
    tokens.close();
    return stk.pop();
}
```

```java
public static void main3(String[] args) {
    String expn = "6 5 2 3 + 8 * + 3 + *";
    int value = postfixEvaluate (expn);
    System.out.println("Given Postfix Expn: "+expn);
    System.out.println("Result after Evaluation: "+value);
}
```

Analysis:

1) Create a stack to store values or operands.
2) Scan through the given expression and do following for each element:
   a) If the element is a number, then push it into the stack.
   b) If the element is an operator, then pop values from the stack. Evaluate the operator over the values and push the result into the stack.
3) When the expression is scanned completely, the number in the stack is the result.

Min stack

Design a stack in which get minimum value in stack should also work in *O(1)* *Time Complexity*.

Hint: Keep two stack one will be general stack, which will just keep the elements. The second will keep the min value.
1. Push: Push an element to the top of stack1. Compare the new value with the value at the top of the stack2. If the new value is smaller, then push the new value into stack2. Or push the value at the top of the stack2 to itself once more.
2. Pop: Pop an element from top of stack1 and return. Pop an element from top of stack2 too.
3. Min: Read from the top of the stack2 this value will be the min.

Palindrome string

Find if given string is a palindrome or not using a stack.
Definition of palindrome: A palindrome is a sequence of characters that is same backward or forward.
Eg. "AAABBBCCCBBBAAA", "ABA" & "ABBA"

Hint: Push characters to the stack till the half length of the string. Then pop these characters and then compare. Make sure you take care of the odd length and even length.

Reverse Stack

Given a stack how to reverse the elements of the stack without using any other data-structure. You cannot use another stack too.
Time Complexity and *Space Complexity* is wrong, it is *O(n)* for both cases.

Hint: Use recursion (system stack.) When you go inside the stack pop elements from stack in each subsequent call until stack is empty. Then push these elements one by one when coming out of the recursion. The elements will be reversed.

Example 8.8:

```
public static <T> void reverseStack (ArrayDeque<T> stk)
{
    if(stk.isEmpty())
    {
        return;
    }
    else
    {
        T value = stk.pop();
        reverseStack(stk);
        insertAtBottom(stk, value);
    }
}
```

Insert At Bottom

Example 8.9:

```
public static <T> void insertAtBottom (ArrayDeque<T> stk, T value) {
    if(stk.isEmpty()) {
        stk.push(value);
    }
    else {

        T out = stk.pop();
        insertAtBottom(stk, value);
        stk.push(out);
    }
}
```

Depth-First Search with a Stack

In a depth-first search, we traverse down a path until we get a dead end; then we *backtrack* by popping a stack to get an alternative path.
- Create a stack
- Create a start point
- Push the start point onto the stack
- While (value searching not found and the stack is not empty)
    - Pop the stack
    - Find all possible points after the one which we just tried
    - Push these points onto the stack

Stack using a queue

How to implement a stack using a queue. Analyze the running time of the stack operations.
See queue chapter for this.

Stock Span Problem

Given a list of daily stock price in an array A[i]. Find the span of the stocks for each day. A span of stock is the maximum number of days for which the price of stock was lower than that day.

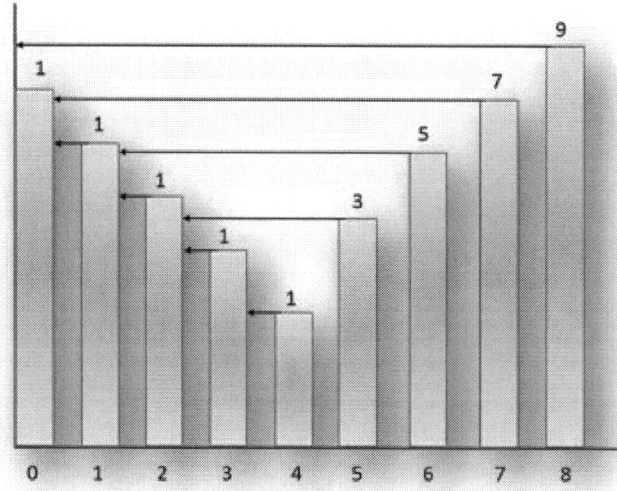

Example 8.10: Approach 1

```
public static int[] StockSpanRange(int[] arr) {
    int[] SR = new int[arr.length];
    SR[0] = 1;
    for (int i = 1; i < arr.length; i++) {
        SR[i] = 1;
        for (int j = i - 1; (j >= 0) && (arr[i] >= arr[j]); j--)
            SR[i]++;
    }
    return SR;
}
```

Example 8.11: Approach 2:

```
int[] StockSpanRange2 (int[] arr) {
    ArrayDeque<Integer> stk = new ArrayDeque<Integer>();

    int[] SR = new int[arr.length];
    stk.push(0);
    SR[0] = 1;
    for (int i = 1; i < arr.length; i++) {
        while (!stk.isEmpty() && arr[stk.peek()]<= arr[i])
            stk.pop();
        SR[i] = (stk.isEmpty()) ? (i + 1) : (i - stk.peek());
        stk.push(i);
    }
    return SR;
}
```

Get Max Rectangular Area in a Histogram

Given a histogram of rectangle bars of each one unit wide. Find the maximum area rectangle in the histogram.

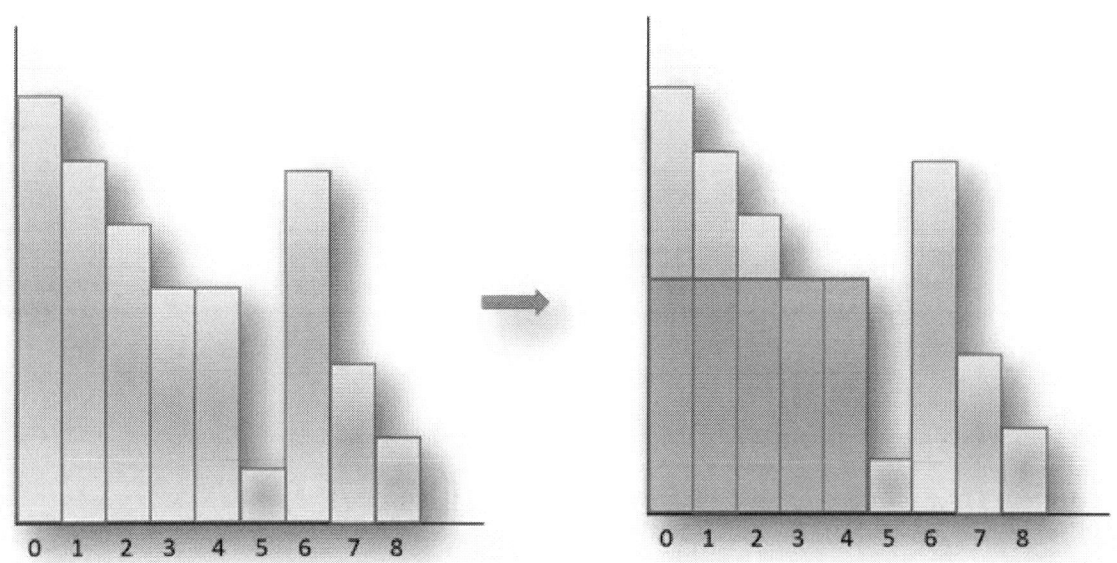

Example 8.12: Approach 1

```java
public static int GetMaxArea(int[] arr)
{
    int size = arr.length;
    int maxArea = -1;
    int currArea;
    int minHeight = 0;
    for (int i = 1; i < size; i++)
    {
        minHeight = arr[i];
        for (int j = i - 1; j >= 0; j--)
        {
            if (minHeight > arr[j])
                minHeight = arr[j];
            currArea = minHeight * (i - j + 1);
            if (maxArea < currArea)
                maxArea = currArea;
        }
    }
    return maxArea;
}
```

Approach 2: Divide and conquer

Example 8.13: Approach 3

```java
public static int GetMaxArea2(int[] arr)
{
    int size=arr.length;
    ArrayDeque<Integer> stk = new ArrayDeque<Integer>();
    int maxArea = 0;
    int top;
    int topArea;
    int i = 0;
    while (i < size)
    {
        while ( (i < size) && (stk.isEmpty() || arr[stk.peek()] <= arr[i]) )
        {
            stk.push(i);
            i++;
        }
        while ( !stk.isEmpty() && ( i == size || arr[stk.peek()] > arr[i]) )
        {
            top = stk.peek();
            stk.pop();
            topArea = arr[top] * (stk.isEmpty() ? i : i - stk.peek() - 1);
            if (maxArea < topArea)
                maxArea = topArea;
        }
    }
    return maxArea;
}
```

Two stacks using single array

Example 8.14: How to implement two stacks using one single array.

```java
public class TwoStack {
        private final int  MAX_SIZE = 50;
        int top1;
        int top2;
        int[] data;

        public TwoStack() {
                top1 = -1;
                top2 = MAX_SIZE;
                data = new int[MAX_SIZE];
        }
        //Other methods.
}
```

```java
public void StackPush1 (int value)
{
        if (top1 < top2 - 1)
        {
                data[++top1] = value;
        }
        else
                System.out.print("Stack is Full!");
}
```

```java
public void StackPush2 (int value)
{
        if (top1 < top2 - 1)
        {
                data[--top2] = value;
        }
        else
                System.out.print("Stack is Full!");
}
```

```java
public int StackPop1 ()
{
        if (top1 >= 0)
        {
                int value = data[top1--];
                return value;
        }
        else
                System.out.print("Stack Empty!");
        return -999;
}
```

```java
public int StackPop2 ()
{
    if (top2 < MAX_SIZE)
    {
        int value = data[top2++];
        return value;
    }
    else
        System.out.print("Stack Empty!");
    return -999;
}
```

```java
public static void main(String[] args)
{
    TwoStack st = new TwoStack();
    for(int i=0;i<10;i++)
        st.StackPush1(i);

    for(int j=0;j<10;j++)
        st.StackPush2(j+10);

    for(int i=0;i<10;i++)
    {
        System.out.println("stack one pop value is : " + st.StackPop1());
        System.out.println("stack two pop value is : " + st.StackPop2());
    }
}
```

Analysis: Same array is used to implement two stack. First stack is filled from the beginning of the array and second stack is filled from the end of the array. Overflow and underflow conditions need to be taken care of carefully.

Pros and cons of array and linked list implementation of stack.

Linked lists: List implementation uses 1 reference extra memory per item. There is no size restriction.

Arrays: Allocated a constant amount of space, when the stack is nearly empty, then lost of space is waste as it's not used. Maximum size is determined when the stack is created.

Uses of Stack

- Recursion can also be done using stack. (In place of the system stack)
- The function call is implemented using stack.
- Some problems when we want to reverse a sequence, we just push everything in stack and pop from it.
- Grammar checking, balance parenthesis, infix to postfix conversion, postfix evaluation of expression etc.

Exercise

Ex 1: Converting Decimal Numbers to Binary Numbers using stack data structure.

Hint: store reminders into the stack and then print the stack.

Ex 2: Convert an infix expression to prefix expression.

Hint: Reverse given expression, Apply infix to postfix, and then reverse the expression again.
Step 1. Reverse the infix expression.
    5^E+D*) C^B+A (
Step 2. Make Every '(' as ')' and every ')' as '('
    5^E+D*(C^B+A)
Step 3. Convert an expression to postfix form.

Step 4. Reverse the expression.
    +*+A^BCD^E5

Ex 3: Write an HTML opening tag and closing tag-matching program.
Hint: parenthesis matching.

Ex 4: Write a function that will do Postfix to Infix Conversion

Ex 5:: Write a function that will do Prefix to Infix Conversion

Ex 6: Write a palindrome matching function, which ignore characters other than English alphabet and digits. String "Madam, I'm Adam." should return true.

Ex 7: In the Growing-Reducing Stack implementation using array. Try to figure out a better algorithm which will work similar to Vector<> or ArrayDeque<>.

Chapter 9: Queue

Introduction

A queue is a basic data structure that organized items in first-in-first-out (FIFO) manner. First element inserted into a queue will be the first to be removed. It is also known as "first-come-first-served".

The real life analogy of queue is typical lines in which we all participate time to time.
- We wait in a line of railway reservation counter.
- We wait in the cafeteria line (to pop a plate from "stack of plates").
- We wait in a queue when we call to some customer case.

The elements, which are at the front of the queue, are the one that stayed in the queue for the longest time.

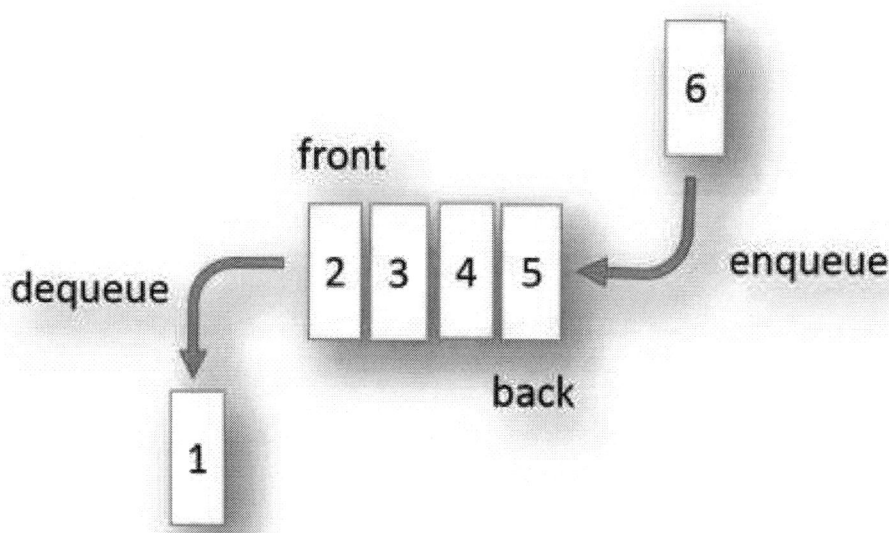

Computer science also has common examples of queues. We issue a print command from our office to a single printer per floor, print task gets lined up in a printer queue. The print command that was issued first will be printed before the next commands in line.

In addition to printing queues, operating system is also using different queues to control process scheduling. Processes are added to processing queue, which is used by an operating system for various scheduling algorithms.

Soon we will be reading about graphs and will come to know about breadth-first traversal, which uses a queue.

The Queue Abstract Data Type

Queue abstract data type is defined as a class whose object, follows FIFO or first-in-first-out for the elements, added to it.
Queue should support the following operation:
1. add(): Which add a single element at the back of a queue
2. remove(): Which remove a single element from the front of a queue.
3. isEmpty(): Returns 1 if the queue is empty
4. size(): Returns the number of elements in a queue.

Queue Using Array

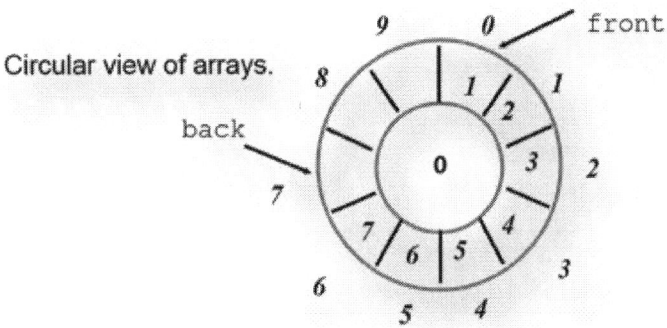

Example 9.1:

```
public class Queue {

    private int size;
    private int Capacity=100;
    private int[] data;
    int front=0;
    int back=0;

    public Queue() {
        size = 0;
        data = new int[100];
    }
}
```

```
boolean isEmpty()
{
    return size == 0;
}
```

```java
int size()
{
    return size;
}
```

```java
public boolean add(int value)
{
    if (size >= Capacity )
    {
        System.out.println("Queue is full.");
        return false;
    }
    else
    {
        size++;
        data[back] = value;
        back = (++back) % (Capacity - 1);
    }
    return true;
}
```

```java
public int remove()
{
    int value;
    if (size <= 0)
    {
        System.out.println("Queue is empty.");
        return -999;
    }
    else
    {
        size--;
        value = data[front];
        front = (++front) % (Capacity - 1);
    }
    return value;
}
```

```java
public static void main(String[] args) {
    Queue que = new Queue();

    for (int i = 0; i < 20; i++)
    {
        que.add(i);
    }
    for (int i = 0; i < 22; i++)
    {
        System.out.println(que.remove());
    }
}
```

Analysis:
1. Hear queue is created from an array of size 100.
2. The number of element in queue to zero. By assigning front, back and size of queue to zero.
3. Add() insert one element at the back of the queue.
4. Remove() delete one element from the front of the queue.

Queue Using linked list

Example 9.2:

```java
public class Queue {
    private static class Node {
        private int value;
        private Node next;

        public Node( int v, Node n){
            value = v;
            next = n;
        }
    }
    private Node head = null;
    private Node tail = null;
    private int size = 0;

    //Other Methods.
}
```

```java
public int size() {
    return size;
}
```

```java
public boolean isEmpty(){
    return size == 0;
}
```

```java
public void print(){
    Node temp = head;
    while(temp != null){
        System.out.print(temp.value +" ");
        temp = temp.next;
    }
}
```

```java
public int peek() throws IllegalStateException {
    if(isEmpty())
        throw new IllegalStateException("QueueEmptyException");
    return head.value
}
```

Add

Enqueue into a queue using linked list. Nodes are added to the end of the linked list. Below diagram indicates how a new node is added to the list. The tail is modified every time when a new value is added to the queue. However, the head is also updated in the case when there is no element in the queue and when that first element is added to the queue both head and tail will be pointing to it.

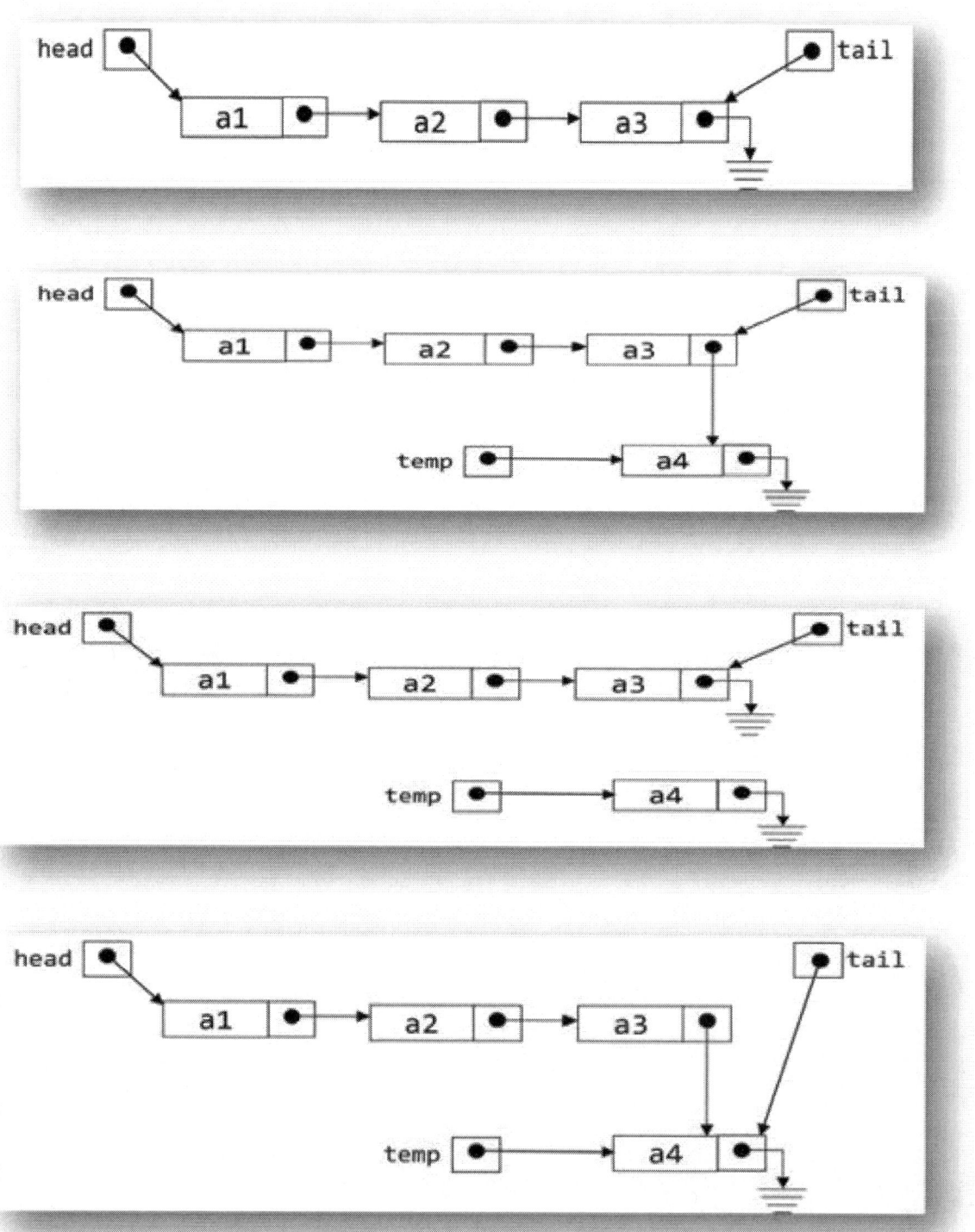

Example 9.3:

```
public void add(int value) {
        Node temp = new Node(value, null);

        if(head == null)
                head = tail = temp;
        else{
                tail.next = temp;
                tail = temp;
        }
        size++;
}
```

Analysis: add operation add one element at the end of the Queue (linked list).

Remove

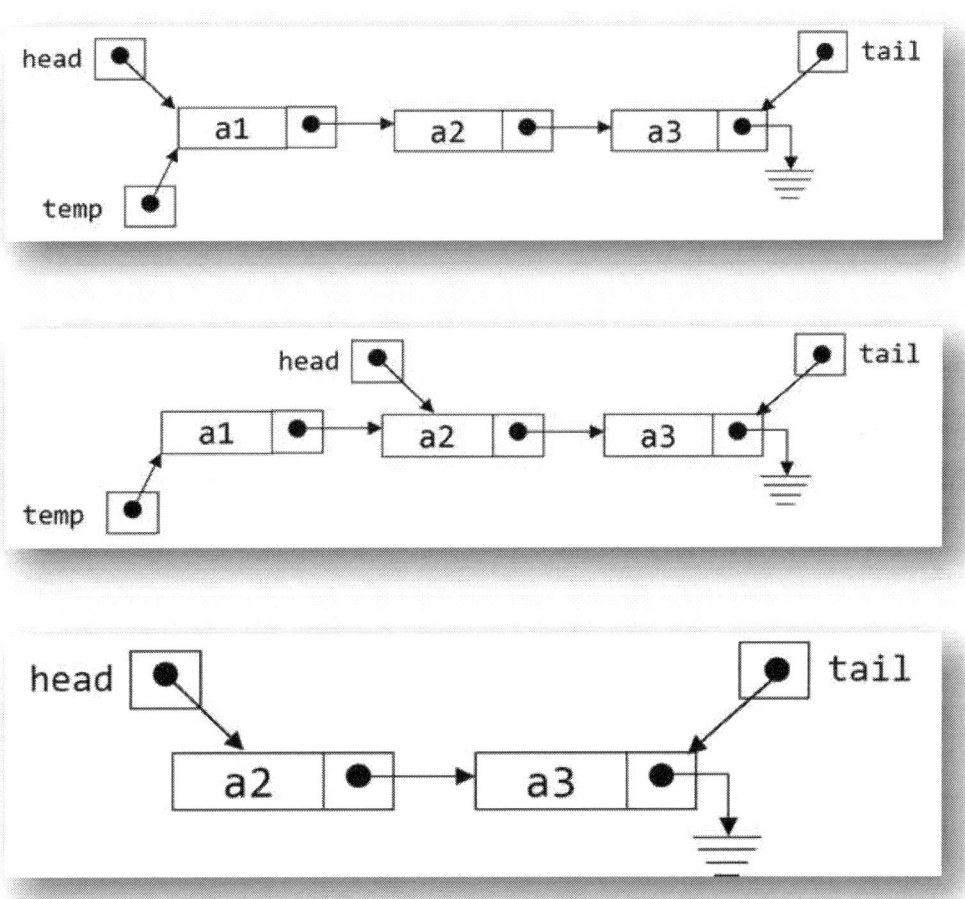

In this we need the tail reference as it may be the case there was only one element in the list and the tail reference will also be modified in case of the remove.

Example 9.4:

```
public int remove() throws IllegalStateException {
    if(isEmpty())
        throw new IllegalStateException("QueueEmptyException");
    int value = head.value;
    head = head.next;
    size--;
    return value;
}
```

```
public static void main(String[] args) {
    Queue q = new Queue();
    for(int i=1;i<=100;i++)
    {
        q.add(i);
    }
    for(int i=1;i<=50;i++)
    {
        q.remove();
    }
    q.print();
}
```

Analysis: Remove operation removes first node from the start of the queue(linked list).

Problems in Queue

Queue using a stack

How to implement a queue using a stack. You can use more than one stack.

Solution: We can use two stack to implement queue. We need to simulate first in first our using stack.
 a) Enqueue Operation: new elements are added to the top of first stack.
 b) Dequeue Operation: elements are popped from the second stack and when second stack is empty then all the elements of first stack are popped one by one and pushed to second stack.

Example 9.5:

```
import java.util.ArrayDeque;

public class QueueUsingStack {

    public QueueUsingStack() {
        stk1 = new ArrayDeque<Integer>();
        stk2 = new ArrayDeque<Integer>();
    }
    private ArrayDeque<Integer> stk1;
    private ArrayDeque<Integer> stk2;
}
```

```java
void add(int value)
{
    stk1.push(value);
}
```

```java
int remove()
{
    int value;
    if(stk2.isEmpty() == false){
        return stk2.pop();
    }
    
    while(stk1.isEmpty() == false){
        value = stk1.pop();
        stk2.push(value);
    }
    return stk2.pop();
}
```

```java
public static void main(String[] args) {
    QueueUsingStack que = new QueueUsingStack();
    
    que.add(1);
    que.add(11);
    que.add(111);
    
    System.out.println(que.remove());
    
    que.add(2);
    que.add(21);
    que.add(211);
    System.out.println(que.remove());
    System.out.println(que.remove());
    System.out.println(que.remove());
    System.out.println(que.remove());
    System.out.println(que.remove());
}
```

Analysis: All add() happens to stack 1. When remove() is called remove happens from stack 2. When the stack 2 is empty then stack 1 is poped and pushed into stack 2. This popping from stack 1 and pushing into stack 2 revert the order of retrival there by making queue behavior out of two stacks.

Stack using a Queue

Implement stack using a queue.

Solution 1: use two queue
Push: add new elements to queue1.
Pop: while size of queue1 is bigger than 1. Push all items from queue 1 to queue 2 except the last item. Switch the name of queue 1 and queue 2. And return the last item.
Push operation is *O(1)* and Pop operation is *O(n)*

Solution 2: This same can be done using just one queue.
Push: add the element to queue.
Pop: find the size of queue. If size is zero then return error. Else, if size is positive then remove size-1 elements from the queue and again add to the same queue. At last, remove the next element and return it.
Push operation is *O(1)* and Pop operation is *O(n)*

Solution 3: In the above solutions the push is efficient and pop is un efficient can we make pop efficient *O(1)* and push inefficient *O(n)*
Push: add new elements to queue2. Then add all the elements of queue 1 to queue 2. Then switch names of queue1 and queue 2.
Pop: remove from queue1
Push operation is *O(n)* and Pop operation is *O(1)*

Reverse a stack

Reverse a stack using queue
Solution:
  a) Pop all the elements of stack and add them into a queue.
  b) Then remove all the elements of the queue into stack
  c) We have the elements of the stack reversed.

Reverse a queue

Reverse a queue using stack
Solution:
  a) Dequeue all the elements of the queue into stack
  b) Then pop all the elements of stack and add them into a queue.
  c) We have the elements of the queue reversed.

Breadth-First Search with a Queue

In breadth-first search, we explore all the nearest nodes first by finding all possible successors and add them to a queue.
  a) Create a queue
  b) Create a start point
  c) Enqueue the start point onto the queue
  d) while (value searching not found and the queue is not empty)
     o Dequeue from the queue
     o Find all possible points after the last one tried
     o Enqueue these points onto the queue

Josephus problem

There are n people standing in a queue waiting to be executed. The counting begins at the front of the queue. In each step, k number of people are removed and again addd one by one from the queue. Then the next person is executed. The execution proceeds around the circle until only the last person remains, who is given freedom.

Find that position where you want to stand and gain your freedom.

Solution:
1) Just insert integer for 1 to k in a queue. (corresponds to k people)
2) Define a Kpop() function such that it will remove and add the queue k-1 times and then remove one more time. (This man is dead.)
3) Repeat second step until size of queue is 1.
4) Print the value in the last element. This is the solution.

Exercise

1) Implement queue using dynamic memory allocation. Such that the implementation should follow the following constraints.
   a) The user should use memory allocation from the heap using new operator. In this, you need to take care of the max value in the queue.

   b) Once you are done with the above exercise and you are able to test your queue. Then you can add some more complexity to your code. In add() function when the queue is full in place of printing "Queue is full" you should allocate more space using new operator.

   c) Once you are done with the above exercise. Now in remove function once you are below half of the capacity of the queue, you need to decrease the size of the queue by half. You should add one more variable "min" to queue so that you can track what is the original value capacity passed at init() function. And the capacity of the queue will not go below the value passed in the initialization.

(If you are not able to solve the above exercise, and then have a look into stack chapter where we have done similar for stack)

2) Implement the below function for the queue:
   a. IsEmpty: This is left as an exercise for the user. Take a variable, which will take care of the size of a queue if the value of that variable is zero, isEmpty should return 1 (true). If the queue is not empty, then it should return 0 (false).

   b. Size: Use the size variable to be used under size function call. Size() function should return the number of elements in the queue.

3) Implement stack using a queue. Write a program for this problem. You can use just one queue.

4) Write a program to Reverse a stack using queue

5) Write a program to Reverse a queue using stack

6) Write a program to solve Josephus problem (algo already discussed.). There are n people standing in a queue waiting to be executed. The counting begins at the front of the queue. In each step, k number of people are removed and again added one by one from the queue. Then the next person is executed. The elimination proceeds around the circle until only the last person remains, who is given freedom. Find that position where you want to stand and gain your freedom.

7) Write a CompStack() function which takes reference to two stack as an argument and return true or false depending upon whether all the elements of the stack are equal or not. You are given isEqual(int, int) which will compare and return 1 if both values are equal and 0 if they are different.

Chapter 10: Tree

Introduction

We have already read about various linear data structures like an array, linked list, stack, queue etc. Both array and linked list have a drawback of linear time required for searching an element.

A tree is a nonlinear data structure, which is used to represent hierarchical relationships (parent-child relationship). Each node is connected by another node by directed edges.

Example 1: Tree in organization

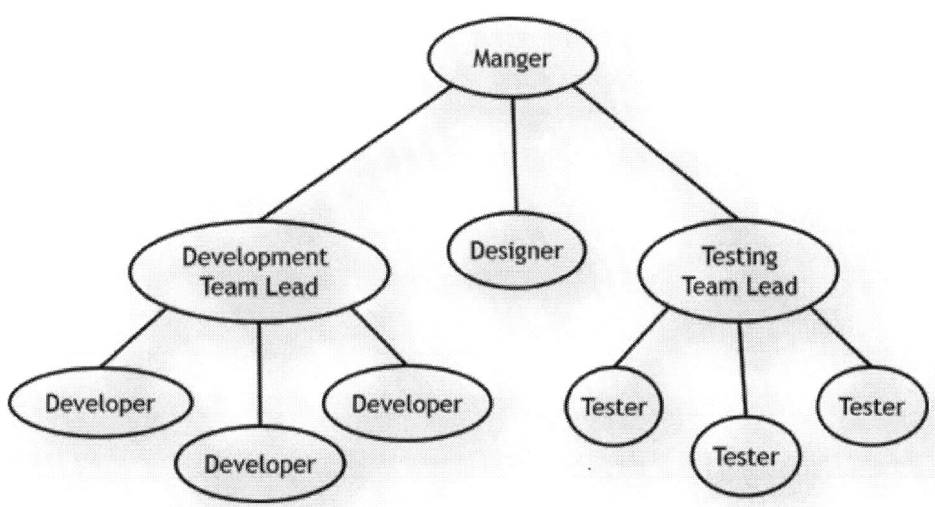

Example 2: Tree in a file system

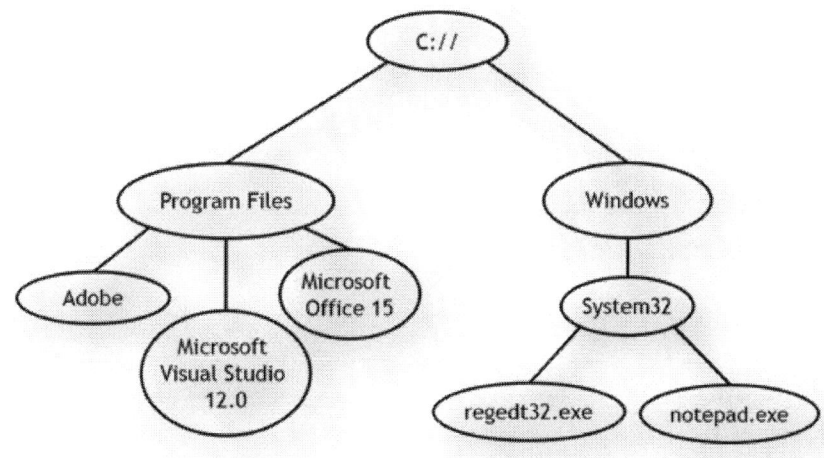

Terminology in tree

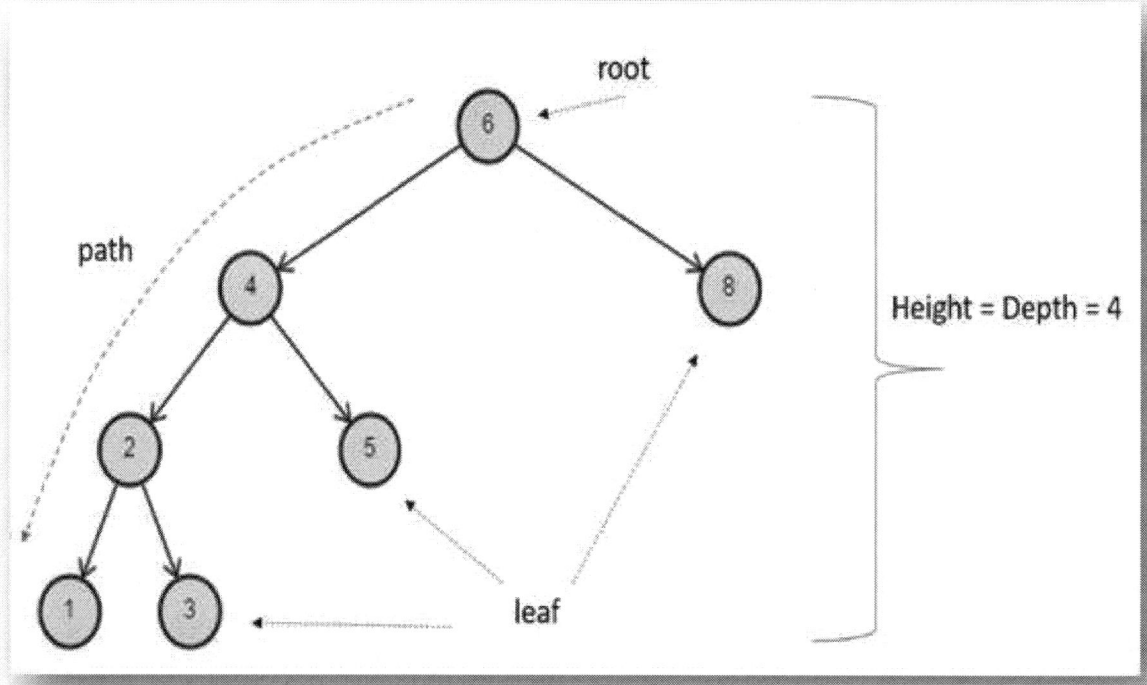

Root: The root of the tree is the only node in the tree that has no incoming edges. It is the top node of a tree.

Node: It is a fundamental element of a tree. Each node has data and two references that may point to null or its child's.

Edge: It is also a fundamental part of a tree, which is used to connect two nodes.

Path: A path is an ordered list of nodes that are connected by edges.

Leaf: A leaf node is a node that has no children.

Height of the tree: The height of a tree is the number of edges on the longest path between the root and a leaf.

The level of node: The level of a node is the number of edges on the path from the root node to that node.

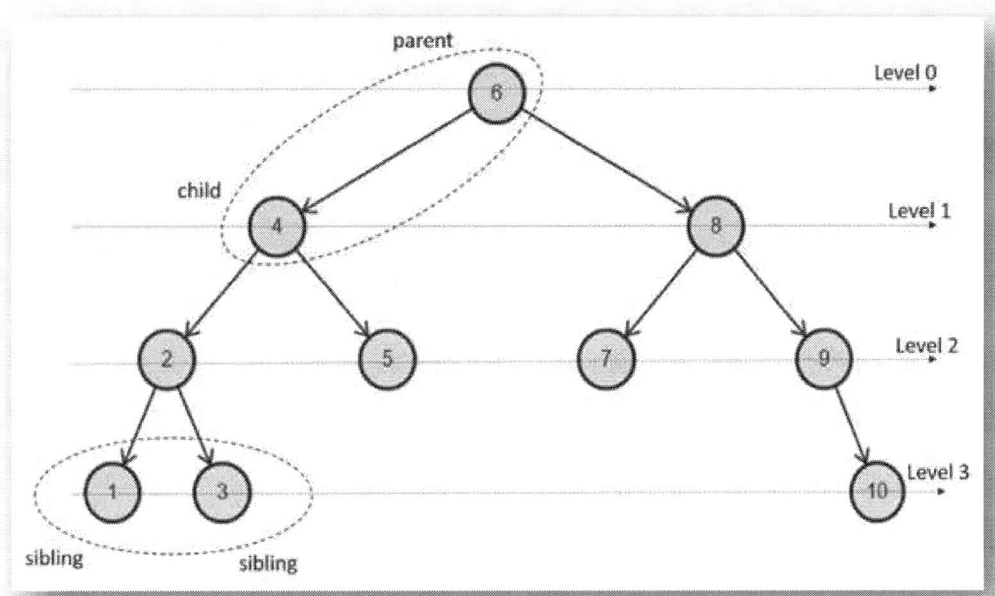

Children: Nodes that have incoming edges from the same node to be said to be the children of that node.

Parent: Node is a parent of all the child nodes that are linked by outgoing edges.

Sibling: Nodes in the tree that are children of the same parent are said to be siblings'

Ancestor: A node reachable by repeated moving from child to parent.

Binary Tree

A binary tree is a type tree in which each node has at most two children (0, 1 or 2), which are referred to as the left child and the right child.

Below is a node of the binary tree with "a" stored as data and whose left child (lChild) and whose right child (rchild) both pointing towards null.

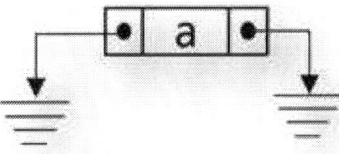

Below is a class definition used to define node.

```
public class Tree {

    private static class Node {
        private int value;
        private Node lChild;
        private Node rChild;

        public Node( int v, Node l, Node r)        {
            value = v;
            lChild = l;
            rChild = r;
        }

        public Node( int v ) {
            value = v;
            lChild = null;
            rChild = null;
        }
    }

    private Node root;

    public Tree() {
        root = null;
    }
}
```

Below is a binary tree whose nodes contains data from 1 to 10

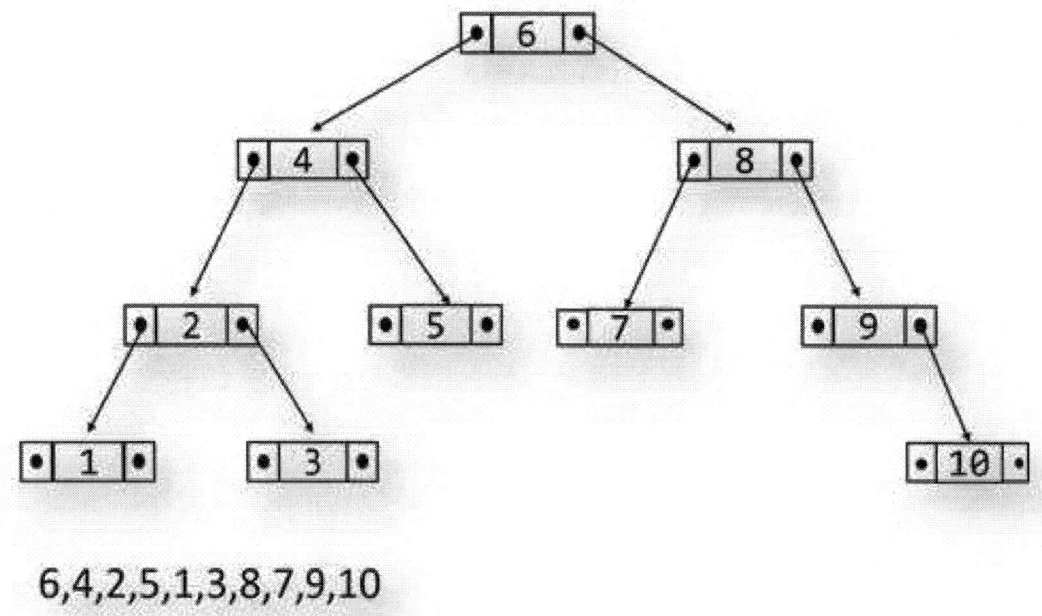

6,4,2,5,1,3,8,7,9,10

In the rest of the book binary tree will be represented as below:

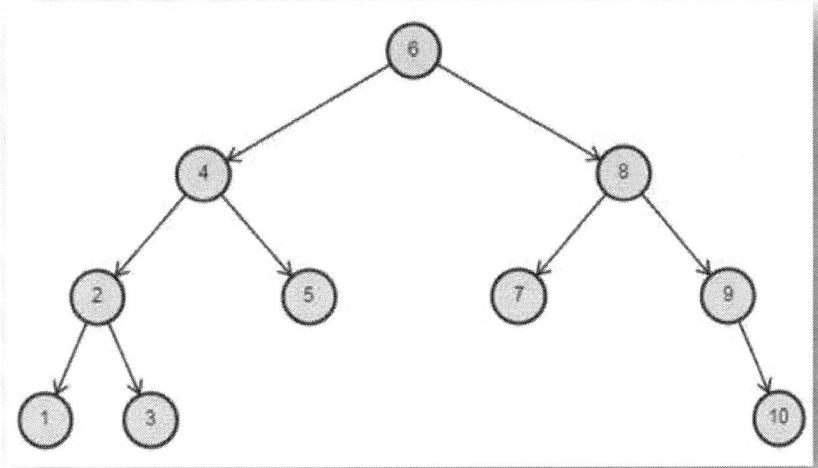

Properties of Binary tree are:
1. The maximum number of nodes on level i of a binary tree is 2^i, where i >= 1
2. The maximum number of nodes in a binary tree of depth k is 2^{k+1}, where k >= 1
3. There is exactly one path from the root to any nodes in a tree.
4. A tree with N nodes have exactly N-1 edges connecting these nodes.
5. The height of a complete binary tree of N nodes is $\log_2 N$.

Types of Binary trees

Complete binary tree

In a complete binary tree, every level except the last one is completely filled. All nodes in the left are filled first, then the right one.
A binary heap is an example of a complete binary tree.

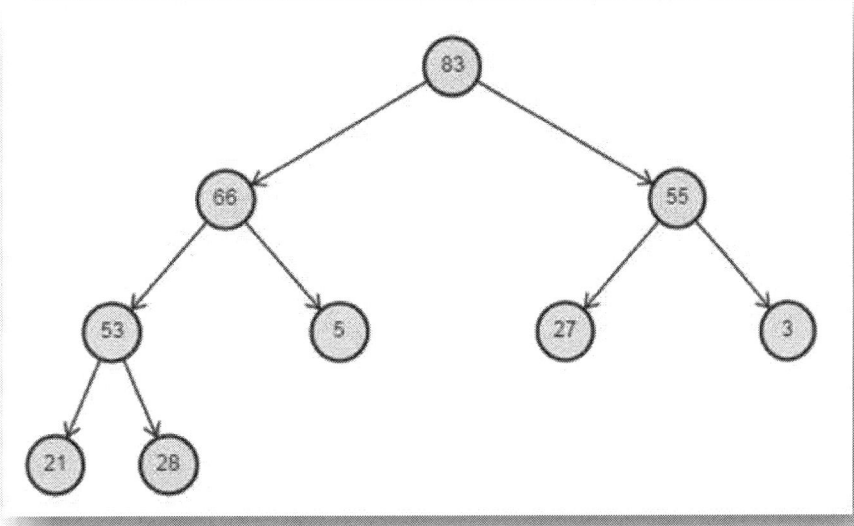

Full/ Strictly binary tree

The full binary tree is a binary tree in which each node has exactly zero or two children.

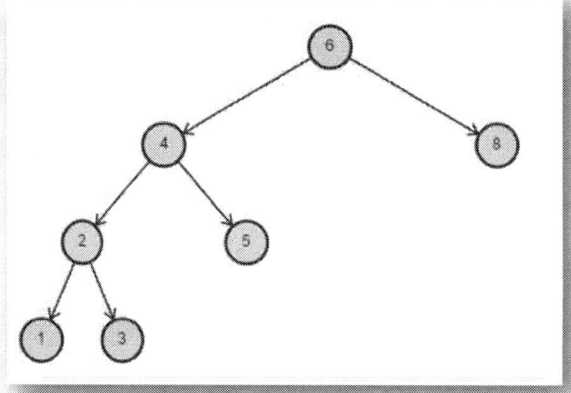

Perfect binary tree

The perfect binary tree is a type of full binary tree in which each non leaf node has exactly two child nodes.
All leaf nodes have identical path length and all possible node slots are occupied

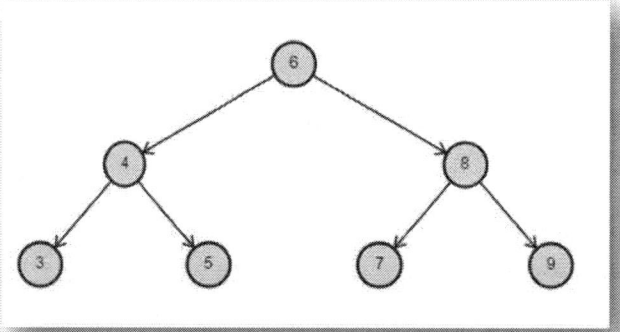

Right skewed binary tree

A binary tree in which each node is having either only a right child or no child (leaf) is called as right skewed binary tree

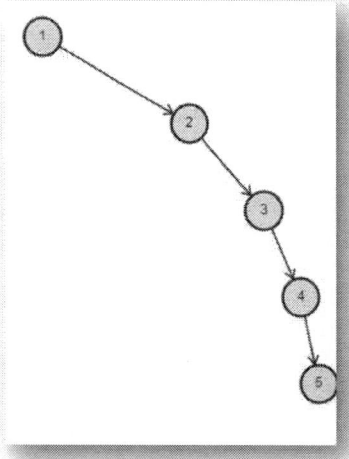

Left skewed binary tree

A binary tree in which each node is having either only a left child or no child (leaf) is called as Left skewed binary tree

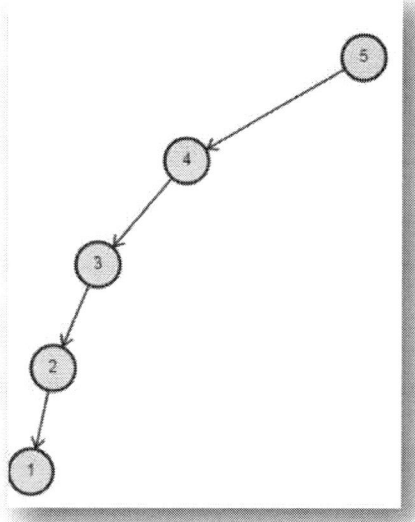

Height-balanced Binary Tree

A height-balanced binary tree is a binary tree such that the left & right subtrees for any given node differ in height by max one.
Note: Each complete binary tree is a height-balanced binary tree

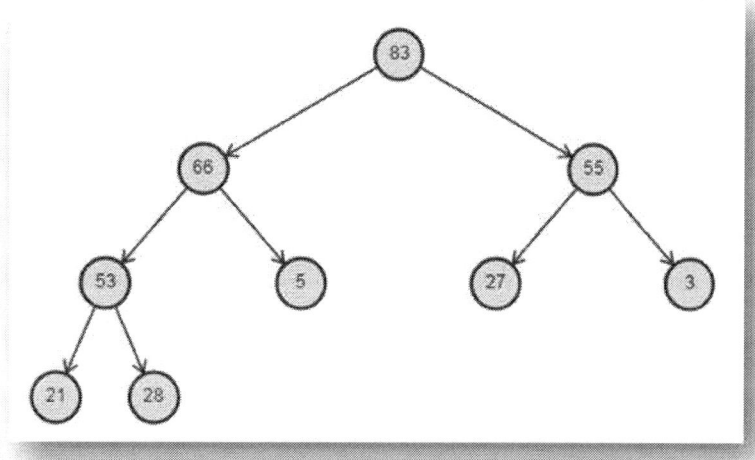

AVL tree and RB tree are an example of height balanced tree we will discuss these trees in advance tree topic.

Problems in Binary Tree

Create a Complete binary tree

Create a binary tree given a list of values in an array.
Solution: Since there is no order defined in a binary tree, so nodes can be inserted in any order so it can be a skewed binary tree. But it is inefficient to do anything in a skewed binary tree so we will create a Complete binary tree. At each node, the middle value stored in the array is assigned to node and left of array is passed to the left child of the node to create left sub-tree. And right portion of array is passed to right child of the node to crate right sub-tree.

Example 10.1:

```java
public void levelOrderBinaryTree(int[] arr){
    root = levelOrderBinaryTree(arr, 0);
}
```

```java
public Node levelOrderBinaryTree(int[] arr, int start)
{
    int size = arr.length;
    Node curr = new Node(arr[start]);
    int left = 2 * start + 1;
    int right = 2 * start + 2;

    if (left < size)
        curr.lChild = levelOrderBinaryTree(arr, left);
    if (right < size)
        curr.rChild = levelOrderBinaryTree(arr, right);

    return curr;
}
```

```java
public static void main(String[] args){
    Tree t = new Tree();
    int[] arr = {1,2,3,4,5,6,7,8,9,10};
    t.levelOrderBinaryTree(arr);
    t.Print();
}
```

Complexity Analysis:
This is an efficient algorithm for creating a complete binary tree.
Time Complexity: $O(n)$ Space Complexity: $O(n)$

Pre-Order Traversal

Traversal is a process of visiting each node of a tree. In Pre-Order Traversal parent is visited/traversed first, then left child and right child. Pre-Order traversal is a type of depth-first traversal.

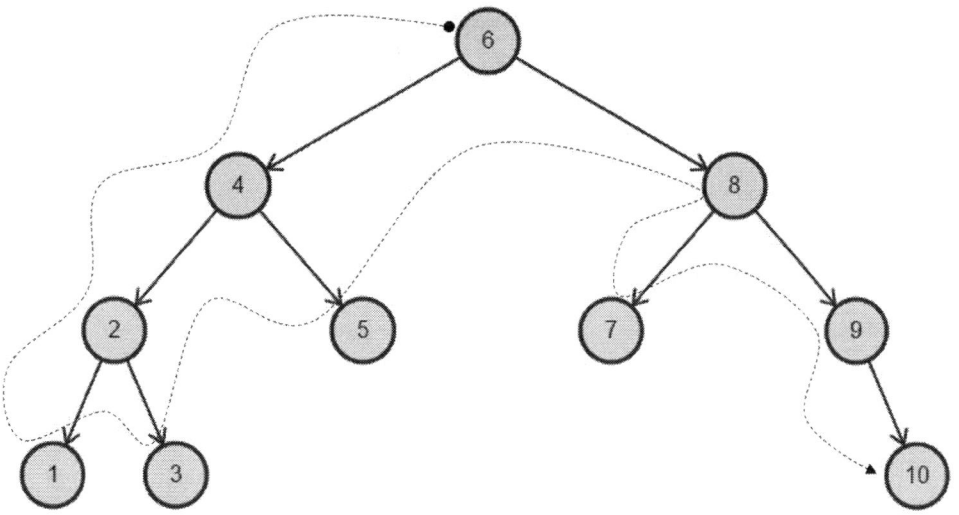

Solution:
Preorder traversal is done using recursion. At each node, first the value stored in it is printed and then followed by the value of left child and right child. At each node its value is printed followed by calling printTree() function to its left and right child to print left and right sub-tree.

Example 10.2:

```java
public void PrintPreOrder(){
    PrintPreOrder(root);
}
```

```java
private void PrintPreOrder(Node node)
{
    if(node != null)
    {
        System.out.print(" "+ node.value);
        PrintPreOrder(node.lChild);
        PrintPreOrder(node.rChild);
    }
}
```

Output:

```
6 4 2 1 3 5 8 7 9 10
```

Complexity Analysis: Time Complexity: $O(n)$ Space Complexity: $O(n)$

Note: When there is an algorithm in which all nodes are traversed, then complexity can't be less then $O(n)$. When there is a large portion of the tree, which is not traversed, then complexity reduces.

Post-Order Traversal

In Post-Order Traversal left child is visited/traversed first, then right child and last parent
Post-Order traversal is a type of depth-first traversal.

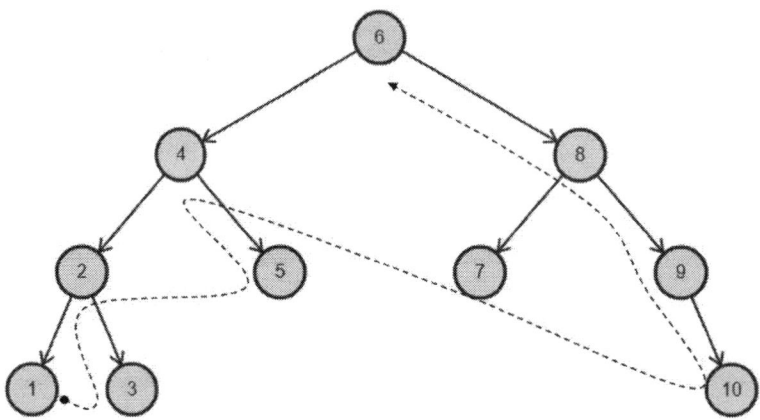

Solution: In post order traversal, first, the left child is traversed then right child and in the end, current node value is printed to the screen.

Example 10.3:

```
public void PrintPostOrder(){
      PrintPostOrder(root);
}
```

```
private void PrintPostOrder(Node node)/*   post order   */
{
      if(node != null)
      {
            PrintPostOrder(node.lChild);
            PrintPostOrder(node.rChild);
            System.out.print(" "+ node.value);
      }
}
```

Output:

```
1 3 2 5 4 7 10 9 8 6
```

Complexity Analysis: Time Complexity: $O(n)$ Space Complexity: $O(n)$

In-Order Traversal

In In-Order Traversal left child is visited/traversed first, then the parent and last right child
In-Order traversal is a type of depth-first traversal. The output of In-Order traversal of BST is a sorted list.

Solution:

In In-Order traversal first, the value of left child is traversed, then the value of node is printed to the screen and then the value of right child is traversed.

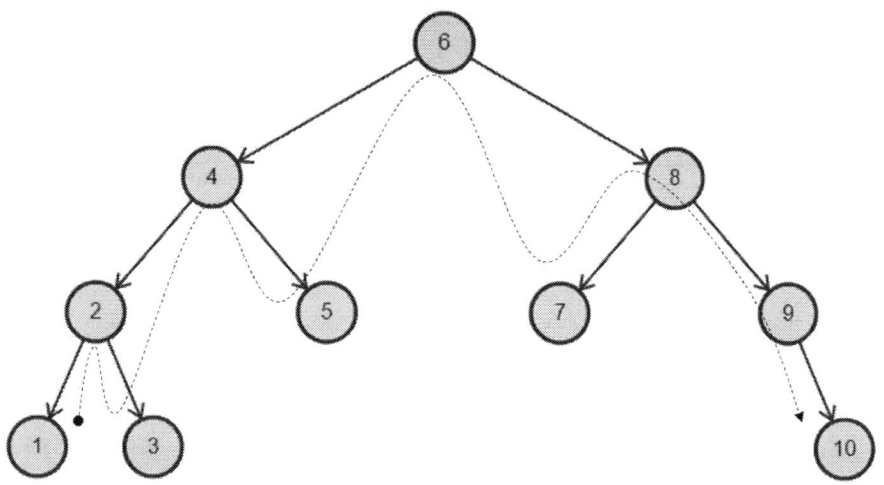

Example 10.4:

```
public void PrintInOrder(){
    PrintPostOrder(root);
}
```

```
private void PrintInOrder(Node node)/*   In order   */
{
    if(node != null)
    {
        PrintPostOrder(node.lChild);
        System.out.print(" "+ node.value);
        PrintPostOrder(node.rChild);
    }
}
```

Output:

1 2 3 4 5 6 7 8 9 10

Complexity Analysis: Time Complexity: $O(n)$ Space Complexity: $O(n)$

Note: Pre-Order, Post-Order, and In-Order traversal are for all binary trees. They can be used to traverse any kind of a binary tree.

Level order traversal / Breadth First traversal

Write code to implement level order traversal of a tree. Such that nodes at depth k is printed before nodes at depth k+1.

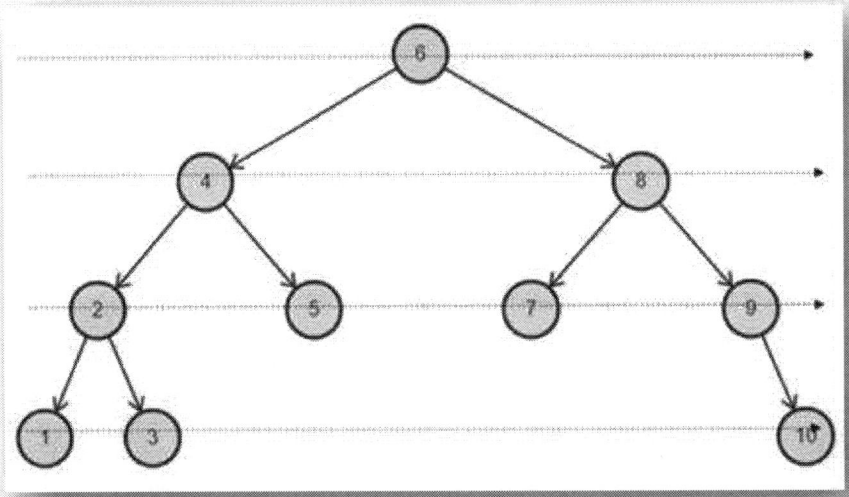

Solution:
Level order traversal or Breadth First traversal of a tree is done using a queue. At the start, the root node reference is added to queue. The traversal of tree happens until its queue is empty. When we traverse the tree, we first remove an element from the queue, print the value stored in that node and then its left child and right child will be added to the queue.

Example 10.5:

```java
public void PrintBredthFirst(){
    ArrayDeque<Node> que = new ArrayDeque<Node>();
    Node temp;

    if(root != null)
        que.add(root);

    while(que.isEmpty() == false)
    {
        temp= que.remove();
        System.out.println(temp.value);

        if(temp.lChild != null)
            que.add(temp.lChild);
        if(temp.rChild != null)
            que.add(temp.rChild);
    }
}
```

Complexity Analysis: Time Complexity: $O(n)$ Space Complexity: $O(n)$

Print Depth First without using the recursion / system stack.

Solution: Depth first traversal of the tree is done using recursion by using system stack. The same can be done using stack. In the start root node reference is added to the stack. The whole tree is traversed until the stack is empty. In each iteration, an element is popped from the stack its value is printed to screen. Then right child and then left child of the node is added to stack.

Example 10.6:

```java
public void PrintDepthFirst(){

    ArrayDeque<Node> stk = new ArrayDeque<Node>();
    Node temp;

    if(root != null)
        stk.push(root);

    while(stk.isEmpty() == false)
    {
        temp= stk.pop();
        System.out.println(temp.value);

        if(temp.lChild != null)
            stk.push(temp.lChild);
        if(temp.rChild != null)
            stk.push(temp.rChild);
    }
}
```

Complexity Analysis: Time Complexity: $O(n)$ Space Complexity: $O(n)$

Tree Depth

Solution: Depth is tree is calculated recursively by traversing left and right child of the root. At each level of traversal depth of left child is calculated and depth of right child is calculated. The greater depth among the left and right child is added by one (which is the depth of the current node) and this value is returned.

Example 10.7:

```java
private int TreeDepth(Node root) {
    if(root == null)
        return 0;
    else
    {
        int lDepth=TreeDepth(root.lChild);
        int rDepth=TreeDepth(root.rChild);
        if(lDepth>rDepth)
            return lDepth+1;
        else
            return rDepth+1;
    }
}
```

```
public int TreeDepth(){
    return TreeDepth(root);
}
```

Complexity Analysis: Time Complexity: $O(n)$ Space Complexity: $O(n)$

Nth Pre-Order

Solution: We want to print the node which will be at the nth index when we print the tree in PreOrder traversal. So we keep a counter to keep track of the index. When the counter is equal to index, then we print the value and return the Nth preorder index node.

Example 10.8:

```
public void NthPreOrder(int index){
    NthPreOrder(root, index, 0);
}

private void NthPreOrder(Node node, int index, int counter)/* pre order */
{
    if(node != null)
    {
        counter++;
        if(counter == index)
        {
            System.out.print(" "+ node.value);
        }
        NthPreOrder(node.lChild, index, counter);
        NthPreOrder(node.rChild, index, counter);
    }
}
```

Complexity Analysis: Time Complexity: $O(n)$ Space Complexity: $O(n)$

Nth Post Order

Solution: We want to print the node that will be at the nth index when we print the tree in post order traversal. So we keep a counter to keep track of the index, but at this time we will increment the counter after left child and right child traversal. When the counter is equal to index, then we print the value and return the nth post order index node.

Example 10.9

```
public void NthPostOrder(int index)
{
    NthPostOrder(root, index, 0);
}
```

```
private void NthPostOrder(Node node, int index, int counter)
{
    if(node != null)
    {
        NthPostOrder(node.lChild,index, counter);
        NthPostOrder(node.rChild,index, counter);
        counter++;
        if(counter == index)
        {
            System.out.print(" "+ node.value);
        }
    }
}
```

Complexity Analysis:
Time Complexity: $O(n)$ Space Complexity: $O(n)$

Nth In Order

Solution:
We want to print the node which will be at the nth index when we print the tree in in-order traversal. So we keep a counter to keep track of the index, but at this time we will increment the counter after left child traversal but before the right child traversal. When the counter is equal to index, then we print the value and return the nth in-order index node.

Example 10.10:

```
public void NthInOrder(int index){
    NthInOrder(root, index, 0);
}
```

```
private void NthInOrder(Node node, int index, int counter)
{
    if(node != null)
    {
        NthInOrder(node.lChild, index, counter);
        counter++;
        if(counter == index)
        {
            System.out.print(" "+ node.value);
        }
        NthInOrder(node.rChild, index, counter);
    }
}
```

Complexity Analysis:
Time Complexity: $O(n)$ Space Complexity: $O(1)$

Copy Tree

Solution: Copy tree is done by copy nodes of the input tree at each level of the traversal of the tree. At each level of the traversal of nodes of tree a new node is created and the value of the input tree node is copied to it. The left child tree is copied recursively and then reference to new subtree is returned will be assigned to the left child of the current new node. Similarly for the right child too. Finally, the tree is copied.

Example 10.11:

```
private Node CopyTree(Node curr)
{
        Node temp;
        if(curr != null)
        {
                temp = new Node(curr.value);
                temp.setLeftChild(CopyTree(curr.lChild));
                temp.setRightChild(CopyTree(curr.rChild));
                return temp;
        }
        else
                return null;
}
```

```
public Tree CopyTree(){
        Tree tree2 = new Tree();
        tree2.root = CopyTree(root);
        return tree2;
}
```

Complexity Analysis: Time Complexity: $O(n)$
Space Complexity: $O(n)$

Copy Mirror Tree

Solution: Copy, mirror image of the tree is done same as copy tree, but in place of left child pointing to the tree formed by left child traversal of input tree, this time left child points to the tree formed by right child traversal. Similarly right child point to the traversal of the left child of the input tree.

Example 10.12:

```
public Tree CopyMirrorTree(){
        Tree tree2 = new Tree();
        tree2.root = CopyMirrorTree(root);
        return tree2;
}
```

```
private Node CopyMirrorTree(Node curr)
{
        Node temp;
        if(curr != null)
        {
                temp = new Node(curr.value);
                temp.setRightChild(CopyTree(curr.lChild));
                temp.setLeftChild(CopyTree(curr.rChild));
                return temp;
        }
        else
                return null;
}
```

Complexity Analysis: Time Complexity: $O(n)$
Space Complexity: $O(n)$

Number of Element

Solution: Number of nodes at the right child and the number of nodes at the left child is added by one and we get the total number of nodes in any tree/sub-tree.

Example 10.13:

```
public int numNodes(){
        return numNodes(root);
}
```

```
public int numNodes(Node curr)
{
        if(curr == null)
                return 0;
        else
                return (1 + numNodes(curr.rChild) + numNodes(curr.lChild) );
}
```

Complexity Analysis: Time Complexity: $O(n)$ Space Complexity: $O(n)$

Number of Leaf nodes

Solution: If we add the number of leaf node in the right child with the number of leaf nodes in the left child, we will get the total number of leaf node in any tree or subtree.

Example 10.14:

```
public int numLeafNodes()
{
        return numLeafNodes(root);
}
```

```
private int numLeafNodes(Node curr)
{
    if(curr == null)
        return 0;
    if( curr.lChild == null && curr.rChild == null )
        return 1;
    else
        return (numLeafNodes(curr.rChild) + numLeafNodes(curr.lChild) );
}
```

Complexity Analysis: Time Complexity: $O(n)$ Space Complexity: $O(n)$

Identical

Solution: Two trees have identical values if at each level the value is equal.

Example 10.15:

```
public boolean isEqual(Tree T2){
    return Identical(root, T2.root);
}
```

```
private boolean Identical(Node node1,Node node2)
{
    if(node1 == null && node2 == null)
        return true;
    else if(node1 == null || node2 == null)
        return false;
    else
        return(Identical(node1.lChild, node2.lChild)
            && Identical(node1.rChild, node2.rChild)
            && (node1.value == node2.value));
}
```

Complexity Analysis: Time Complexity: $O(n)$ Space Complexity: $O(n)$

Free Tree

Solution: The tree is traversed and nodes of tree are freed in such a manner such that all child nodes are freed before it.

Example 10.16:

```
public void Free(){
    root = null;
}
```

Complexity Analysis:
Time Complexity: $O(1)$ Space Complexity: $O(1)$
System will do garbage collection so for user action the time complexity is 0.

Tree to List Rec

Solution: Tree to the list is done recursively. At each node we will suppose that the tree to list function will do its job for the left child and right child. Then we will combine the result of the left child and right child traversal. We need a head and tail reference of the left list and right list to combine them with the current node. In the process of integration the current node will be added to the tail of the left list and current node will be added to the head to the right list. Head of the left list will become the head of the newly formed list and tail of the right list will become the tail of the newly created list.

Example 10.17:

```
private Node treeToListRec(Node curr)
{
        Node Head=null, Tail=null;
        if(curr == null)
                return null;

        if(curr.lChild == null && curr.rChild == null)
        {
                curr.lChild = curr;
                curr.rChild = curr;
                return curr;
        }

        if(curr.lChild != null)
        {
                Head = treeToListRec(curr.lChild);
                Tail = Head.lChild;

                curr.lChild = Tail;
                Tail.rChild = curr;
        }
        else
                Head=curr;

        if(curr.rChild != null)
        {
                Node tempHead = treeToListRec(curr.rChild);
                Tail = tempHead.lChild;

                curr.rChild = tempHead;
                tempHead.lChild = curr;
        }
        else
                Tail=curr;

        Head.lChild=Tail;
        Tail.rChild=Head;
        return Head;
}
```

```java
public Node treeToListRec(){
    Node head = treeToListRec(root);
    Node temp = head;
    return temp;
}
```

Complexity Analysis:
Time Complexity: $O(n)$ Space Complexity: $O(n)$

Print all the paths

Print all the paths from the roots to the leaf

Solution: Whenever we traverse a node we add that node to the list. When we reach a leaf we print the whole list. When we return from a function, then we remove the element that was added to the list when we entered this function.

Example 10.18:

```java
public void printAllPath()
{
    ArrayDeque<Integer> stk = new ArrayDeque<Integer>();
    printAllPath(root,stk);
}
```

```java
private void printAllPath(Node curr, ArrayDeque<Integer> stk)
{
    if(curr == null)
        return;

    stk.push(curr.value);

    if(curr.lChild == null && curr.rChild == null)
    {
        System.out.println(stk);
        stk.pop();
        return;
    }

    printAllPath(curr.rChild,stk);
    printAllPath(curr.lChild,stk);
    stk.pop();
}
```

Complexity Analysis:
Time Complexity: $O(n)$, Space Complexity: $O(n)$

Least Common Ancestor

Solution: We recursively traverse the nodes of a binary tree. And we find any one of the node we are searching for then we return that node. And when we get both the left and right as some valid reference location other than null we will return that node as the common ancestor.

Example 10.19:

```
public int LCA(int first, int second){
        Node ans = LCA(root, first, second);
        if(ans != null)
                return ans.value;
        else
                return Integer.MIN_VALUE;
}

private Node LCA(Node curr, int first, int second)
{
        Node left, right;

        if (curr == null)
                return null;

        if (curr.value == first || curr.value == second)
                return curr;

        left = LCA(curr.lChild, first, second);
        right = LCA(curr.rChild, first, second);

        if (left != null && right != null)
                return curr;
        else if (left != null)
                return left;
        else
                return right;
}
```

Complexity Analysis: Time Complexity: $O(n)$ Space Complexity: $O(n)$

Find Max in Binary Tree

Solution: We recursively traverse the nodes of a binary tree. We will find the maximum value in the left and right subtree of any node then will compare the value with the value of the current node and finally return the largest of the three values.

Example 10.20:

```
public int findMaxBT(){
        int ans = findMaxBT(root);
        return ans;
}
```

```
private int findMaxBT(Node curr)
{
    int left, right;

    if (curr == null)
        return Integer.MIN_VALUE;

    int max = curr.value;

    left = findMaxBT(curr.lChild);
    right = findMaxBT(curr.rChild);

    if (left > max)
        max = left;
    if (right > max)
        max = right;

    return max;
}
```

Search value in a Binary Tree

Solution: To find if some value is there in a binary tree or not is done using exhaustive search of the binary tree. First, the value of current node is compared with the value which we are looking for. Then it is compared recursively inside the left child and right child.

Example 10.21:

```
public boolean searchBT(Node root, int value)
{
    int max;
    boolean left, right;

    if (root == null)
        return false;

    if(root.value== value)
        return true;

    left = searchBT(root.lChild, value);
    if (left)
        return true;

    right = searchBT(root.rChild, value);
    if (right)
        return true;

    return false;
}
```

Maximum Depth in a Binary Tree

Solution: To find the maximum depth of a binary tree we need to find the depth of the left tree and depth of right tree then we need to store the value and increment it by one so that we get depth of the given node.

Example 10.22:

```
public int TreeDepth(){
        return TreeDepth(root);
}

private int TreeDepth(Node root)
{
        if(root == null)
                return 0;
        else
        {
                int lDepth=TreeDepth(root.lChild);
                int rDepth=TreeDepth(root.rChild);

                if(lDepth>rDepth)
                        return lDepth+1;
                else
                        return rDepth+1;
        }
}
```

Number of Full Nodes in a BT

Solution: A full node is a node which have both left and right child. We will recursively travers the whole tree and will increase the count of full node as we find them.

Example 10.23:

```
public int numFullNodesBT(){
        return numNodes(root);
}

public int numFullNodesBT(Node curr)
{
        int count;
        if(curr == null)
                return 0;

        count = numFullNodesBT(curr.rChild) + numFullNodesBT(curr.lChild);
        if(curr.rChild != null && curr.lChild != null)
                count++;

        return count;
}
```

Maximum Length Path in a BT/ Diameter of BT

Solution: To find the diameter of BT we need to find the depth of left child and right child then will add these two values and increment it by one so that we will get the maximum length path (diameter candidate) which contains the current node. Then we will find max length path in the left child sub-tree. And will also find the max length path in the right child sub-tree. Finally, we will compare the three values and return the maximum value out of these this will be the diameter of the Binary tree.

Example 10.24:

```java
public int maxLengthPathBT()
{
    return maxLengthPathBT(root);
}
```

```java
private int maxLengthPathBT(Node curr)//diameter
{
    int max;
    int leftPath, rightPath;
    int leftMax, rightMax;

    if (curr == null)
        return 0;

    leftPath = TreeDepth(curr.lChild);
    rightPath = TreeDepth(curr.rChild);

    max = leftPath + rightPath + 1;

    leftMax = maxLengthPathBT(curr.lChild);
    rightMax = maxLengthPathBT(curr.rChild);

    if (leftMax > max)
        max = leftMax;

    if (rightMax > max)
        max = rightMax;

    return max;
}
```

Sum of All nodes in a BT

Solution: We will find the sum of all the nodes recursively. sumAllBT() will return the sum of all the node of left and right subtree then will add the value of current node and will return the final sum.

Example 10.25:

```java
public int sumAllBT(){
    return sumAllBT(root);
}
```

```java
private int sumAllBT(Node curr)
{
    int sum, leftSum, rightSum;

    if(curr == null)
        return 0;

    rightSum = sumAllBT(curr.rChild);
    leftSum = sumAllBT(curr.lChild);

    sum = rightSum + leftSum + curr.value;

    return sum;
}
```

Iterative Pre-order

Solution: In place of using system stack in recursion, we can traverse the tree using stack data structure.

Example 10.26:

```java
public void iterativePreOrder(){
    ArrayDeque<Node> stk = new ArrayDeque<Node>();
    Node curr;

    if(root != null)
        stk.add(root);

    while(stk.isEmpty() == false){
        curr = stk.pop();
        System.out.print(curr.value + " ");

        if(curr.rChild != null)
            stk.push(curr.rChild);

        if(curr.lChild != null)
            stk.push(curr.lChild);
    }
}
```

Complexity Analysis:
Time Complexity: $O(n)$ Space Complexity: $O(n)$

Iterative Post-order

Solution: In place of using system stack in recursion, we can traverse the tree using stack data structure.

Example 10.27:

```java
public void iterativePostOrder() {
        ArrayDeque<Node> stk = new ArrayDeque<Node>();
        ArrayDeque<Integer> visited=new ArrayDeque<Integer>();
        Node curr;
        int vtd;

        if (root != null) {
                stk.add(root);
                visited.add(0);
        }

        while (stk.isEmpty() == false) {
                curr = stk.pop();
                vtd = visited.pop();
                if (vtd == 1) {
                        System.out.print(curr.value + " ");
                } else {
                        stk.push(curr);
                        visited.push(1);
                        if (curr.rChild != null) {
                                stk.push(curr.rChild);
                                visited.push(0);
                        }
                        if (curr.lChild != null) {
                                stk.push(curr.lChild);
                                visited.push(0);
                        }
                }
        }
}
```

Complexity Analysis: Time Complexity: $O(n)$ Space Complexity: $O(n)$

Iterative In-order

Solution: In place of using system stack in recursion, we can traverse the tree using stack data structure.

Example 10.28:

```java
public void iterativeInOrder() {
        ArrayDeque<Node> stk = new ArrayDeque<Node>();
        ArrayDeque<Integer> visited = new ArrayDeque<Integer>();
        Node curr;
        int vtd;

        if (root != null) {
                stk.add(root);
                visited.add(0);
        }
```

```
            while (stk.isEmpty() == false) {
                curr = stk.pop();
                vtd = visited.pop();
                if (vtd == 1) {
                    System.out.print(curr.value + " ");
                } else {
                    if (curr.rChild != null) {
                        stk.push(curr.rChild);
                        visited.push(0);
                    }
                    stk.push(curr);
                    visited.push(1);
                    if (curr.lChild != null) {
                        stk.push(curr.lChild);
                        visited.push(0);
                    }
                }
            }
        }
    }
```

Complexity Analysis: Time Complexity: $O(n)$ Space Complexity: $O(n)$

Binary Search Tree (BST)

A binary search tree (BST) is a binary tree on which nodes are ordered in the following way:
- The key in the left subtree is less than the key in its parent node.
- The key in the right subtree is greater the key in its parent node.
- No duplicate key allowed.

Note: there can be two separate key and value fields in the tree node. But for simplicity, we are considering value as the key. All problems in the binary search tree are solved using this supposition that the value in the node is key for the tree.

Note: Since binary search tree is a binary tree to all the above algorithm of a binary tree are applicable to a binary search tree.

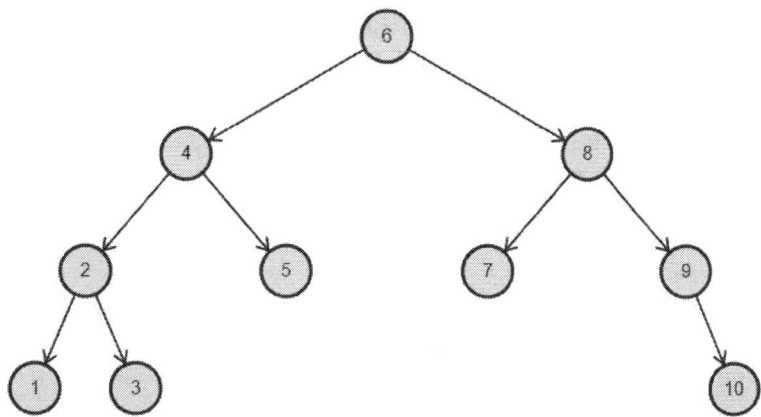

Problems in Binary Search Tree (BST)

All binary tree algorithms are valid for binary search tree too.

Create a binary search tree from sorted array

Create a binary tree given list of values in an array in sorted order

Since the elements in the array are in sorted order and we want to create a binary search tree in which left subtree nodes are having values less than the current node and right subtree nodes have value greater than the value of the current node.

Solution: We have to find the middle node to create a current node and send the rest of the array to construct left and right subtree.
Example 10.29:

```
public void CreateBinaryTree (int[] arr)
{
        root = CreateBinaryTree (arr, 0, arr.length - 1);
}
```

```
private Node CreateBinaryTree (int[] arr, int start, int end)
{
        Node curr= null;
        if (start > end)
                return null;

        int mid = ( start + end ) / 2;
        curr = new Node(arr[mid]);
        curr.lChild = CreateBinaryTree (arr, start, mid-1);
        curr.rChild = CreateBinaryTree (arr, mid+1, end);
        return curr;
}
```

```
public static void main(String[] args){
        Tree t = new Tree();
        int[] arr = {1,2,3,4,5,6,7,8,9,10};
        t.CreateBinaryTree(arr);
        t.PrintInOrder();
}
```

Insertion

Nodes with key 6,4,2,5,1,3,8,7,9,10 are inserted in a tree. Below is step by step tree after inserting nodes in the order.

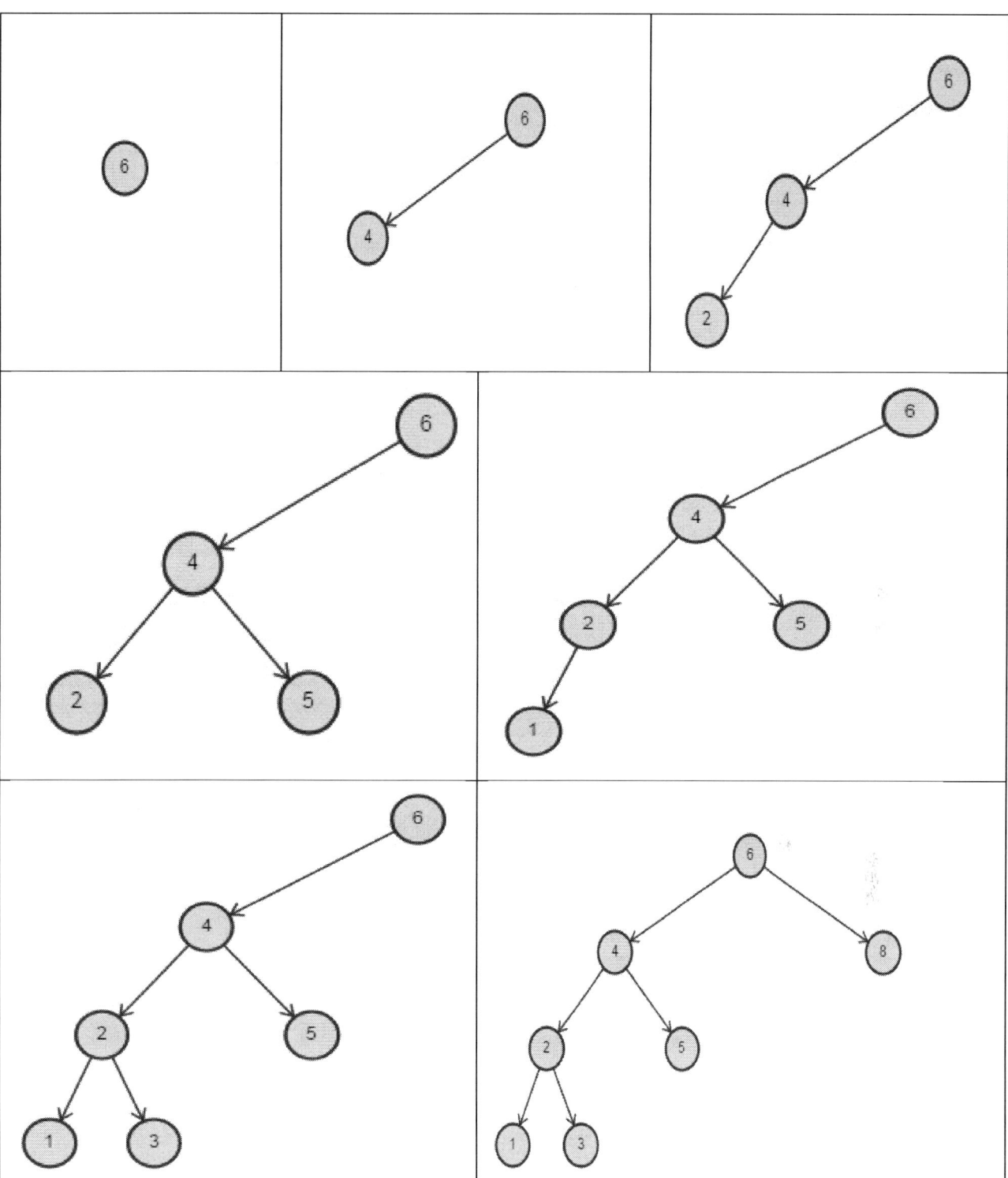

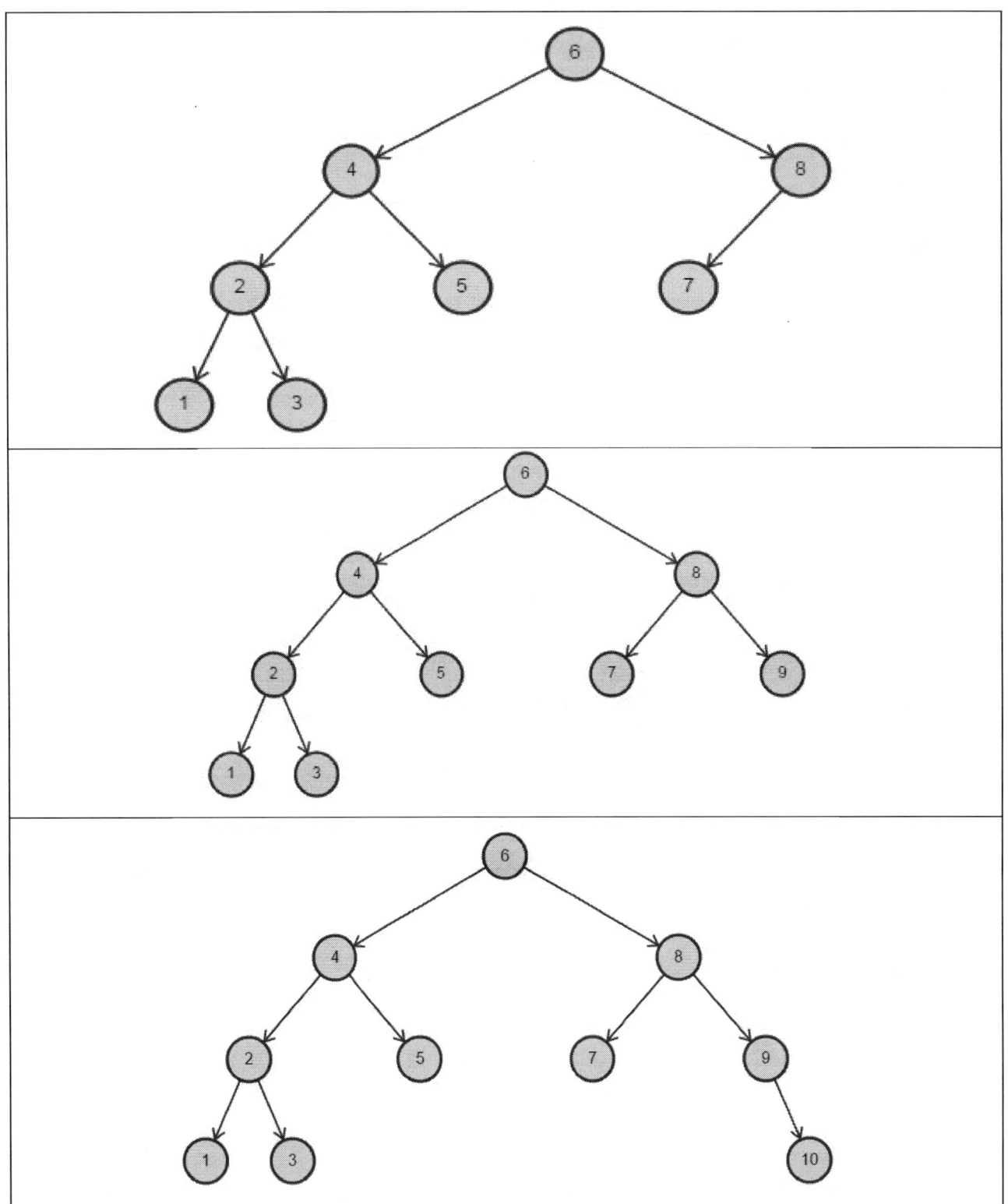

Solution: Smaller values will be added to the left child sub-tree of a node and greater value will be added to the right child sub-tree of the current node.

Example 10.30:

```java
public void InsertNode(int value) {
    root = InsertNode(value, root);
}

private Node InsertNode(int value, Node node) {
    if(node==null)
    {
        node=new Node(value,null,null);
    }
    else
    {
        if(node.value > value)
            node.lChild = InsertNode(value,node.lChild);
        else
            node.rChild = InsertNode(value, node.rChild);
    }
    return node;
}
```

Complexity Analysis: Time Complexity: $O(n)$ Space Complexity: $O(n)$

Find Node

Solution: The value grater then the current node value will be in the right child sub-tree and the value smaller than the current node is in the left child sub-tree. We can find a value by traversing the left or right subtree iteratively.

Example 10.31: Find the node with the value given.

```java
public boolean Find(int value){
    Node curr=root;

    while(curr != null)
    {
        if(curr.value == value)
            return true;
        else if(curr.value > value)
            curr = curr.lChild;
        else
            curr = curr.rChild;
    }
    return false;
}
```

Complexity Analysis: Time Complexity: $O(n)$ Space Complexity: $O(1)$

Example 10.32: Operators are generally read from left to right

```
public boolean Find2(int value)
{
        Node curr = root;
        while(curr != null && curr.value != value)
                curr = (curr.value > value)? curr.lChild : curr.rChild;
        return curr != null;
}
```

Complexity Analysis:
Time Complexity: $O(n)$ Space Complexity: $O(n)$

Find Min

Find the node with the minimum value.

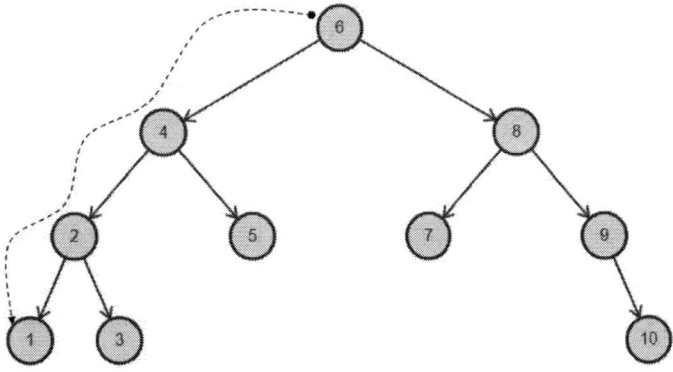

Solution: left most child of the tree will be the node with the minimum value.

Example 10.33:

```
public int FindMin()
{
        Node node=root;
        if(node == null)
        {
                return Integer.MAX_VALUE;
        }

        while(node.lChild != null)
        {
                node = node.lChild;
        }
        return node.value;
}
```

Complexity Analysis:
Time Complexity: $O(n)$ Space Complexity: $O(1)$

Find Max

Find the node in the tree with the maximum value.

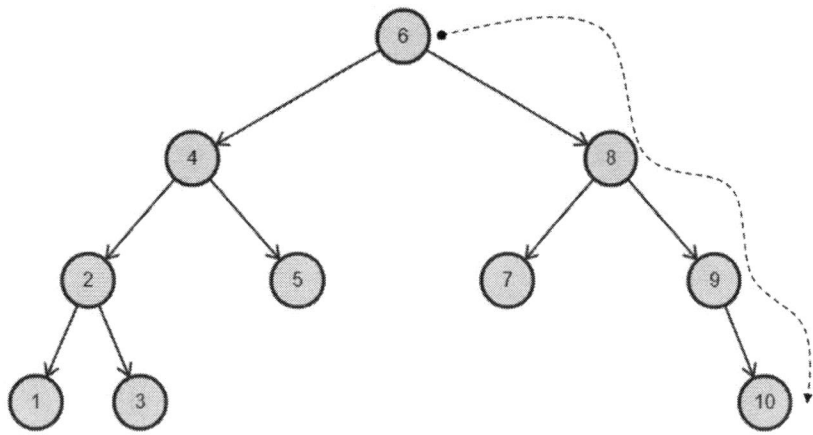

Solution: Right most node of the tree will be the node with the maximum value.

Example 10.34:

```java
public int FindMax() {
    Node node=root;
    if(node == null) {
        return Integer.MIN_VALUE;
    }
    while(node.rChild != null) {
        node = node.rChild;
    }
    return node.value;
}
```

Complexity Analysis: Time Complexity: $O(n)$ Space Complexity: $O(1)$

Is tree a BST

Solution: At each node we check, max value of left subtree is smaller than the value of current node and min value of right subtree is grater then the current node.

Example 10.35:

```java
public boolean isBST(Node root) {
    if(root == null)
        return true;
    if(root.lChild != null && FindMax(root.lChild).value > root.value )
        return false;
    if(root.rChild != null && FindMin(root.rChild).value <= root.value )
        return false;
    return (isBST(root.lChild) && isBST(root.rChild));
}
```

Complexity Analysis:
Time Complexity: $O(n)$ Space Complexity: $O(n)$

The above solution is correct but it is not efficient as same tree nodes are traversed many times.

Solution: A better solution will be the one in which we will look into each node only once. This is done by narrowing the range. We will be using a isBSTUtil() function which will take the max and min range of the values of the nodes. The initial value of min and max will be INT_MIN and INT_MAX.

Example 10.36:

```
public boolean isBST(){
    return isBST(root, Integer.MIN_VALUE, Integer.MAX_VALUE);
}
```

```
public boolean isBST(Node curr, int min, int max)
{
    if(curr == null)
        return true;

    if( curr.value < min || curr.value > max)
        return false;

    return isBST(curr.lChild, min, curr.value)
            && isBST(curr.rChild, curr.value, max);
}
```

Complexity Analysis:
Time Complexity: $O(n)$ Space Complexity: $O(n)$ for stack

Solution: Above method is correct and efficient but there is an easy method to do the same. We can do in-order traversal of nodes and see if we are getting a strictly increasing sequence

Example 10.37:

```
private boolean isBST2(Node root, counter count)/* in order traversal */
{
    boolean ret;
    if(root != null)
    {
        ret = isBST2(root.lChild,count);
        if(!ret)
            return false;

        if(count.value > root.value)
            return false;

        count.value = root.value;

        ret = isBST2(root.rChild,count);
        if(!ret)
            return false;
    }
    return true;
}
```

```
class counter{
      int value;
}
public boolean isBST2(){

      counter c = new counter();
      return isBST2(root,c);
}
```

Complexity Analysis:
Time Complexity: $O(n)$ Space Complexity: $O(n)$ for stack

Delete Node

Description: Remove the node x from the binary search tree, making the necessary, reorganize nodes of binary search tree to maintain its properties.

There are three cases in delete node, let's call the node that need to be deleted as x.
Case 1: node x has no children. Just delete it (i.e. Change parent node so that it does not point to x)
Case 2: node x has one child. Splice out x by linking x's parent to x's child
Case 3: node x has two children. Splice out the x's successor and replace x with x's successor

When the node to be deleted have no children
This is a trivial case in this case we directly delete the node and return null.

When the node to be deleted have only one child.
In this case we save the child in a temp variable, then delete current node, and finally return the child.

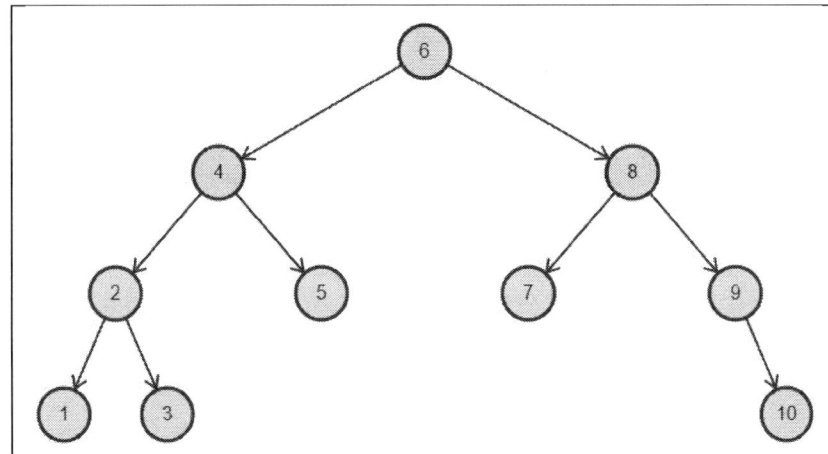

We want to remove node with value 9. The node has only one child.

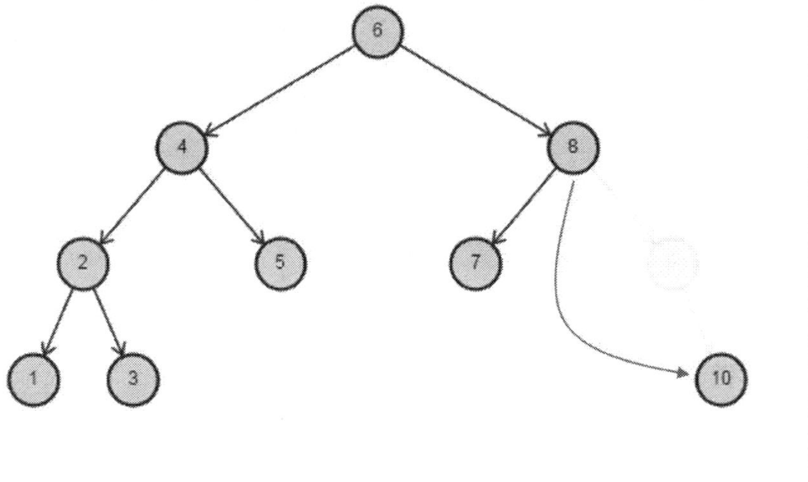	Right child of the parent of node with value 9 that is node with value 8 will point to the node with value 10.
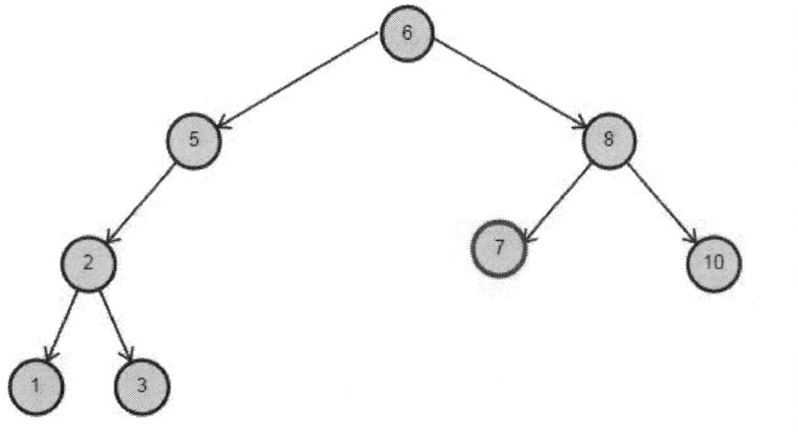	Finally, node with value 9 is removed from the tree.

When the node to be deleted has two children.

	We want to delete node with value 6. Which have two children.

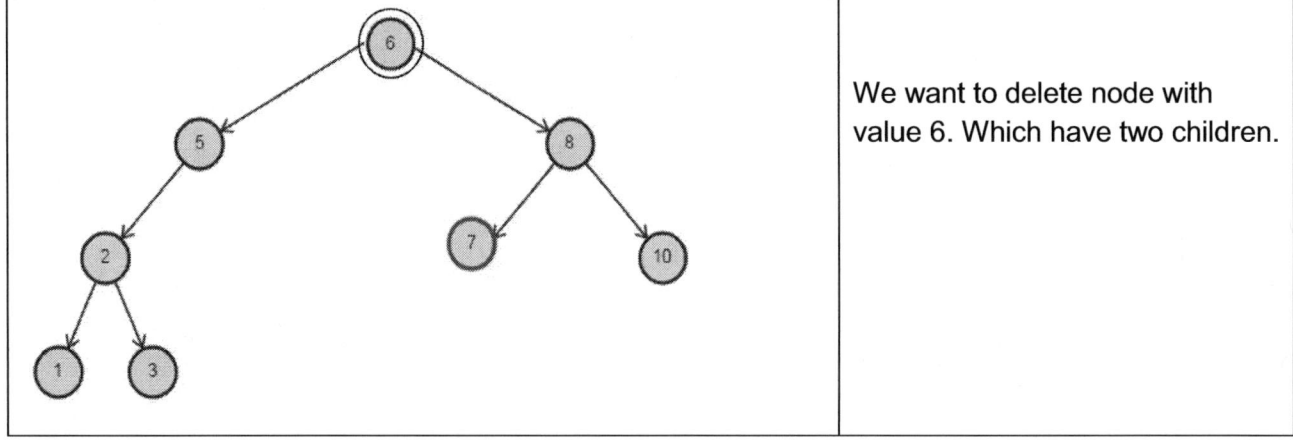

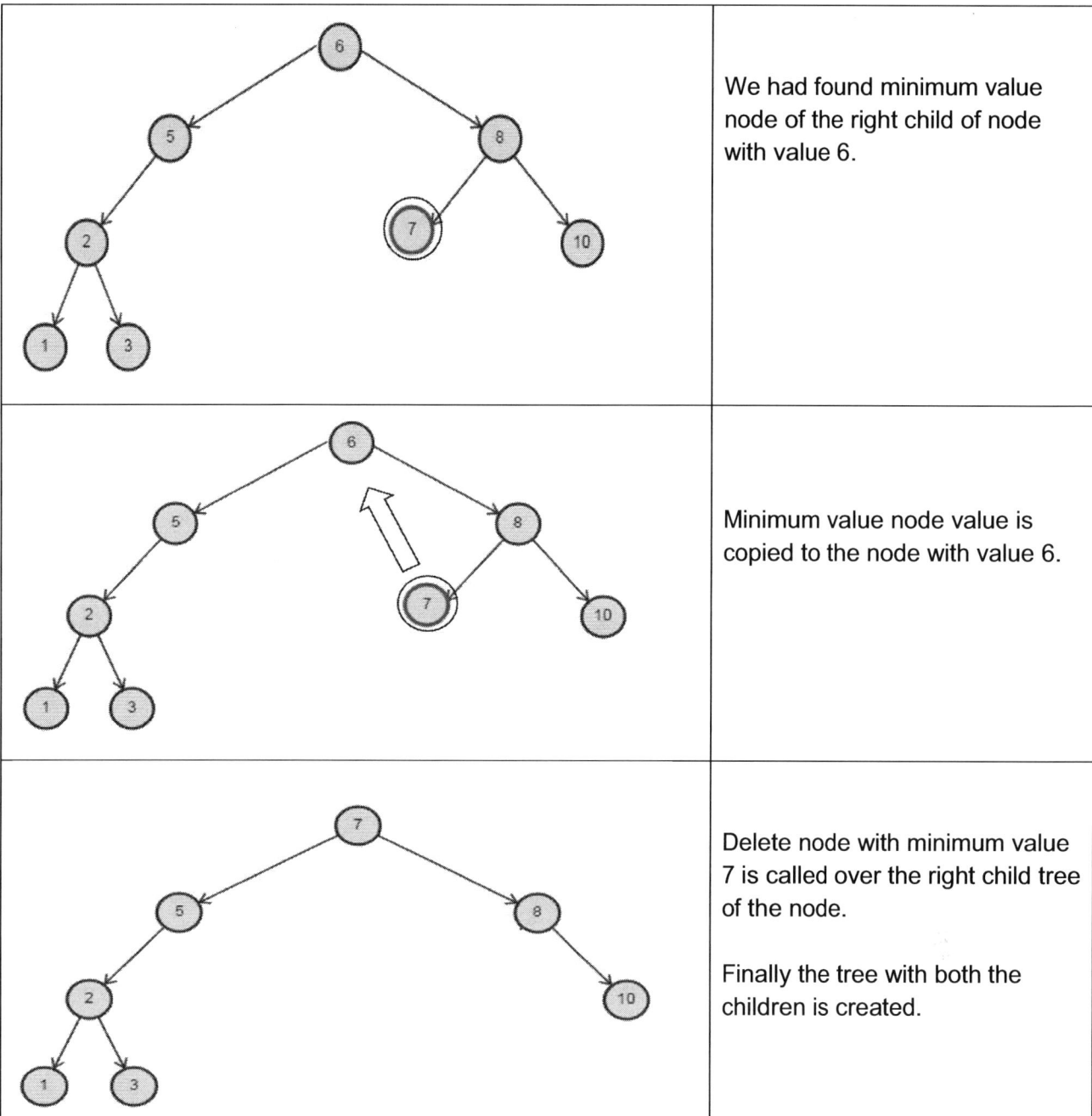

	We had found minimum value node of the right child of node with value 6.
	Minimum value node value is copied to the node with value 6.
	Delete node with minimum value 7 is called over the right child tree of the node. Finally the tree with both the children is created.

Example 10.38:

```
public void DeleteNode(int value)
{
        root = DeleteNode( root, value);
}
```

```
private Node DeleteNode(Node node, int value)
{
    Node temp=null;
    if(node != null)
    {
        if(node.value==value)
        {
            if(node.lChild==null && node.rChild==null)
            {
                return null;
            }
            else
            {
                if(node.lChild==null)
                {
                    temp=node.rChild;
                    return temp;
                }

                if(node.rChild==null)
                {
                    temp=node.lChild;
                    return temp;
                }

                Node maxNode = FindMax(node.lChild);
                int maxValue=maxNode.value;
                node.value = maxValue;
                node.lChild=DeleteNode(node.lChild,maxValue);
            }
        }
        else
        {
            if(node.value > value)
            {
                node.lChild = DeleteNode(node.lChild, value);
            }
            else
            {
                node.rChild = DeleteNode(node.rChild, value);
            }
        }
    }
    return node;
}
```

Analysis: Time Complexity: $O(n)$ Space Complexity: $O(n)$

Least Common Ancestor

In a tree T. The least common ancestor between two nodes n1 and n2 is defined as the lowest node in T that has both n1 and n2 as descendants.

Example 10.39:

```
public int LcaBST(int first, int second)
{
    return LcaBST(root, first, second);
}
```

```
private int LcaBST(Node curr, int first, int second)
{
    if(curr == null)
    {
        return Integer.MAX_VALUE;
    }

    if(curr.value > first &&
            curr.value > second)
    {
        return LcaBST(curr.lChild, first, second);
    }
    if(curr.value < first &&
            curr.value < second)
    {
        return LcaBST(curr.rChild, first, second);
    }
    return curr.value;
}
```

Trim the Tree nodes which are Outside Range

Given a range as min, max. We need to delete all the nodes of the tree that are out of this range.

Solution: Traverse the tree and each node that is having value outside the range will delete itself. All the deletion will happen from inside out so we do not have to care about the children of a node as if they are out of range then they already had deleted themselves.

Example 10.40:

```
public void trimOutsideRange(int min, int max)
{
    trimOutsideRange(root, min, max);
}
```

```
private Node trimOutsideRange(Node curr, int min, int max)
{
    if (curr == null)
        return null;

    curr.lChild=trimOutsideRange(curr.lChild, min, max);
    curr.rChild=trimOutsideRange(curr.rChild, min, max);

    if (curr.value < min)
    {
        return curr.rChild;
    }

    if (curr.value > max)
    {
        return curr.lChild;
    }

    return curr;
}
```

Print Tree nodes which are in Range

Print only those nodes of the tree whose value is in the range given.

Solution: Just normal inorder traversal and at the time of printing we will check if the value is inside the range provided.

Example 10.41:

```
public void printInRange(int min, int max)
{
    printInRange(root, min, max);
}
```

```
private void printInRange(Node root, int min, int max)
{
    if(root == null)
        return;

    printInRange(root.lChild, min, max);

    if(root.value >= min && root.value <= max)
        System.out.print(root.value + " ");

    printInRange(root.rChild, min, max);
}
```

Find Ceil and Floor value inside BST given key

Given a tree and a value we need to find the ceil value of node in tree which is smaller than the given value and need to find the floor value of node in tree which is bigger. Our aim is to find ceil and floor value as close as possible then the given value.

Example 10.42:

```
public int FloorBST(int val)
{
        Node curr = root;
        int floor=Integer.MAX_VALUE;
        while (curr != null)
        {
                if (curr.value == val)
                {
                        floor = curr.value;
                        break;
                }
                else if (curr.value > val)
                {
                        curr = curr.lChild;
                }
                else
                {
                        floor = curr.value;
                        curr = curr.rChild;
                }
        }
        return floor;
}
```

```
public int CeilBST(int val)
{
        Node curr = root;
        int ceil=Integer.MIN_VALUE;
        while (curr != null)
        {
                if (curr.value == val)
                {
                        ceil = curr.value;
                        break;
                }
                else if (curr.value > val)
                {
                        ceil = curr.value;
                        curr = curr.lChild;
                }
                else
                {
                        curr = curr.rChild;
                }
        }
        return ceil;
}
```

Exercise

1. Construct a tree given its in-order and pre-order traversal strings.
   - inorder: 1 2 3 4 5 6 7 8 9 10
   - pre-order: 6 4 2 1 3 5 8 7 9 10

2. Construct a tree given its in-order and post-order traversal strings.
   - inorder: 1 2 3 4 5 6 7 8 9 10
   - post-order: 1 3 2 5 4 7 10 9 8 6

3. Write a delete node function in Binary tree.

4. Write a function print depth first in a binary tree without using system stack (use STL queue or stack etc.)

   Hint: you may want to keep another element to tree node like visited flag.

5. Check whether a given Binary Tree is Complete or not
   - In a complete binary tree, every level except the last one is completely filled. All nodes in the left are filled first, then the right one.

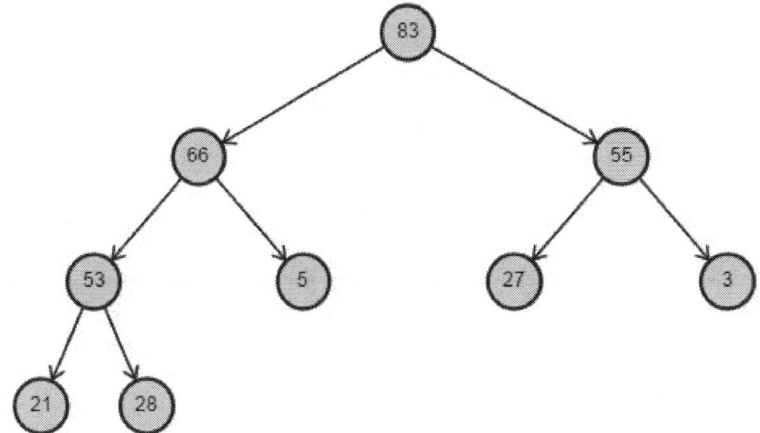

6. Check whether a given Binary Tree is Full/ Strictly binary tree or not
   - The full binary tree is a binary tree in which each node has zero or two children.

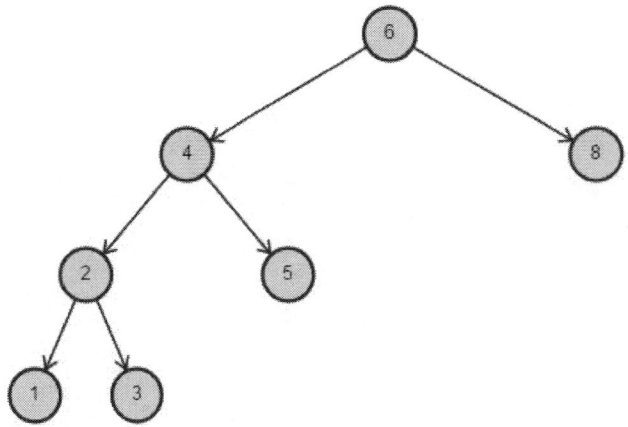

7. Check whether a given Binary Tree is a Perfect binary tree or not
    - The perfect binary tree- is a type of full binary trees in which each non-leaf node has exactly two child nodes.

8. Check whether a given Binary Tree is Height-balanced Binary Tree or not
    - A height-balanced binary tree is a binary tree such that the left & right subtrees for any given node differ in height by no more than one

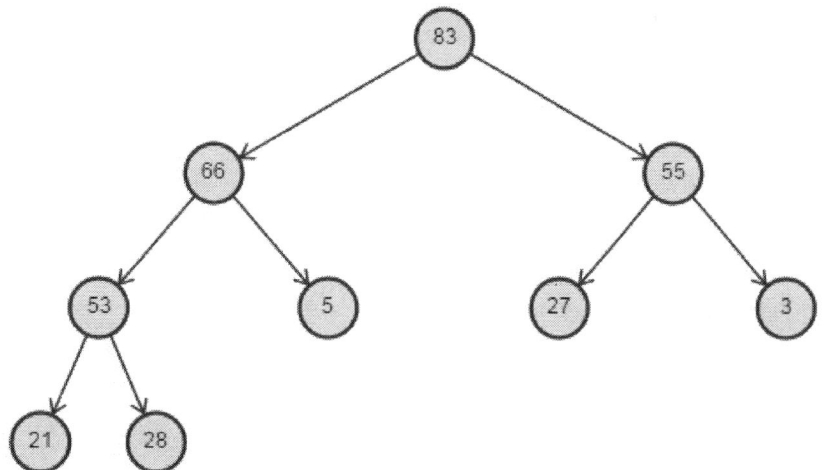

9. Isomorphic: two trees are isomorphic if they have the same shape, it does not matter what the value is. Write a program to find if two given tree are isomorphic or not.

10. The worst-case run*time* Complexity of building a BST with n nodes
    - $O(n^2)$
    - $O(n * \log n)$
    - $O(n)$
    - $O(\log n)$

11. The worst-case runtime Complexity of insertion into a BST with n nodes is
    - $O(n2)$
    - $O(n * \log n)$
    - $O(n)$
    - $O(\log n)$

12. The worst-case runtime *Complexity* of a search of a value in a BST with n nodes.
    - $O(n^2)$
    - $O(n * \log n)$
    - $O(n)$
    - $O(\log n)$

13. Which of the following traversals always gives the sorted sequence of the elements in a BST?
    - Preorder
    - Ignored
    - Postorder
    - Undefined

14. The height of a Binary Search Tree with n nodes in the worst case?
    - O(n * log n)
    - O(n)
    - O(logn)
    - O(1)

15. Try to optimize the above solution to give a DFS traversal without using recursion use some stack or queue.

16. This is an open exercise for the readers. Every algorithm that is solved using recursion (system stack) can also be solved using user defined or library defined (STL) stack. So try to figure out what all algorithms that are using recursion and try to figure out how you will do this same issue using user layer stack.

17. In a binary tree, print the nodes in zigzag order. In the first level, nodes are printed in the left to right order. In the second level, nodes are printed in right to left and in the third level again in the order left to right.
    - *Hint:* Use two stacks. Pop from first stack and push into another stack. Swap the stacks alternatively.

18. Find nth smallest element in a binary search tree.
    - *Hint:* Nth inorder in a binary tree.

19. Find the floor value of key that is inside a BST.

20. Find the Ceil value of key, which is inside a BST.

21. What is *Time Complexity* of the below code:

```java
void DFS(Node head) {
    Node curr = head, prev;
    int count = 0;
    while (curr && ! curr.visited) {
        count++;
        if (curr.lChild && ! curr.lChild.visited) {
            curr= curr.lChild;
        }
        else if (curr.rChild && ! curr.rChild.visited) {
            curr= curr.rChild;
        }
        else {
            System.out.print((" " + curr.value);
            curr.visited = 1;
            curr = head;
        }
    }
    System.out.print(("count is : " + count);
}
```

Chapter 11: Priority Queue

Introduction

A Priority-Queue also knows as Binary-Heap, is a variant of queue. Items are removed from the start of the queue. However, in a Priority-Queue the logical ordering of objects is determined by their priority. The highest priority item are at the front of the Priority-Queue. When you add an item to Priority-Queue the new item can more to the front of the queue. A Priority-Queue is a very important data structure. Priority-Queue is used in various Graph algorithms like Prim's Algorithm and Dijkstra's algorithm. Priority-Queue is also used in the timer implementation etc.

A Priority-Queue is implemented using a Heap (Binary Heap). A Heap data structure is an array of elements that can be observed as a complete binary tree. The tree is completely filled on all levels except possibly the lowest. And heap satisfies the heap ordering property. A heap is a complete binary tree so the height of tree with N nodes is always *O(logn)*.

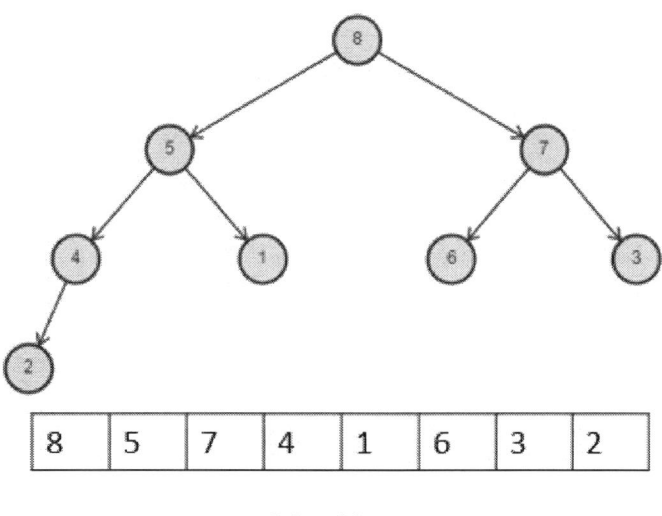

Max Heap

A heap is not a sorted data structure and can be regarded as partially ordered. As you see from the picture, there is no relationship among nodes at any given level, even among the siblings.

Heap is implemented using an array. And because heap is a complete binary tree, the left child of a parent (at position x) is the node that is found in position 2x in the array. Similarly, the right child of the parent is at position 2x+1 in the array. To find the parent of any node in the heap, we can simply division. Given the index y of a node, the parent index will by y/2.

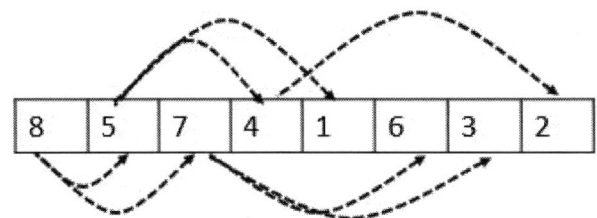

Types of Heap

There are two types of heap and the type depends on the ordering of the elements. The ordering can be done in two ways: Min-Heap and Max-Heap

Max Heap

Max-Heap: the value of each node is less than or equal to the value of its parent, with the largest-value element at the root.

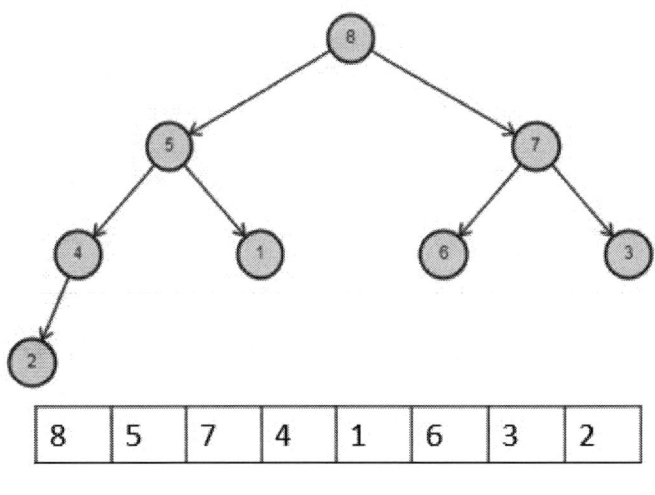

Max Heap

Max Heap Operations

Operation	Complexity
Insert	O(logn)
DeleteMax	O(logn)
Remove	O(logn)
FindMax	O(1)

Min Heap

Min-Heap: the value of each node is greater than or equal to the value of its parent, with the minimum-value element at the root.

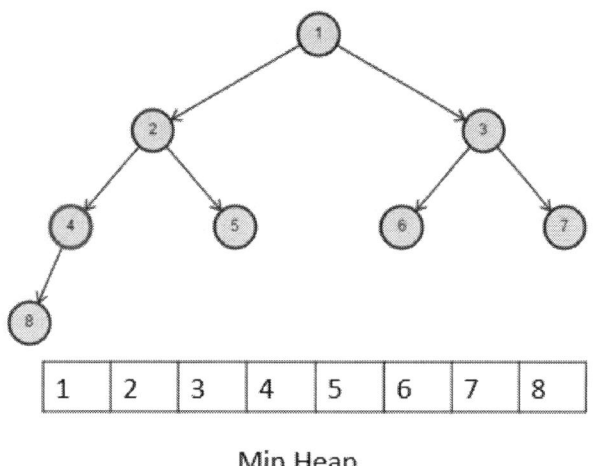

Min Heap

Use it whenever you need quick access to the smallest item, because that item will always be at the root of the tree or the first element in the array. However, the remainder of the array is kept partially sorted. Thus, instant access is only possible for the smallest item.

Min Heap Operations

Insert	O(logn)
DeleteMin	O(logn)
Remove	O(logn)
FindMin	O(1)

Throughout this chapter, the word "heap" will always refer to a max-heap. The implementation of min-heap is left for the user to do it as an exercise.

Heap ADT Operations

The basic operations of binary heap are as follows:

Binary Heap	Create a new empty binary heap	O(1)
Insert	Adding a new element to the heap	O(logn)
DeleteMax	Delete the maximum element form the heap.	O(logn)
FindMax	Find the maximum element in the heap.	O(1)
isEmpty	return true if the heap is empty else return false	O(1)
Size	Return the number of elements in the heap.	O(1)
BuildHeap	Build a new heap from the array of elements	O(logn)

Operation on Heap

Create Heap from an array

1. Starts by putting the elements to an array.
2. Starting from the middle of the array move downward towards the start of the array. At each step, compare parent value with its left child and right child. And restore the heap property by shifting the parent value with its greatest-value child. Such that the parent value will always be greater than or equal to left child and right child.
3. For all elements from middle of the array to the start of the array. We are doing comparisons and shift till we reach the leaf nodes of the heap. The *Time Complexity* of build heap is **O(N)**.

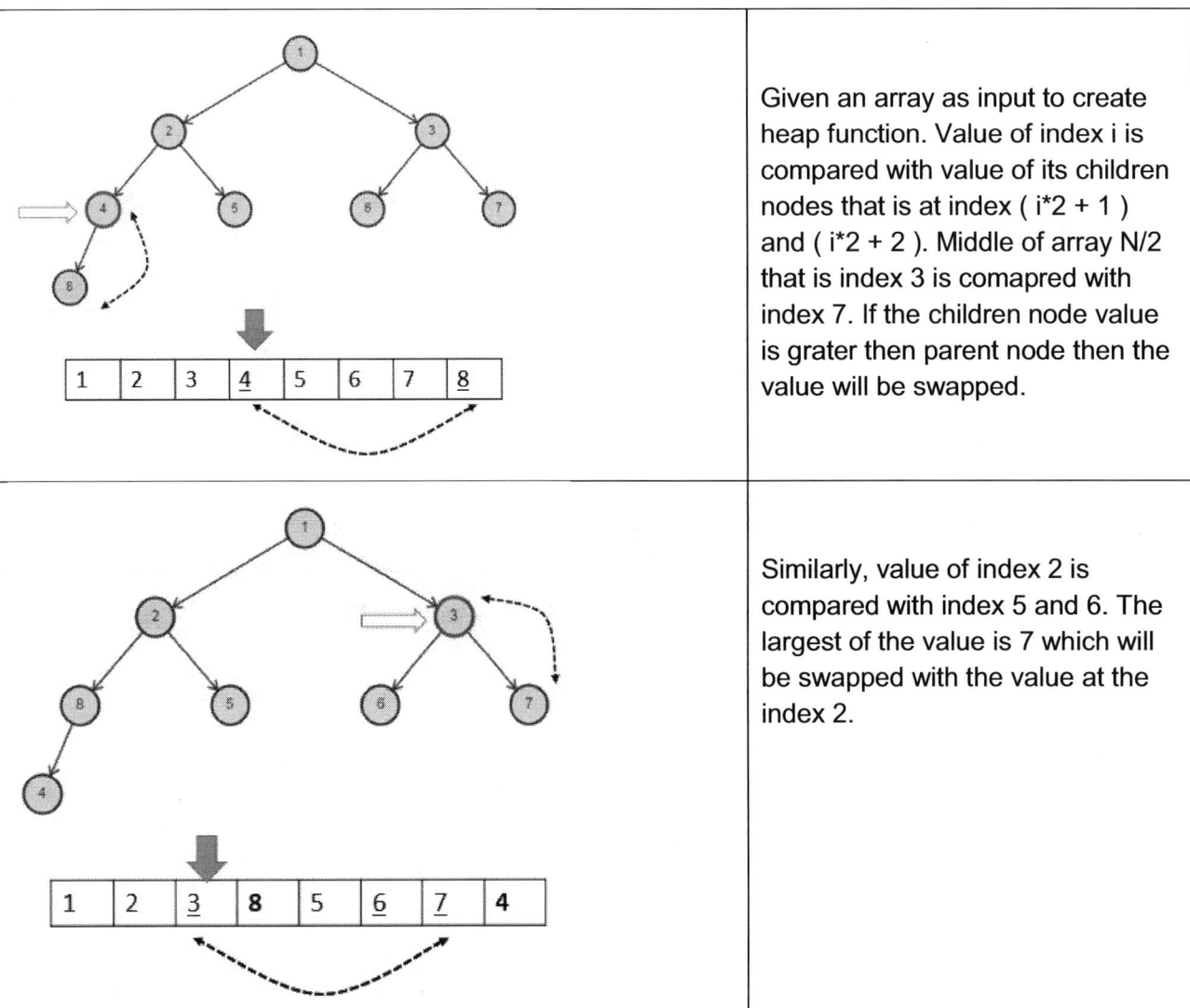

Given an array as input to create heap function. Value of index i is compared with value of its children nodes that is at index (i*2 + 1) and (i*2 + 2). Middle of array N/2 that is index 3 is comapred with index 7. If the children node value is grater then parent node then the value will be swapped.

Similarly, value of index 2 is compared with index 5 and 6. The largest of the value is 7 which will be swapped with the value at the index 2.

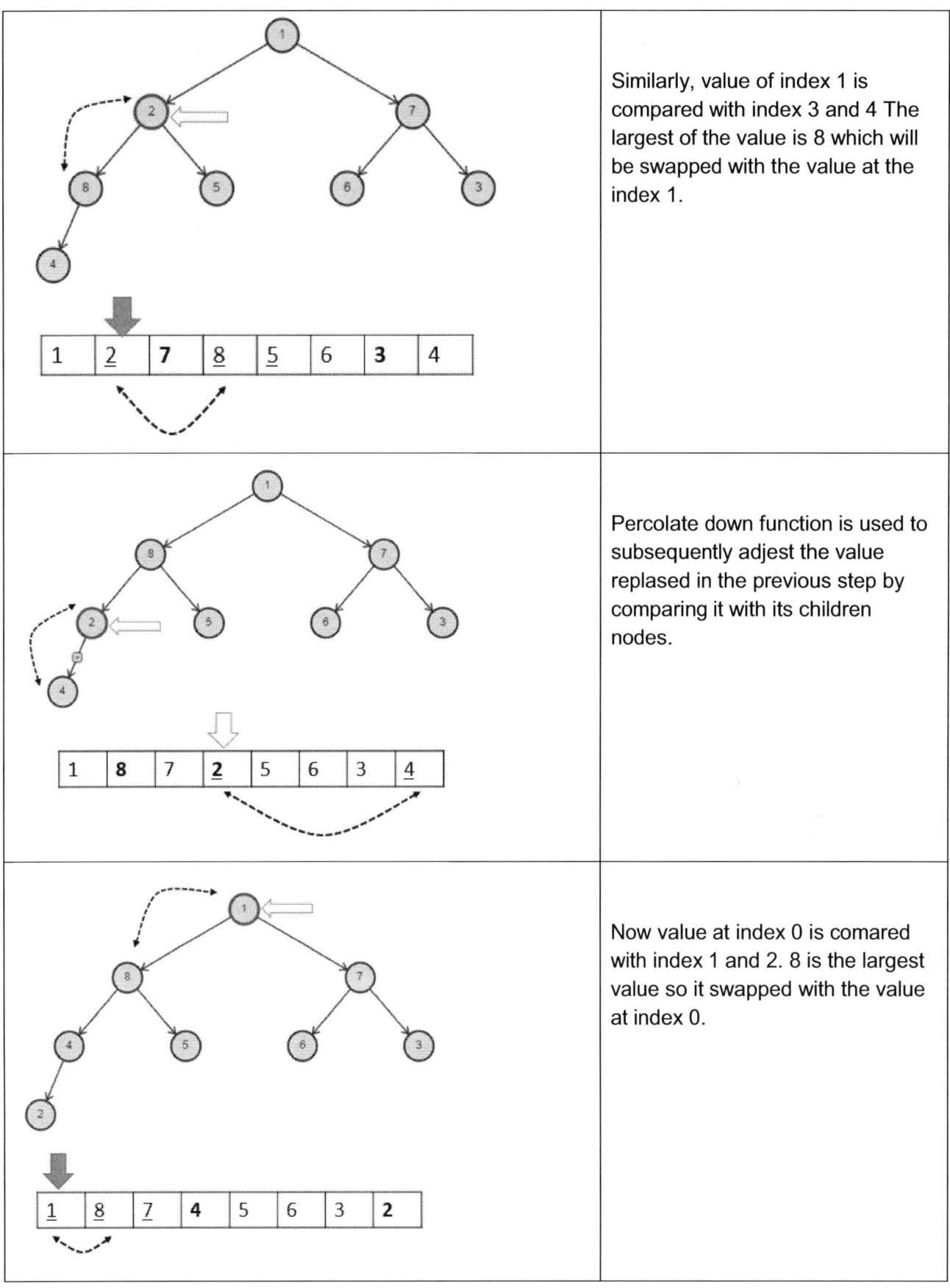

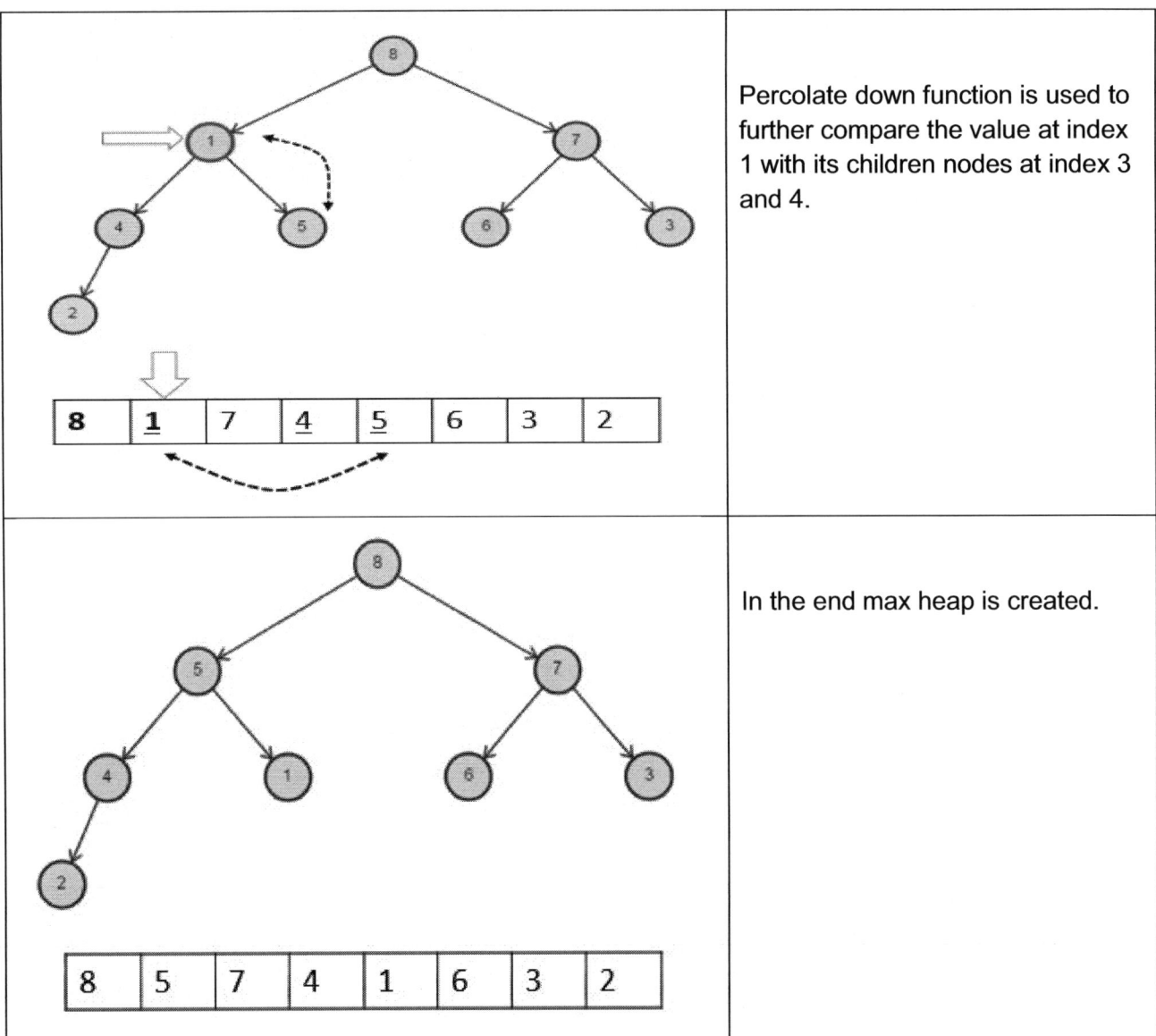

	Percolate down function is used to further compare the value at index 1 with its children nodes at index 3 and 4.
	In the end max heap is created.

Example 11.1:

```
public class Heap {

    private static final int CAPACITY = 16;
    private int size;      // Number of elements in heap
    private int[] arr;     // The heap array

    public Heap() {
        arr = new int[CAPACITY];
        size = 0;
    }
```

```java
public Heap(int[] array) {
        size = array.length;
        arr = new int[array.length+1];
        //we do not use 0 index
        System.arraycopy(array, 0, arr, 1, array.length);

        //Build Heap operation over array
        for(int i=(size/2);i>0;i--)
        {
                percolateDown(i);
        }
}
//Other Methods.
}
```

```java
private void percolateDown(int position) {
      int lChild=2*position;
      int rChild=lChild+1;
      int small=-1;
      int temp;

      if( lChild <= size )
            small =lChild;

      if( rChild <= size && (arr[rChild] - arr[lChild])<0 )
            small =rChild;

      if( small!=-1 && (arr[small] - arr[position])<0 ) {
            temp=arr[position];
            arr[position]=arr[small];
            arr[small]=temp;
            percolateDown(small);
      }
}
```

Initializing an empty Heap

Example 11.2:

```java
private void percolateUp( int position ) {
      int parent=position/2;
      int temp;
      if(parent==0)
            return;

      if((arr[parent] - arr[position])<0) {
            temp =      arr[position];
            arr[position]=arr[parent];
            arr[parent] = temp;
            percolateUp(parent);
      }
}
```

Enqueue / Insert

1. Add the new element at the end of the array. This keeps the structure as a complete binary tree, but it might no longer be a heap since the new element might have a value greater than its parent.
2. Swap the new element with its parent until it has value greater than its parents.
3. Step 2 will terminate when the new element reaches the root or when the new element's parent have a value greater than or equal to the new element's value.

Let's take an example of the Max heap created in the above example.

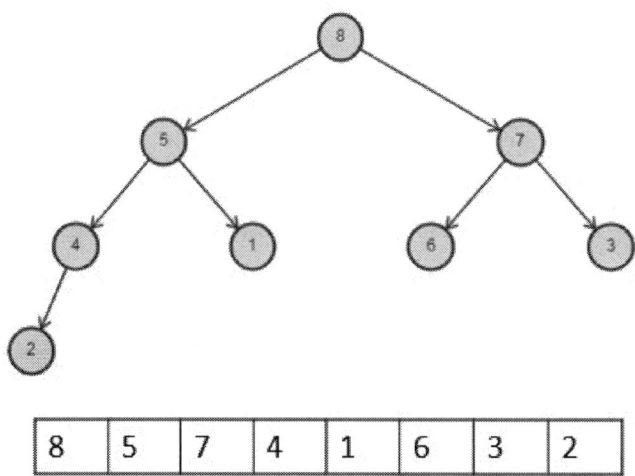

Let's take an example by inserting element with value 9 to the heap. The element is added to the end of the heap array. Now the value will be percolate up by comparing it with the parent. The value is added to index 8 and its parent will be (N-1)/2 = index 3.

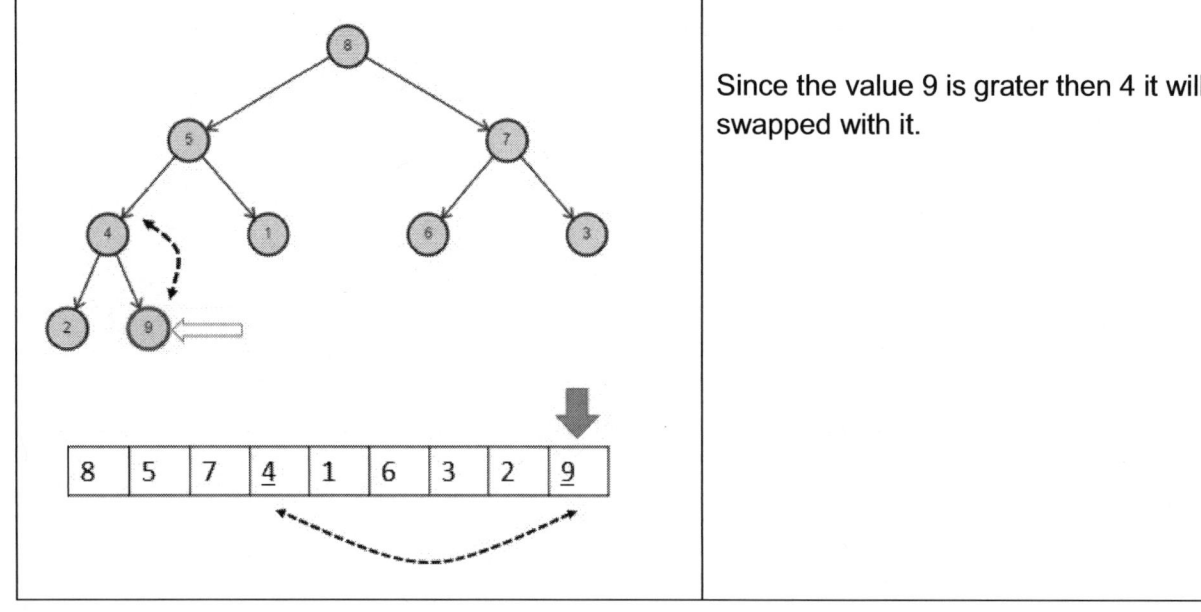

Since the value 9 is grater then 4 it will be swapped with it.

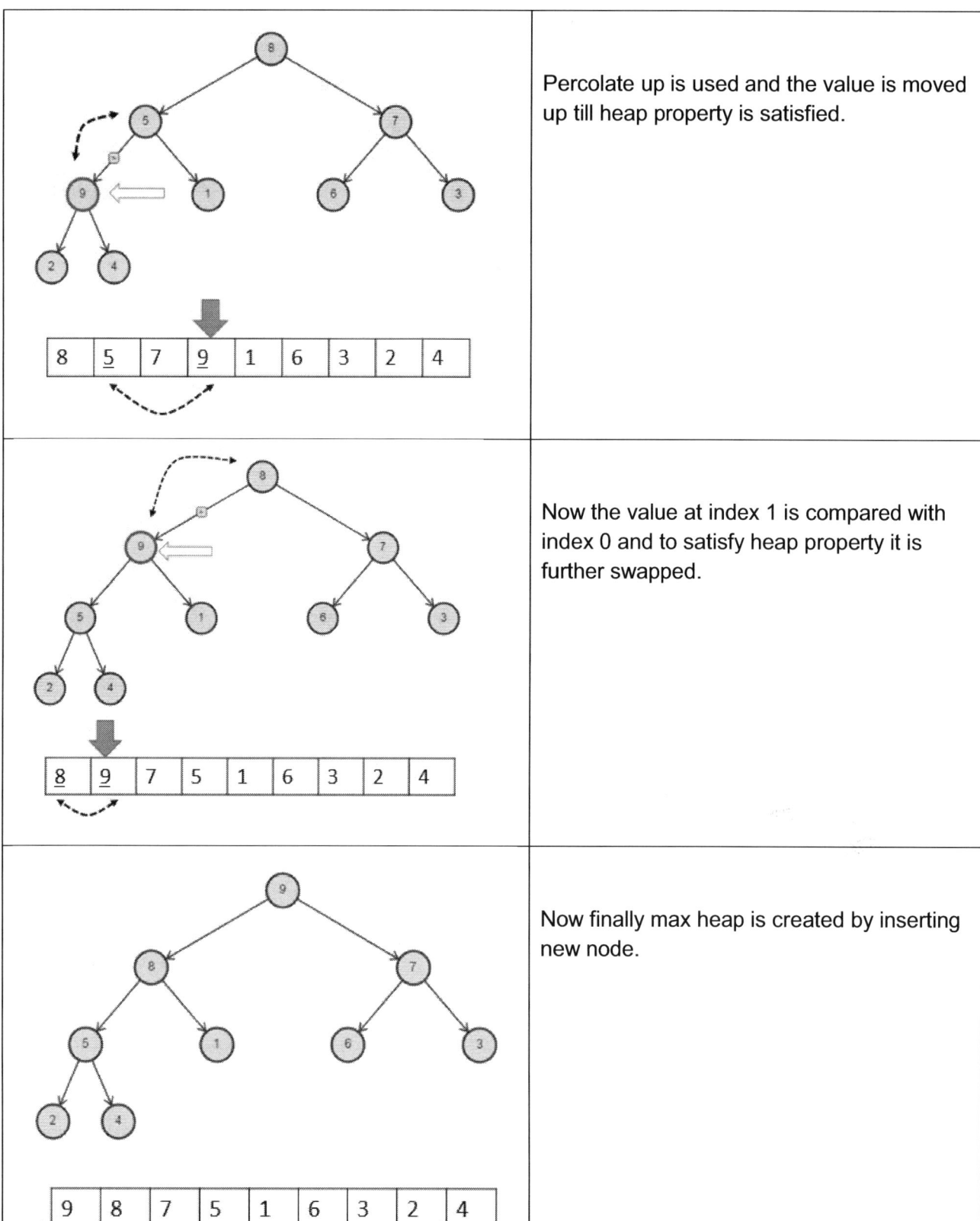

Example 11.3:

```
public void add( int value )
{
        if(size == arr.length - 1)
            doubleSize();

        arr[++size]=value;
        percolateUp(size);
}
```

```
private void doubleSize()
{
        int[] old = arr;
        arr = new int[arr.length * 2];
        System.arraycopy(old, 1, arr, 1, size);
}
```

Dequeue / Delete

1. Copy the value at the root of the heap to the variable used to return a value.
2. Copy the last element of the heap to the root, and then reduce the size of heap by 1. This element is called the "out-of-place" element.
3. Restore heap property by swapping the out-of-place element with its greatest-value child. Repeat this process until the out-of-place element reaches a leaf or it has a value that is greater or equal to all its children.
4. Return the answer that was saved in Step 1.

To remove an element from heap its top value is swapped to the end of the heap array and size fo heap is reduced by 1.

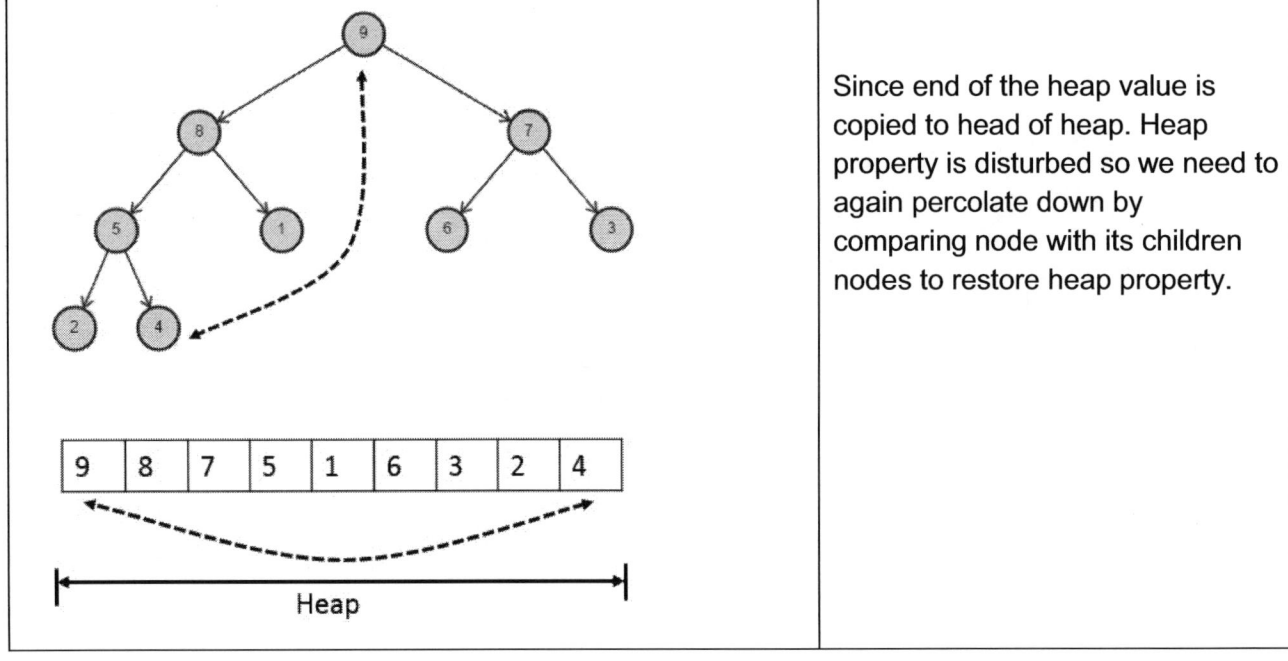

Since end of the heap value is copied to head of heap. Heap property is disturbed so we need to again percolate down by comparing node with its children nodes to restore heap property.

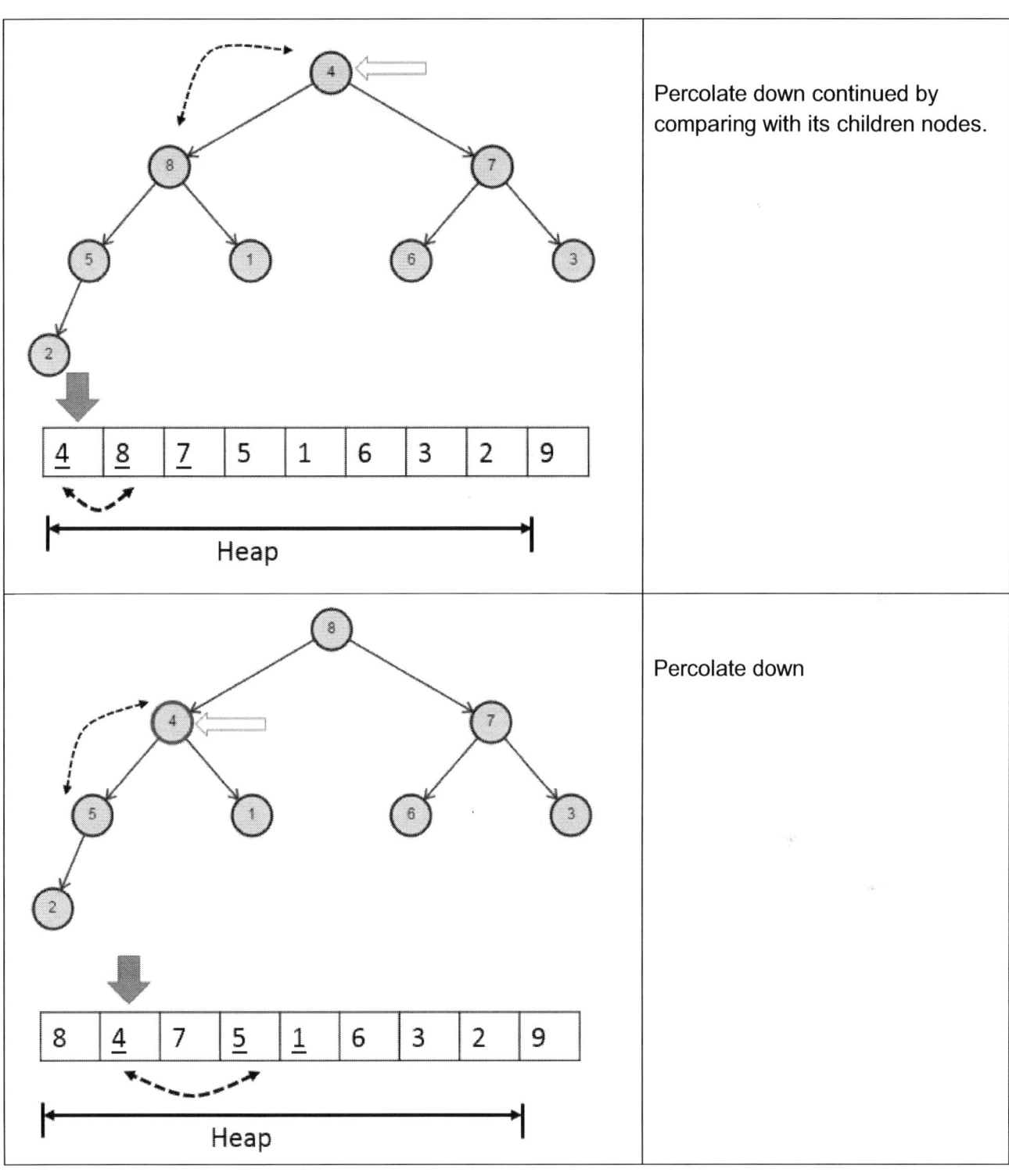

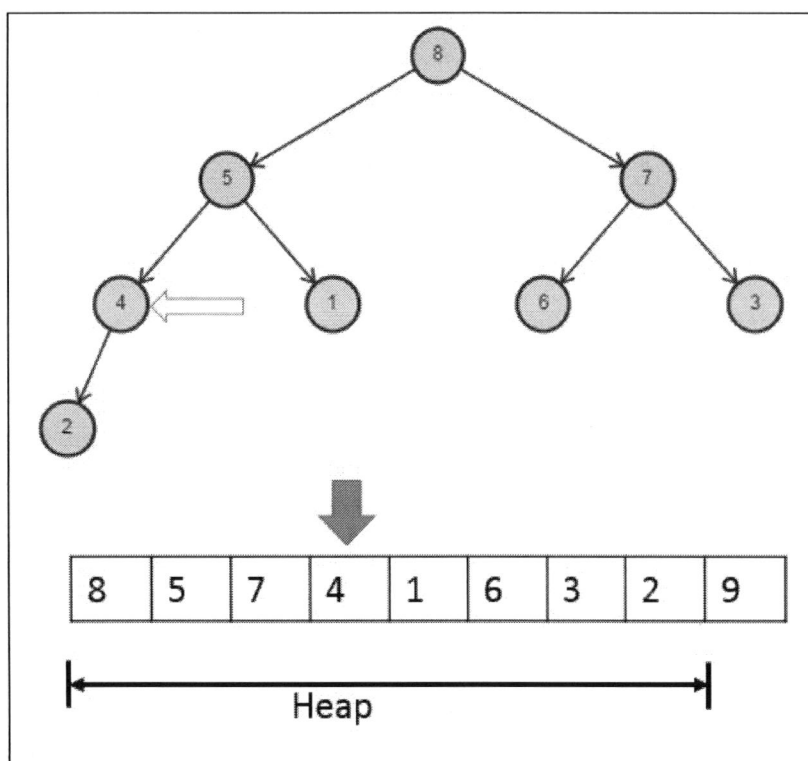

Percolate down Complete

Example 11.4:

```
public int remove( )
{
        if (isEmpty())
                throw new IllegalStateException();

        int value=arr[1];
        arr[1]=arr[size];
        size--;
        percolateDown(1);
        return value;
}
```

Heap-Sort

1. Use create heap function to build a max heap from the given array of elements. This operation will take *O(N)* time.
2. Dequeue the max value from the heap and store this value to the end of the array at location arr[size-1]
    a) Copy the value at the root of the heap to end of the array.
    b) Copy the last element of the heap to the root, and then reduce the size of heap by 1. This element is called the "out-of-place" element.
    c) Restore heap property by swapping the out-of-place element with its greatest-value child. Repeat this process until the out-of-place element reaches a leaf or it has a value that is greater or equal to all its children
3. Repeat this operation till there is just one element in the heap.

Let's take example of the heap which we had created at the start of the chapter. Heap sort is algorithm starts by creating a heap of the given array which is done in linear time. Then at each step head of the heap is swapped with the end of the heap and the heap size is reduced by 1. Then percolate down is used to restore the heap property. And this same is done multiple times till the heap contain just one element.

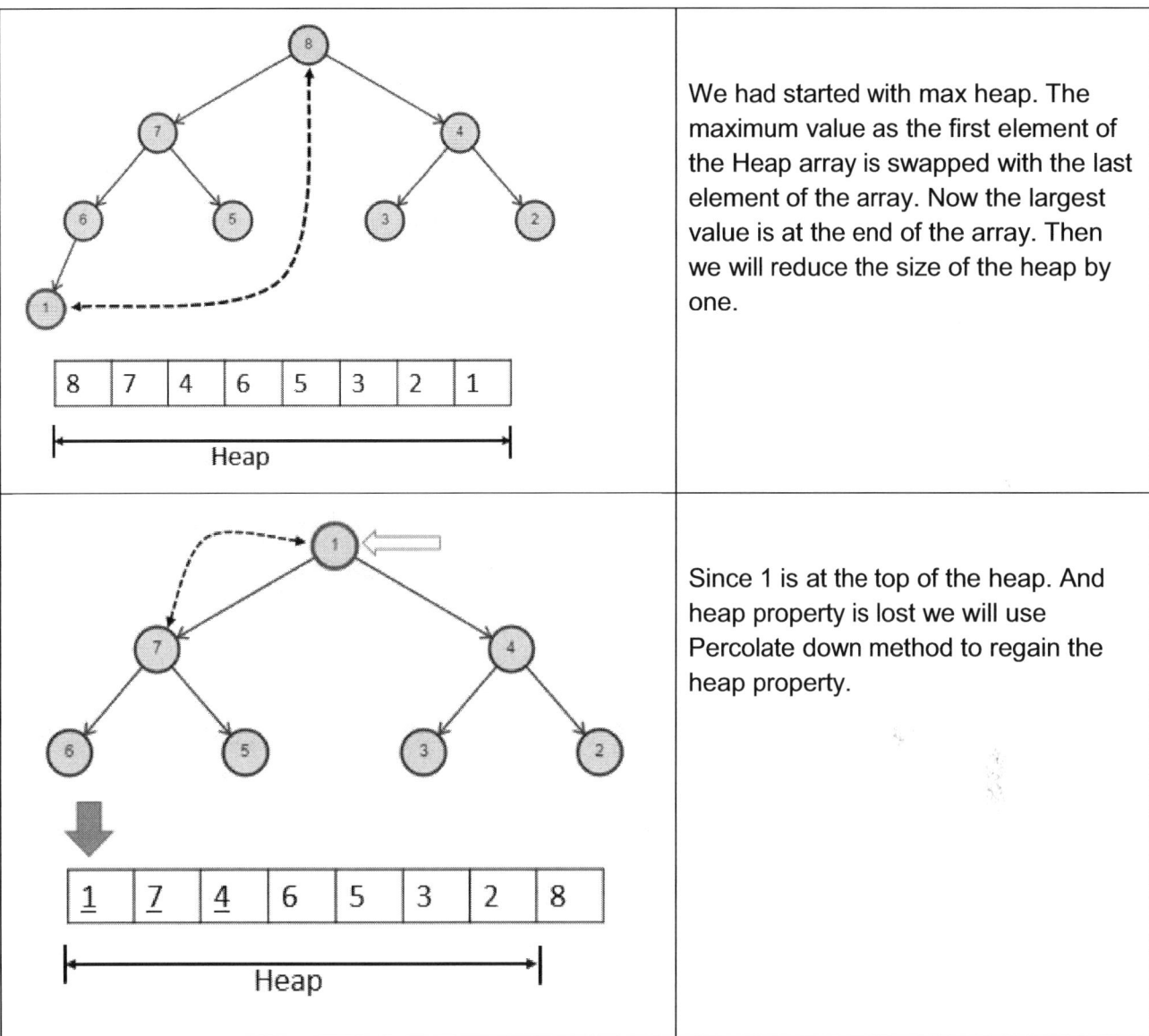

We had started with max heap. The maximum value as the first element of the Heap array is swapped with the last element of the array. Now the largest value is at the end of the array. Then we will reduce the size of the heap by one.

Since 1 is at the top of the heap. And heap property is lost we will use Percolate down method to regain the heap property.

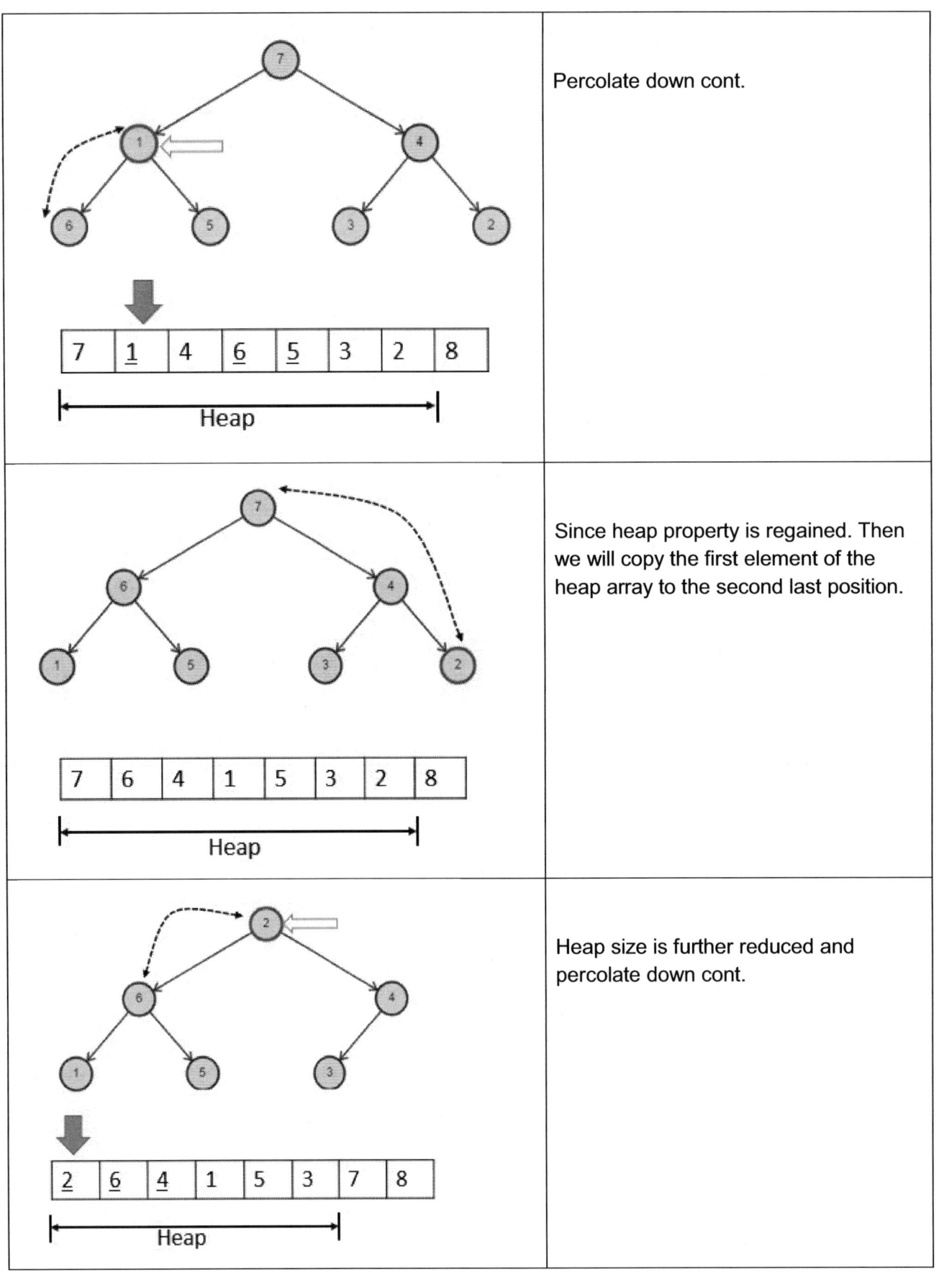

Percolate down cont.

Since heap property is regained. Then we will copy the first element of the heap array to the second last position.

Heap size is further reduced and percolate down cont.

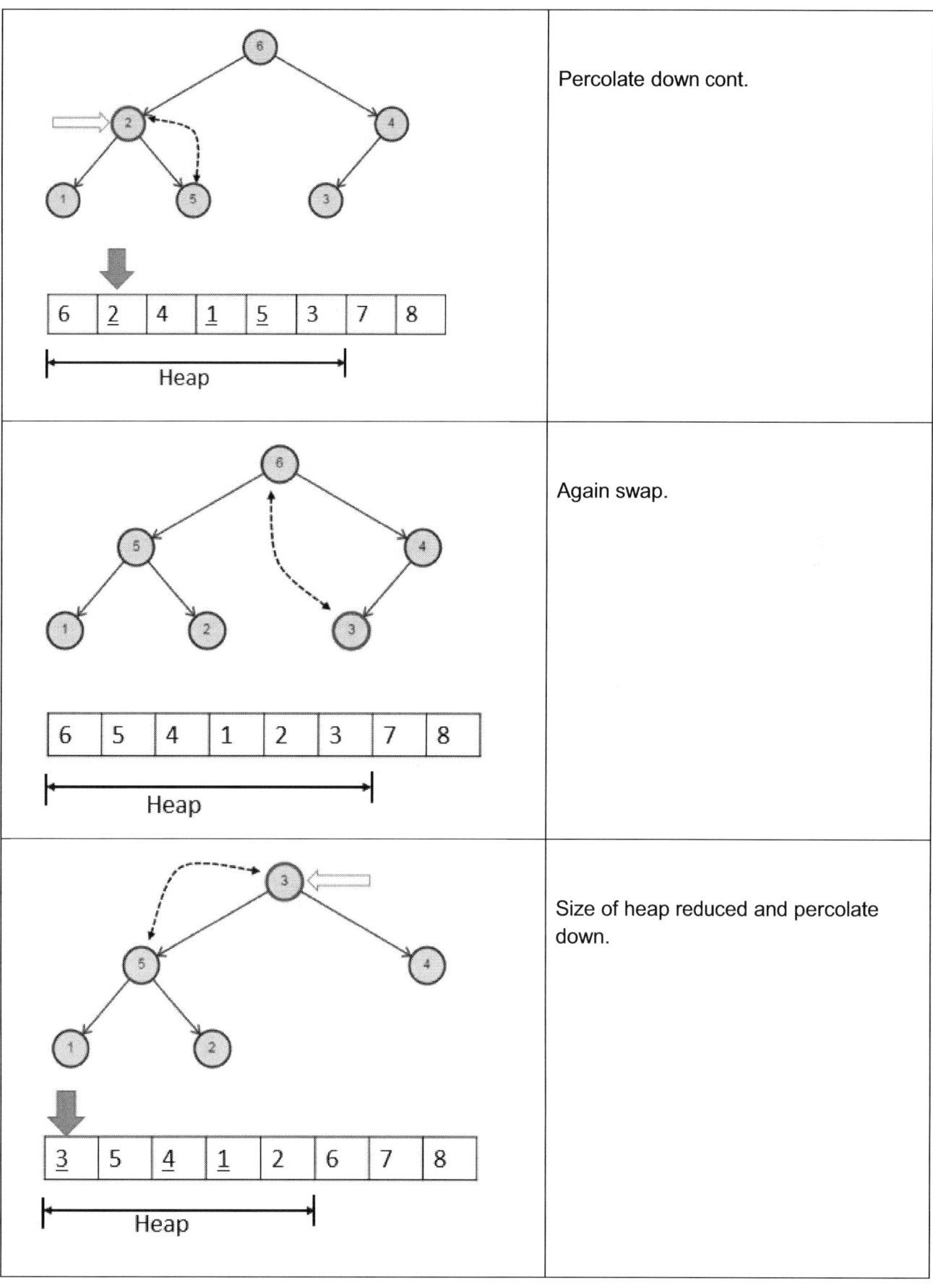

	Percolate down cont.
	Again swap.
	Size of heap reduced and percolate down.

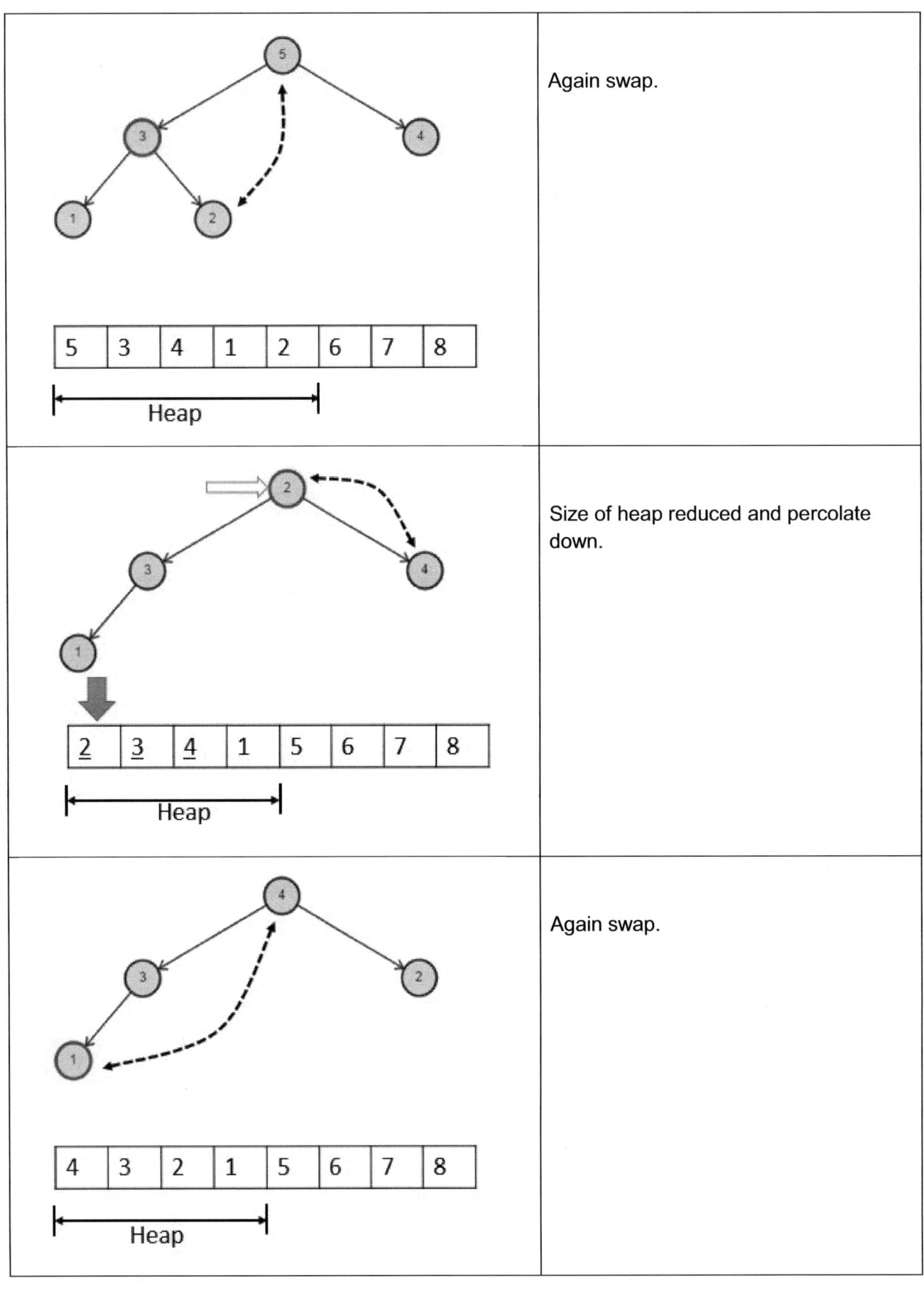

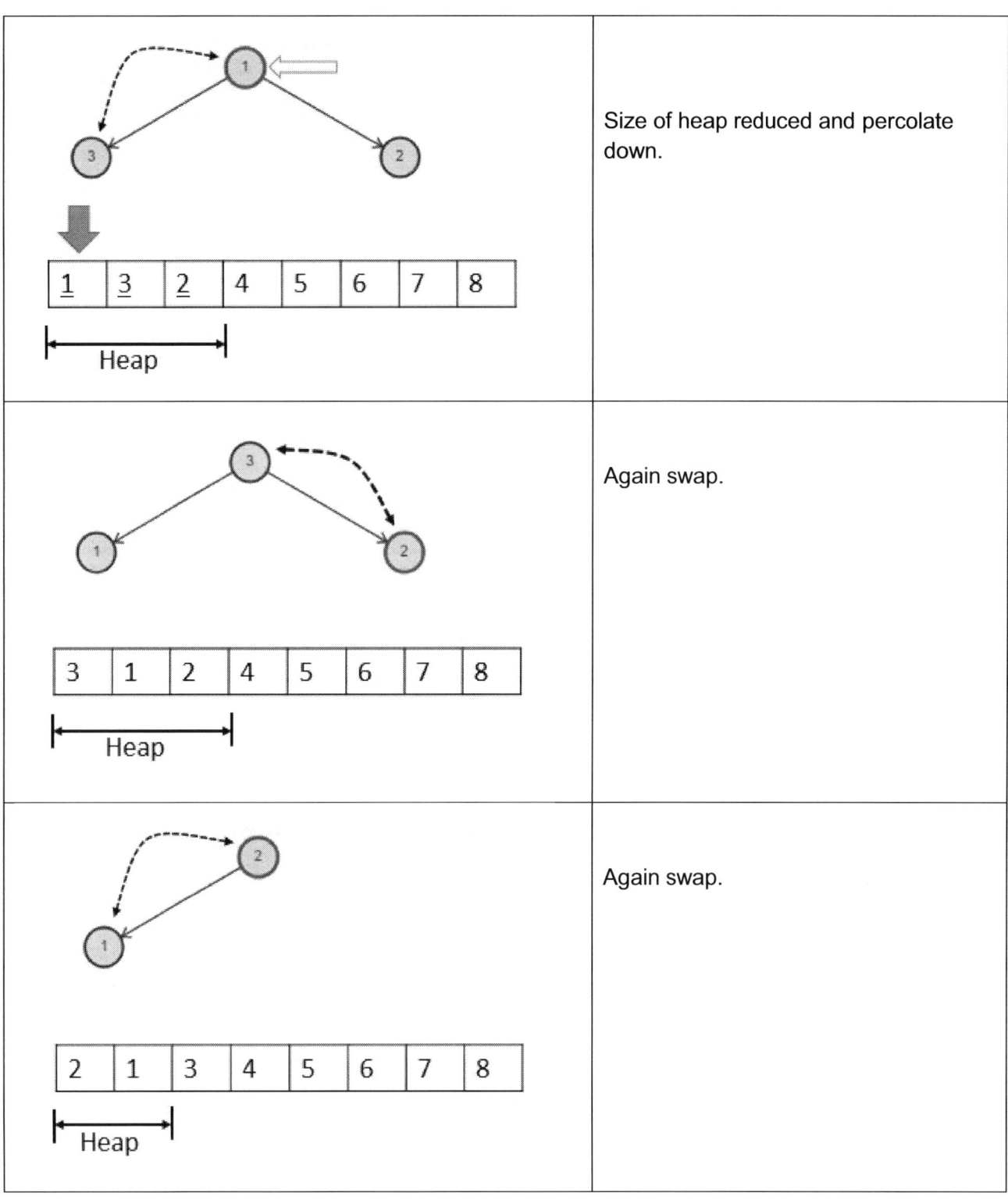

	Size of heap reduced and percolate down.
	Again swap.
	Again swap.

273

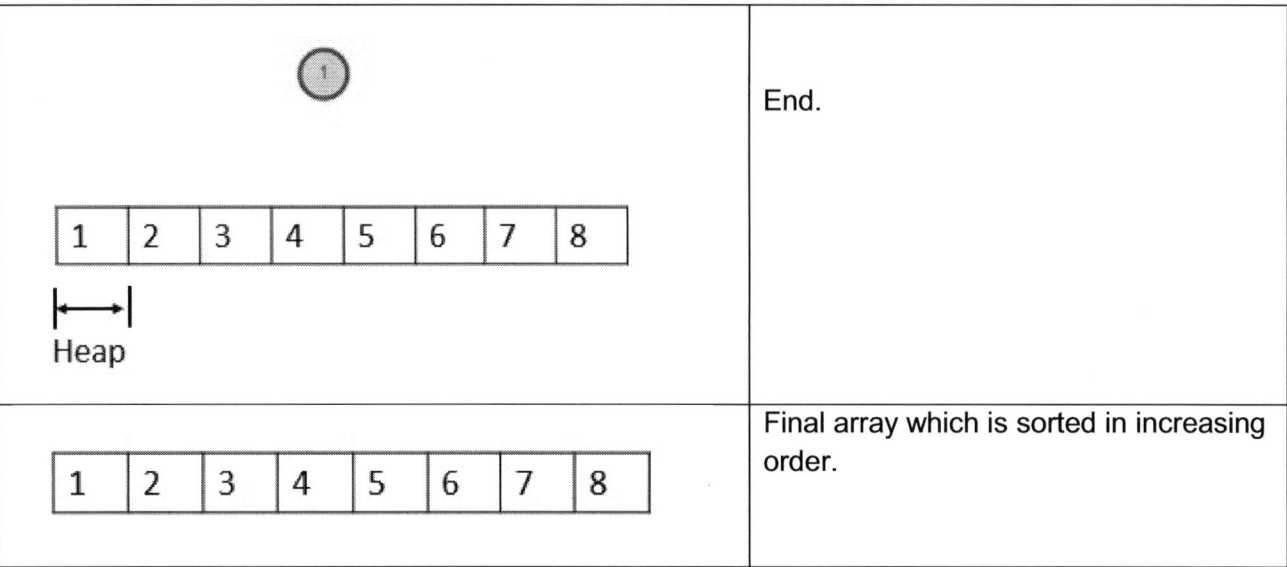

	Final array which is sorted in increasing order.

Example 11.5:

```java
public void print( )
{
    for(int i=1 ; i<=size ; i++ )
        System.out.println("value is :: " + arr[i] );
}
```

```java
public boolean isEmpty(){
    return (size==0) ;
}
```

```java
public int peek() {
    if (isEmpty()) {
        throw new IllegalStateException();
    }
    return arr[1];
}
```

```java
public static void heapSort(int[] array){
    Heap hp = new Heap(array);
    for(int i=0; i < array.length; i++)
        array[i] = hp.remove();
}
```

```java
public static void main(String[] args) {
    int[] a = {1,9,6,7,8,0,2,4,5,3};
    Heap hp = new Heap(a);
    hp.print();
    for(int i=0;i<a.length;i++)
        System.out.println("pop value :: " + hp.remove() );

    Heap.heapSort(a);
    for(int i=0 ; i<a.length ; i++ )
        System.out.println("value is :: " + a[i] );
}
```

Data structure	Array
Worst Case Time Complexity	O(nlogn)
Best Case Time Complexity	O(nlogn)
Average Time Complexity	O(nlogn)
Space Complexity	O(1)

Note: Heap-Sort is not a Stable sort and do not require any extra space for sorting a list.

Uses of Heap

1. **Heapsort**: One of the best sorting methods being in-place and log(N) time complexity in all scenarios.

2. **Selection algorithms**: Finding the min, max, both the min and max, median, or even the kth largest element can be done in linear time (often constant time) using heaps.

3. **Priority Queues**: Heap Implemented priority queues are used in Graph algorithms like Prim's Algorithm and Dijkstra's algorithm. A heap is a useful data structure when you need to remove the object with the highest (or lowest) priority. Schedulers, timers

4. **Graph algorithms**: By using heaps as internal traversal data structures, run time will be reduced by polynomial order. Examples of such problems are Prim's minimal

5. Because of the lack of references, the operations are faster than a binary tree. Also, some more complicated heaps (such as binomial) can be merged efficiently, which isn't easy to do for a binary tree.

Problems in Heap

Kth Smallest in a Min Heap

Just call DeleteMin() operation K-1 times and then again call DeleteMin() this last operation will give Kth smallest value. *Time Complexity* O(KlogN)

Kth Largest in a Max Heap

Just call DeleteMax() operation K-1 times and then again call DeleteMax () this last operation will give Kth smallest value. *Time Complexity* O(KlogN)

100 Largest in a Stream

There are billions of *integers* coming out of a stream some getInt() function is providing *integers* one by one. How would you determine the largest 100 numbers?

> *Solution:* Large hundred (or smallest hundred etc.) such problems are solved very easily using a Heap. In our case, we will create a min heap.

1. First from 100 first integers builds a min heap.
2. Then for each coming integer compare if it is greater than the top of the min heap.
3. If not, then look for next integer. If yes, then remove the top min value from the min heap, insert the new value at the top of the heap, use procolateDown, and move it to its proper position down the heap.
4. Every time you have largest 100 values stored in your head

Merge two Heap

How can we merge two heaps?

Solution: There is no single solution for this. Let us suppose the size of the bigger heap is N and the size of the smaller heap is M.
1. If both heaps are comparable size, then put both heap arrays in same bigger arrays. Or in one of the arrays if they are big enough. Then apply CreateHeap() function which will take theta(N+M) time.
2. If M is much smaller then N then add() each element of M array one by one to N heap. This will take O(MlogN) the worst case or O(M) best case.

Get Median function

Example 11.6: Give a data structure that will provide median of given values in constant time.
Solution: We will be using two heap one min heap and other max heap. First, there will be a max heap which will contain the first half of data and there will be an min heap which will contain the second half of the data. Max heap will contain the smaller half of the data and its max value that is at the top of the heap will be the median contender. Similarly, the Min heap will contain the larger values of the data and its min value that is at its top will contain the median contender. We will keep track of the size of heaps. Whenever we insert a value to heap, we will make sure that the size of two heaps

differs by max one element, otherwise we will pop one element from one and insert into another to keep them balanced.

```java
import java.util.Collections;
import java.util.PriorityQueue;

public class MedianHeap
{
    PriorityQueue<Integer> minHeap;
    PriorityQueue<Integer> maxHeap;

    public MedianHeap()
    {
        minHeap = new PriorityQueue<Integer>();
        maxHeap = new PriorityQueue<Integer>
                        (Collections.reverseOrder());
    }
    //Other Methods.
}
```

```java
public void insert(int value)
{
    if (maxHeap.size() == 0 || maxHeap.peek() >= value )
        maxHeap.add(value);
    else
        minHeap.add(value);

    //size balancing
    if (maxHeap.size() > minHeap.size() + 1) {
        value = maxHeap.remove();
        minHeap.add(value);
    }

    if (minHeap.size() > maxHeap.size() + 1) {
        value = minHeap.remove();
        maxHeap.add(value);
    }
}
```

```java
public int getMedian()
{
    if (maxHeap.size() == 0 && minHeap.size() == 0)
        return Integer.MAX_VALUE;

    if (maxHeap.size() ==  minHeap.size() )
        return (maxHeap.peek() +  minHeap.peek())/2;
    else if (maxHeap.size() >  minHeap.size())
        return maxHeap.peek();
    else
        return minHeap.peek();
}
```

```java
public static void main(String[] args) {
    int arr[] = { 1, 9, 2, 8, 3, 7, 4, 6, 5, 1, 9, 2, 8, 3, 7, 4, 6, 5, 10, 10 };
    MedianHeap hp = new MedianHeap();

    for (int i = 0; i < 20; i++)
    {
        hp.insert(arr[i]);
        System.out.println("Median after insertion of " + arr[i] + " is  "
                            +hp.getMedian());
    }
}
```

Is Min Heap

Example 11.7: Given an array, find if it is a binary Heap is Min Heap

```java
boolean IsMinHeap(int[] arr, int size)
{
    for (int i = 0; i <= (size - 2) / 2; i++)
    {
        if (2 * i + 1 < size)
        {
            if (arr[i] > arr[2 * i + 1])
                return false;
        }
        if (2 * i + 2<size)
        {
            if (arr[i] > arr[2 * i + 2])
                return false;
        }
    }
    return true;
}
```

Is Max Heap

Example 11.8: Given an array find if it is a binary Heap Max heap

```java
boolean IsMaxHeap(int[] arr, int size)
{
    for (int i = 0; i <= (size - 2) / 2; i++)
    {
        if (2 * i + 1 < size && arr[i] < arr[2 * i + 1])
            return false;

        if (2 * i + 2 < size && arr[i] < arr[2 * i + 2])
            return false;
    }
    return true;
}
```

Analysis: If each parent is grater then its children then heap property is true. We will start from half of the array and will reduce the size one by one thereby comparing the value of index node with its left child and right child node.

Traversal in Heap

Heaps are not designed to traverse to find some element they are made to get min or max element fast. Still if you want to traverse a heap just traverse the array sequentially. This traversal will be level order traversal. This traversal will have linear *Time Complexity*.

Deleting Arbiter element from Min Heap

Again, heap is not designed to delete an arbitrary element, but still if you want to do so. Find the element by linear search in the heap array. Replace it with the value stored at the end of the Heap value. Reduce the size of the heap by one. Compare the new inserted value with its parent. If its value is smaller than the parent value, then percolate up. Else if its value is greater than its left and right child then percolate down. *Time Complexity* is **O(logn)**

Deleting Kth element from Min Heap

Again, heap is not designed to delete an arbitrary element, but still if you want to do so. Replace the kth value with the value stored at the end of the Heap value. Reduce the size of the heap by one. Compare the new inserted value with its parent. If its value is smaller than the parent value, then percolate up. Else if its value is greater than its left and right child then percolate down. *Time Complexity* is **O(logn)**

Print value in Range in Min Heap

Linearly traverse through the heap and print the value that are in the given range.

Exercise

1. What is the worst-case runtime Complexity of finding the smallest item in a min-heap?

2. Find max in a min heap.
   Hint: normal search in the complete array. There is one more optimization you can search from the mid of the array at index N/2

3. What is the worst-case time Complexity of finding the largest item in a min-heap?

4. What is the worst-case time Complexity of deleteMin in a min-heap?

5. What is the worst-case time Complexity of building a heap by insertion?

6. Is a heap full or complete binary tree?

7. What is the worst time runtime Complexity of sorting an array of N elements using heapsort?

8. Given a sequence of numbers: 1, 2, 3, 4, 5, 6, 7, 8, 9
   a. Draw a binary Min-heap by inserting the above numbers one by one
   b. Also draw the tree that will be formed after calling Dequeue() on this heap

9. Given a sequence of numbers: 1, 2, 3, 4, 5, 6, 7, 8, 9
    a. Draw a binary Max-heap by inserting the above numbers one by one
    b. Also draw the tree that will be formed after calling Dequeue() on this heap

10. Given a sequence of numbers: 3, 9, 5, 4, 8, 1, 5, 2, 7, 6. Construct a Min-heap by calling CreateHeap function.

11. Show an array that would be the result after the call to deleteMin() on this heap

12. Given an array: [3, 9, 5, 4, 8, 1, 5, 2, 7, 6]. Apply heapify over this to make a min heap and sort the elements in decreasing order?

13. In Heap-Sort once a root element has been put in its final position, how much time, does it take to re-heapify the structure so that the next removal can take place? In other words, what is the *Time Complexity* of a single element removal from the heap of size N?

14. What do you think the overall *Time Complexity* for heapsort is? Why do you feel this way?

CHAPTER 12: HASH-TABLE

Introduction

In the previous chapter, we have looked into various searching techniques. Consider a problem of searching a value in an array. If the array is not sorted then we have no other option but to look into each and every element one by one so the searching *Time Complexity* will be **O(n)**. If the array is sorted then we can search the value we are looking for in **O(logn)** logarithmic time using binary search.

What if we have a function that can tell us the location/index of the value we are looking for in the array? We can directly go into that location and tell whether our object we are searching for is present or not in just **O(1)** constant time. Such a function is called a Hash function.

In real life when a letter is handed over to a postman, by looking at the address on the letter, postman precisely knows to which house this letter needs to be delivered. He is not going to ask for a person door to door.

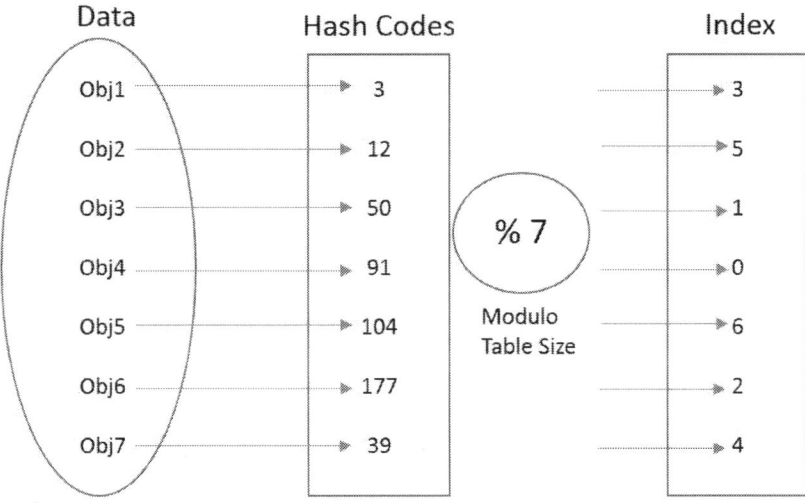

The process of storing objects using a hash function is as follows:
1. Create an array of size M to store objects; this array is called Hash-Table.
2. Find a hash code of an object by passing it through the hash function.
3. Take module of hash code by the size of Hashtable to get the index of the table where objects will be stored.
4. Finally store these objects in the designated index.

The process of searching objects in Hash-Table using a hash function is as follows:
1. Find a hash code of the object we are searching for by passing it through the hash function.
2. Take module of hash code by the size of Hashtable to get the index of the table where objects are stored.
3. Finally, retrieve the object from the designated index.

Hash-Table

A Hash-Table is a data structure that maps keys to values. Each position of the Hash-Table is called a slot. The Hash-Table uses a hash function to calculate an index of an array of slots. We use the Hash-Table when the number of keys actually stored is small relatively to the number of possible keys.

Hash-Table Abstract Data Type (ADT)

ADT of Hash-Table contains the following functions:
1. Insert(x), add object x to the data set.
2. Delete(x), delete object x from the data set.
3. Search(x), search object x in data set.

Hash Function

A hash function is a function that generates an index in a table for a given object.
An ideal hash function should generate a unique index for each and every object is called the perfect hash function.

Example 12.1: Most simple hash function

```
private int ComputeHash(int key)//division method
{
    int hashValue = 0;
    hashValue = key;
    return hashValue % tableSize;
}
```

There are many hash functions, but this is the minimum that a hash function should do. Various hash generation logics will be added to this function to generate a better hash.

Properties of good hash function:

1. It should provide a uniform distribution of hash values. A non-uniform distribution increased the number of collisions and the cost of resolving them.
2. Choose a hash function which can be computed quickly and returns values within the range of the Hash-Table.
3. Chose a hash function with a good collision resolution algorithm which can be used to compute alternative index if the collision occurs.
4. Choose a hash function which uses the necessary information provided in the key.
5. It should have high load factor for a given set of keys.

Load Factor

Load factor = Number of elements in Hash-Table / Hash-Table size

Based on the above definition, Load factor tells whether the hash function is distributing the keys uniformly or not. So it helps in determining the efficiency of the hashing function. It also works as decision parameter when we want to expand or rehash the existing Hash-Table entries.

Collisions

When a hash function generates the same index for the two or more different objects, the problem known as the collision. Ideally, hash function should return a unique address for each key, but practically it is not possible.

Collision Resolution Techniques

Hash collisions are practically unavoidable when hashing large number of objects. Techniques that are used to find the alternate location in the Hash-Table is called collision resolution. There are a number of collision resolution techniques to handle the collision in hashing.

Most common and widely used techniques are:
- Open addressing
- Separate chaining

Hashing with Open Addressing

When using linear open addressing, the Hash-Table is represented by a one-dimensional array with indices that range from 0 to the desired table size-1.

One method of resolving collision is the look into a Hash-Table and find another free slot the hold the object that have caused the collision. A simple way is to move from one slot to another in some sequential order until we find a free space. This collision resolution process is called Open Addressing.

Linear Probing

In Linear Probing, we try to resolve the collision of an index of a Hash-Table by sequentially searching the Hash-Table free location. Let us suppose, if k is the index retrieved from the hash function. If the kth index is already filled then we will look for (k+1) %M, then (k+2) %M and so on. When we get a free slot, we will insert the object into that free slot.

Example 12.2: The resolver function of linear probing

```
int resolverFun(int i)
{
    return i;
}
```

Quadratic Probing

In Quadratic Probing, we try to resolve the collision of the index of a Hash-Table by quadratic ally increasing the search index free location. Let us suppose, if k is the index retrieved from the hash function. If the kth index is already filled then we will look for (k+1^2) %M, then (k+2^2) %M and so on. When we get a free slot, we will insert the object into that free slot.

Example 12.3: The resolver function of quadratic probing

```
int resolverFun(int i)
{
    return i * i;
}
```

Table size should be a prime number to prevent early looping should not be too close to 2powN

Linear Probing implementation

Example 12.4: Below is a linear probing collision resolution Hash-Table implementation.

```
public class HashTable {

    private static int EMPTY_NODE = -1;
    private static int LAZY_DELETED = -2;
    private static int FILLED_NODE = 0;


    private int tableSize;
    int[] Arr;
    int[] Flag;

    public HashTable(int tSize) {
        tableSize = tSize;
        Arr = new int[tSize + 1];
        Flag = new int[tSize + 1];
        for(int i=0;i<=tSize;i++)
            Flag[i]=EMPTY_NODE;
    }
}
```

Table array size will be 50 and we have defined two constant values EMPTY_NODE and LAZY_DELETED.

```
int ComputeHash(int key)
{
    return key%tableSize;
}
```

This is the most simple hash generation function which does nothing but just take the modulus of the key.

```
int resolverFun(int index)
{
        return index;
}
```

When the hash index is already occupied by some element the value will be placed in some other location to find that new location resolver function is used.

Hash-Table has two component one is table size and other is reference to array.

Example 12.5:

```
boolean InsertNode(int value)
{
        int hashValue = ComputeHash(value);
        for (int i = 0; i < tableSize; i++)
        {
                if (Flag[hashValue] == EMPTY_NODE ||
                        Flag[hashValue] == LAZY_DELETED)
                {
                        Arr[hashValue] = value;
                        Flag[hashValue] = FILLED_NODE;
                        return true;
                }
                hashValue += resolverFun(i);
                hashValue %= tableSize;
        }
        return false;
}
```

An insert node function is used to add values to the array. First hash is calculated. Then we try to place that value in the Hash-Table. We look for empty node or lazy deleted node to insert value. In case insert did not success, we try new location using a resolver function.

Example 12.6:

```
boolean FindNode(int value)
{
        int hashValue = ComputeHash(value);
        for (int i = 0; i < tableSize; i++)
        {
                if (Flag[hashValue] == EMPTY_NODE)
                        return false;

                if (Flag[hashValue] == FILLED_NODE && Arr[hashValue] == value)
                        return true;

                hashValue += resolverFun(i);
                hashValue %= tableSize;
        }
        return false;
}
```

Find node function is used to search values in the array. First hash is calculated. Then we try to find that value in the Hash-Table. We look for over desired value or empty node. In case we find the value we are looking for then we return that value or in case we don't we return -1. We use a resolver function to find the next probable index to search.

Example 12.7:

```
boolean DeleteNode(int value)
{
    int hashValue = ComputeHash(value);
    for (int i = 0; i < tableSize; i++)
    {
        if (Flag[hashValue] == EMPTY_NODE)
            return false;

        if (Flag[hashValue] == FILLED_NODE && Arr[hashValue] == value)
        {
            Flag[hashValue] = LAZY_DELETED;
            return true;
        }
        hashValue += resolverFun(i);
        hashValue %= tableSize;
    }
    return false;
}
```

Delete node function is used to delete values from a Hashtable. We do not actually delete the value we just mark that value as LAZY_DELETED. Same as the insert and search we use resolverFun to find the next probable location of the key.

Example 12.8:

```
void Print()
{
    for (int i = 0; i < tableSize; i++)
    {
        if (Flag[i] == FILLED_NODE)
            System.out.println("Node at index [" +i+ " ] :: "+Arr[i] );
    }
}
```

Print method print the content of hash table.

```
public static void main(String[] args) {
    HashTable ht = new HashTable(1000);
    ht.InsertNode(89);
    ht.InsertNode(18);
    ht.InsertNode(49);
    ht.InsertNode(58);
    ht.InsertNode(69);
    ht.InsertNode(89);
    ht.InsertNode(18);
```

```
        ht.InsertNode(49);
        ht.InsertNode(58);
        ht.InsertNode(69);

        ht.Print();
        System.out.println("");

        ht.DeleteNode(89);
        ht.DeleteNode(18);
        ht.DeleteNode(49);
        ht.DeleteNode(58);
        ht.DeleteNode(69);

        ht.Print();
}
```

Main function demonstrating how to use hash table.

Quadratic Probing implementation.

Everything will be same as linear probing implementation only resolver function will be changed.

```
int resolverFun(int index)
{
        return index * index;
}
```

Hashing with separate chaining

Another method for collision resolution is based on an idea of putting the keys that collide in a linked list. This method is called separate chaining. To speed up search we use Insertion-Sort or keeping the linked list sorted.

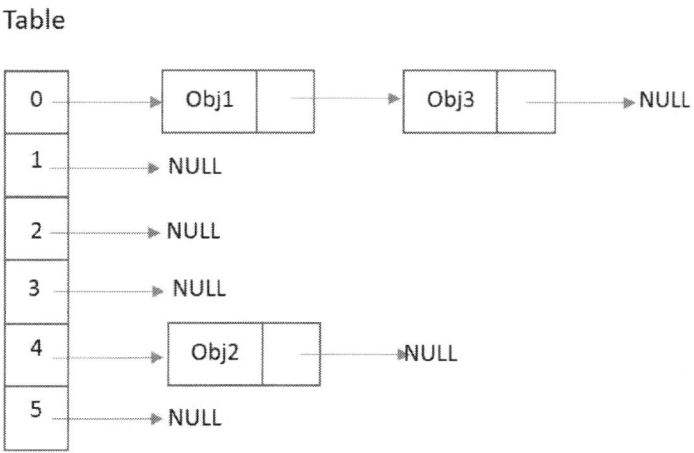

Separate Chaining implementation

Example 12.9: Below is separate chaining implementation of hash tables.

```java
public class HashTableSC {

    private class Node{
        private int value;
        private Node next;

        public Node(int v, Node n){
            value = v;
            next = n;
        }
    };

    private int tableSize;
    Node[] listArray;      //double pointer

    public HashTableSC() {
        tableSize=512;
        listArray = new Node[tableSize];
        for (int i = 0; i<tableSize; i++)
            listArray[i] = null;
    }
}
```

```java
private int ComputeHash(int key)        //division method
{
    int hashValue = 0;
    hashValue = key;
    return hashValue % tableSize;
}
```

```java
public void insert(int value)
{
    int index = ComputeHash(value);
    listArray[index]=new Node(value,listArray[index]);
}
```

```java
public boolean find(int value)
{
    int index = ComputeHash(value);
    Node head = listArray[index];
    while (head!=null){
        if(head.value == value)
            return true;
        head = head.next;
    }
    return false;
}
```

```java
public boolean delete(int value)
{
    int index = ComputeHash(value);
    Node nextNode, head = listArray[index];
    if (head != null && head.value == value)
    {
        listArray[index] = head.next;
        return true;
    }
    while (head!=null)
    {
        nextNode = head.next;
        if (nextNode != null&& nextNode.value == value)
        {
            head.next = nextNode.next;
            return true;
        }
        else
        {
            head = nextNode;
        }
    }
    return false;
}
```

```java
public void print()
{
    for (int i = 0; i<tableSize; i++)
    {
        System.out.println("Printing for index value :: " + i + "List of value printing :: ");
        Node head = listArray[i];
        while (head!=null)
        {
            System.out.println(head.value);
            head = head.next;
        }
    }
}
```

```java
public static void main(String[] args) {
    HashTableSC ht = new HashTableSC();

    for (int i = 100; i < 110; i++)
        ht.insert(i);
    System.out.println("search 100 :: "+ ht.find(100));
    System.out.println("remove 100 :: "+ ht.delete( 100));
    System.out.println("search 100 :: "+ ht.find( 100));
    System.out.println("remove 100 :: "+ ht.delete( 100));
}
```

Note: It is important to note that the size of the "skip" must be such that all the slots in the table will eventually be occupied. Otherwise, part of the table will be unused. To ensure this, it is often suggested that the table size being a prime number. This is the reason we have been using 11 in our examples.

Count Map

Below is CountMap<T> implementation over java collection HashMap<T>, this interface is important if you have repeted values can come and you want to keep track of there count. CountMap<T> is our implementation that we will use to solve many problems.

Example 12.10:

```java
import java.util.HashMap;

public class CountMap<T>
{
        HashMap<T, Integer> hm = new HashMap<T, Integer>();

        public void add(T key)
        {
                if(hm.containsKey(key))
                {
                        int count = hm.get(key);
                        hm.put(key, count+1);
                }
                else
                {
                        hm.put(key, 1);
                }
        }
}
```

```java
public void remove(T key)
{
        if(hm.containsKey(key))
        {
                if(hm.get(key)==1)
                        hm.remove(key);
                else{
                        int count = hm.get(key);
                        hm.put(key, count-1);
                }
        }
}
```

```java
public int get(T key)
{
        if(hm.containsKey(key))
                return hm.get(key);
        return 0;
}
```

```java
public boolean containsKey(T key)
{
        return hm.containsKey(key);
}
```

```java
public int size()
{
        return hm.size();
}
```

```java
public static void main(String[] args)
{
        CountMap<Integer> cm = new CountMap<Integer>();
        cm.add(2);cm.add(2);
        cm.remove(2);
        System.out.println("count is : " + cm.get(2));
        System.out.println("count is : " + cm.get(3));
}
```

```
count is : 1
count is : 0
```

Problems in Hashing

Anagram solver

An anagram is a word or phrase formed by reordering the letters of another word or phrase.

Example 12.11: Two words are anagram if they are of same size and there chracters are same.

```java
public static boolean isAnagram(char[] str1, char[] str2) {
        int size1 = str1.length;
        int size2 = str2.length;

        if (size1 != size2)
                return false;

        CountMap<Character> cm = new CountMap<Character>();

        for (char ch : str1) {
                cm.add(ch);
        }

        for (char ch : str2)
        {
                cm.remove(ch);
        }

        return (cm.size() == 0);
}
```

Remove Duplicate

Remove duplicates in an array of numbers.
Solution: We can use a second array or the same array, as the output array. In the below example Hash-Table is used to solve this problem.

Example 12.12:

```java
public static void removeDuplicate(char[] str)
{
    int index = 0;
    HashSet<Character> hs= new HashSet<Character>();

    for(char ch : str)
    {
        if (hs.contains(ch) == false)
        {
            str[index++] = ch;
            hs.add(ch);
        }
    }
    str[index] = '\0';
}
```

Find Missing

Example 12.13: There is a list of integers we need to find the missing number in the list.

```java
public static int findMissing(int[] arr, int start, int end)
{
    HashSet<Integer> hs = new HashSet<Integer>();
    for (int i : arr)
    {
        hs.add(i);
    }

    for(int curr = start; curr <= end; curr++)
    {
        if (hs.contains(curr) == false)
            return curr;
    }

    return Integer.MAX_VALUE;
}
```

All the element in the list is added to hashtable and then the missing element is found by searching into hashtable and final missing value is returned.

Print Repeating

Example 12.14: Print the repeating integer in a list of integers.

```
public static void printRepeating(int[] arr)
{
        HashSet<Integer> hs = new HashSet<Integer>();

        System.out.print("Repeating elements are:");
        for (int val : arr)
        {
                if (hs.contains(val))
                        System.out.print(" " + val);
                else
                        hs.add(val);
        }
}
```

All the values to the hash table when some value came which is already in the hash table then that is the repeted value.

Print First Repeating

Example 12.15: Same as the above problem in this we need to print the first repeating number. Caution should be taken to find the first repeating number. It should be the one number that is repeating. For example, 1, 2, 3, 2,1. The answer should be 1 as it is the first number which is repeating.

```
public static void printFirstRepeating(int[] arr)
{
        int i;
        int size = arr.length;
        CountMap<Integer> hs = new CountMap<Integer>();

        for (i = 0; i < size; i++)
        {
                hs.add(arr[i]);
        }
        for (i = 0; i < size; i++)
        {
                hs.remove(arr[i]);
                if (hs.containsKey(arr[i]))
                {
                        System.out.println("First Repeating number is : " + arr[i]);
                        return;
                }
        }
}
```

Add values to the count map the one that is repeting will have multiple count. Now traverse the array again and see if the count is more then one. So that is the first repeting.

Exercise

1. Design a number (ID) generator system that generate numbers between 0-99999999 (8-digits). The system should support two functions:
   a. int getNumber();
   b. int requestNumber();

   getNumber() function should find out a number that is not assigned, than marks it as assigned and return that number.
   requestNumber() function checks the number is assigned or not. If it is assigned returns 0, else marks it as assigned and return 1.

   Hint: You can keep a counter for assigning numbers. Whenever there is a getNumber() call you will check if that number is already assigned in a Hash-Table. If it is already assigned, then increase the counter and check again. If you find a number not in the Hash-Table then add it to Hashtable and increase the counter.
   requestNumber() will look in the Hash-Table if the number is already taken, then it will return 0 else it will return 1 and mark that number as taken inside the Hash-Table.

2. Given a large string, find the most occurring words in the string. What is the *Time Complexity* of the above solution?

Hint:-
   a. Create a Hashtable which will keep track of <word, frequency>
   b. Iterate through the string and keep track of word frequency by inserting into Hash-Table.
   c. When we have a new word, we will insert it into the Hashtable with frequency 1. For all repetition of the word, we will increase the frequency.
   d. We can keep track of the most occurring words whenever we are increasing the frequency we can see if this is the most occurring word or not.
   e. The *Time Complexity* is **O(n)** where n is the number of words in the string and *Space Complexity* is the O(m) where m is the unique words in the string.

3. In the above question, What if you are given whole work of OSCAR WILDE, most popular playwrights in the early 1890s.

Hint:-
   a. Who knows how many books are there, let's assume there is a lot and we can't put everything in memory. First, we need a Streaming Library so that we can read section by section in each document. Then we need a tokenizer that will give words to our program. And we need some sort of dictionary let's say we will use HashTable.
   b. What you need is - 1. A streaming library tokenizer, 2. A tokenizer 3. A hashmap

Method:
1. Use streamers to find a stream of the given words
2. Tokenize the input text
3. If the stemmed word is in hash map, increment its frequency count else adds a word to hash map with frequency 1

c. We can improve the performance by looking into parallel computing. We can use the map-reduce to solve this problem. Multiple nodes will read and process multiple documents. Once they are done with their processing, then we can use reduce to merge them.

4. In the above question, What if we wanted to find the most common PHRASE in his writings.
Hint:- We can keep <phrase, frequency> Hash-Table and do the same process of the 2nd and 3rd problems.

5. Write a hashing algorithm for strings.
Hint: Use Horner's method

```
int hornerHash (char[] key, int tableSize )
{
        int size = key.length;
        int h = 0;
        int i;
        for (i=0; i < size ; i++)
        {
                h = (32 * h + key[i]) % tableSize;
        }
        return h;
}
```

6. Pick two data structures to use in implementing a Map. Describe lookup, insert, & delete operations. Give time & *Space Complexity* for each. Give pros & cons for each.
Hint:-
   a) Linked List
      I. Insert is *O(1)*
      II. Delete is *O(1)*
      III. Lookup is *O(1)* auxiliary and *O(N)* worst case.
      IV. Pros: Fast inserts and deletes, can use for any data type.
      V. Cons: Slow lookups.
   b) Balanced Search Tree (RB Tree)
      I. Insert is *O(logn)*
      II. Delete is *O(logn)*
      III. Lookup is *O(logn)*
      IV. Pros: Reasonably fast inserts/deletes and lookups.
      V. Cons: Data needs to have order defined on it.

Chapter 13: Graphs

Introduction

In this chapter, we will study about Graphs. Graphs can be used to represent many interesting things in the real world. Flights from cities to cities, rods connecting various town and cities. Even the sequence of steps that we take to become ready for jobs daily, or even a sequence of classes that we take to become a graduate in computer science. Once we have a good representation of the map, then we use a standard graph algorithms to solve many interesting problems of real life.

The flight connection between major cities of India can also be represented by the below graph. Each node is a city and each edge is a straight flight path from one city to another. You may want to go from Delhi to Chennai, if given this data in good representation to a computer, through graph algorithms the computer may propose shortest, quickest or cheapest path from soured to destination.

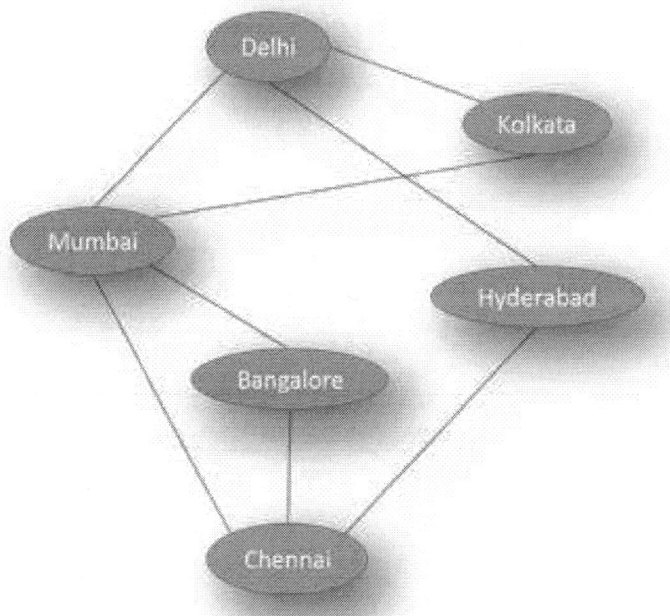

Google map that we use is also a big graph of lots of nodes and edges. And suggest shortest and quickest path to the user.

Graph Definitions

A Graph is represented by G where G = (V, E), where V is a finite set of points called **Vertices** and E is a finite set of **Edges**.

Each **edge** is a tuple (u, v) where u, v ∈ V. There can be a third component weight to the tuple. **Weight** is cost to go from one vertex to another.

Edge in a graph can be directed or undirected. If the edges of graph are one way, it is called **Directed graph** or **Digraph**. The graph whose edges are two ways are called **Undirected graph** or just graph.

A **Path** is a sequence of edges between two vertices. The length of a path is defined as the sum of the weight of all the edges in the path.

Two vertices u and v are **adjacent** if there is an edge whose endpoints are u and v.

In the below graph:
V = { V1, V2, V3, V4, V5, V6, V7, V8, V9 } ,

$$E = \begin{Bmatrix} (V1,V0,1), (V2,V1,2), (V3,V2,3), (V3,V4,4), (V5,V4,5), \\ (V1,V5,6), (V2,V5,7), (V3,V5,8), (V4,V5,9) \end{Bmatrix}$$

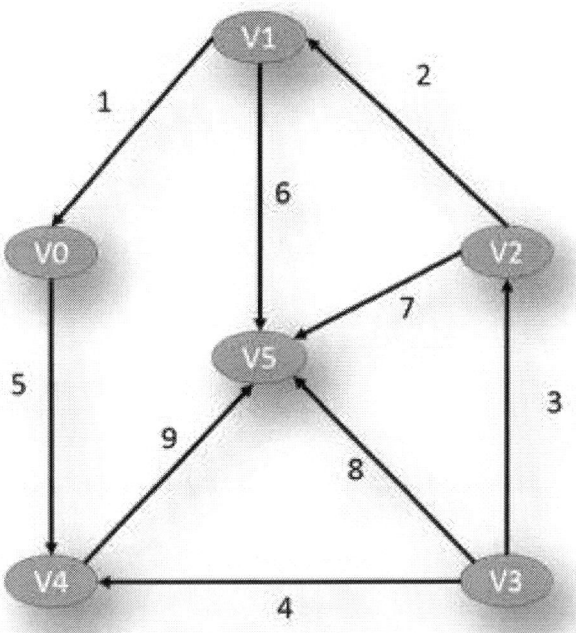

The **in-degree** of a vertex v, denoted by indeg(v) is the number of incoming edges to the vertex v.
The **out-degree** of a vertex v, denoted by outdeg(v) is the number of outgoing edges of a vertex v.
The **degree** of a vertex v, denoted by deg(v) is the total number of edges whose one endpoint is v.

deg(v) = Indeg (v) + outdeg (v)

In the above graph
deg(V4)=3, indeg(V4)=2 and outdeg(V4)=1

A **Cycle** is a path that starts and ends at the same vertex and include at least one vertex.

An edge is a **Self-Loop** if two if its two endpoints coincide. This is a form of a cycle.

A vertex v is **Reachable** from vertex u or "u reaches v" if there is a path from u to v. In an undirected graph if v is reachable from u then u is reachable from v. But in a directed graph it is possible that u reaches v but there is no path from v to u.

A graph is **Connected** if for any two vertices there is a path between them.

A **Forest** is a graph without cycles.

A **Sub-Graph** of a graph G is a graph whose vertices and edges are a subset of the vertices and edges of G.

A **Spanning Sub-Graph** of G is a graph that connects all the vertices of G.

A **tree** is an acyclic connected graph.

A **Spanning tree** of a graph is a spanning sub-graph that is also a tree that means, a connected graph which connects all the vertices of graph and that does not have a cycle.

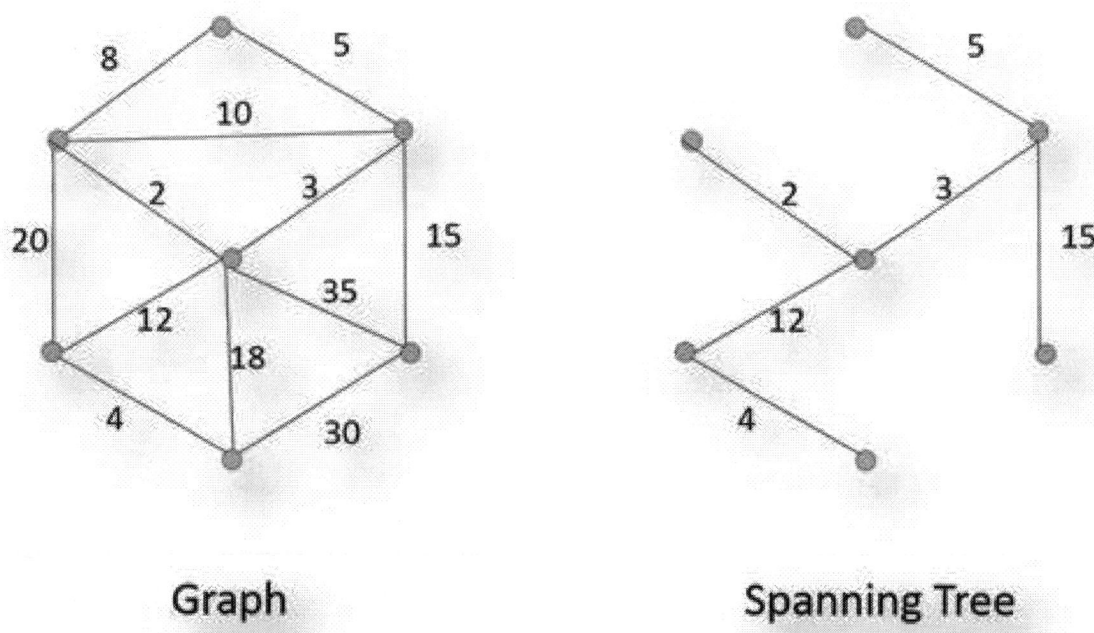

Graph   Spanning Tree

Graph Representation

In this section, we introduce the data structure for representing a graph. In the below representations we maintain a collection to store edges and vertices of the graph.

Adjacency Matrix

One of the ways to represent a graph is to use two-dimensional matrix. Each combination of row and column represent a vertex in the graph. The value stored at the location row v and column w is the edge from vertex v to vertex w. The nodes that are connected by an edge are called adjacent nodes. This matrix is used to store adjacent relation so it is called the Adjacency Matrix. In the below diagram, we have a graph and its Adjacency matrix.

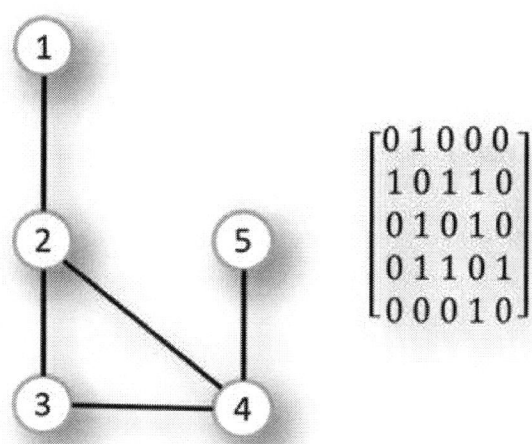

In the above graph, each node has weight 1 so the adjacency matrix has just 1s or 0s. If the edges are of different, weights that that weight will be filled in the matrix.

Pros: Adjacency matrix implementation is simple. Adding/Removing an edge between two vertices is just *O(1)*. Query if there is an edge between two vertices is also *O(1)*

Cons: It always consumes O(V^2) space, which is an inefficient way to store when a graph is a sparse.
Sparse Matrix: In a huge graph, each node is connected with fewer nodes. So most of the places in adjacency matrix are empty. Such matrix is called sparse matrix. In most of the real world problems adjacency matrix is not a good choice for sore graph data.

Adjacency List

A more space efficient way of storing graph is adjacency list. In adjacency list of references to a linked list node. Each reference corresponds to vertices in a graph. Each reference will then point to the vertices that are connected to it and store this as a list.
In the below diagram node 2 is connected to 1, 3 and 4. So the reference at location 2 is pointing to a list which contain 1, 3 and 4.

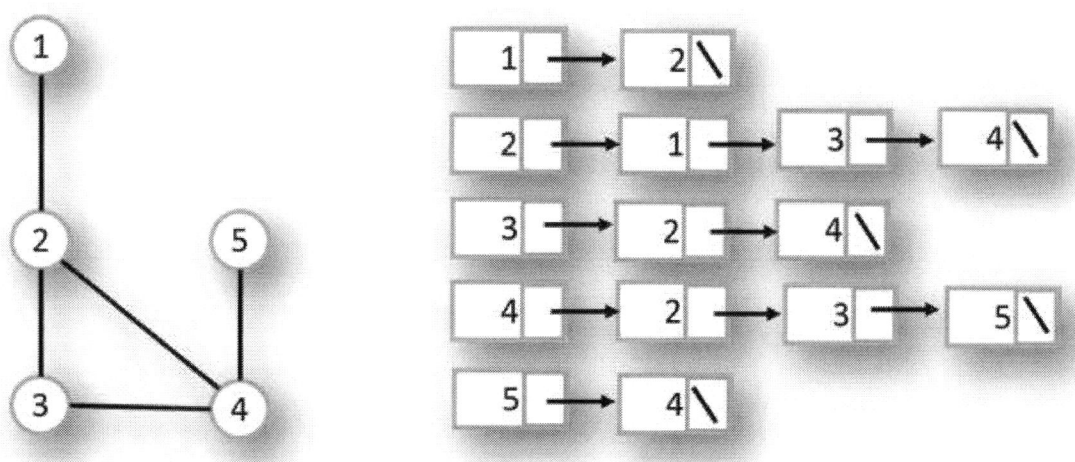

The adjacency list helps us to compactly represent a sparse graph. An adjacency list representation also allows us to find all the vertices that are directly connected to any vertices by just one link list scan. In all our programs, we are going to use the adjacency list to store the graph.

Below is the code for adjacency list representation of an undirected graph:

Example 13.1:

```java
public class Graph {
    private static class AdjNode {
        private int source;
        private int destination;
        private int cost;
        private AdjNode next;

        public AdjNode(int src, int dst, int cst) {
            source = src;
            destination = dst;
            cost = cst;
            next = null;
        }

        public AdjNode(int src, int dst) {
            this(src, dst, 1);
        }
    }

    private class static AdjList {
        private AdjNode head;
    }

    int count;
    AdjList[] array;

    public Graph(int cnt) {
        count = cnt;
        array = new AdjList[cnt];
        for(int i=0; i<cnt; i++){
            array[i]=new AdjList();
            array[i].head = null;
        }
    }

    public void AddEdge(int source, int destination, int cost){
        AdjNode node = new AdjNode(source, destination, cost);
        node.next = array[source].head;
        array[source].head =node;
    }

    public void AddEdge(int source, int destination){
        AddEdge(source, destination, 1);
    }
```

```java
        //bi directional edge
        public void AddBiEdge(int source, int destination, int cost)
        {
            AddEdge(source,destination,cost);
            AddEdge(destination,source,cost);
        }

        public void AddBiEdge(int source, int destination)//bi directional edge
        {
            AddBiEdge(source,destination,1);
        }
        public void Print(){
            AdjNode ad;
            for(int i=0;i<count;i++){
                ad = array[i].head;
                if(ad != null){
                    System.out.print("Vertex " + i+ " is connected to:");
                    while(ad != null)
                    {
                        System.out.print(ad.destination+ " ");
                        ad = ad.next;
                    }
                    System.out.println("");
                }
            }
        }
}
```

Graph traversals

The **Depth first search (DFS)** and **Breadth first search (BFS)** are the two algorithms used to traverse a graph. These same algorithms can also be used to find some node in the graph, find if a node is reachable etc.

Traversal is the process of exploring a graph by examining all its edges and vertices.

A list of some of the problems that are solved using graph traversal are:
1. Determining a path from vertex u to vertex v, or report an error if there is no such path.
2. Given a starting vertex s, finding the minimum number of edges from vertex s to all the other vertices of the graph.
3. Testing of a graph G is connected.
4. Finding a spanning tree of a Graph.
5. Finding if there is some cycle in the graph.

Depth First Traversal

The DFS algorithm we start from starting point and go into depth of graph until we reach a dead end and then move up to parent node (Backtrack). In DFS we use stack to get the next vertex to start a search. Or we can use recursion (system stack) to do the same.

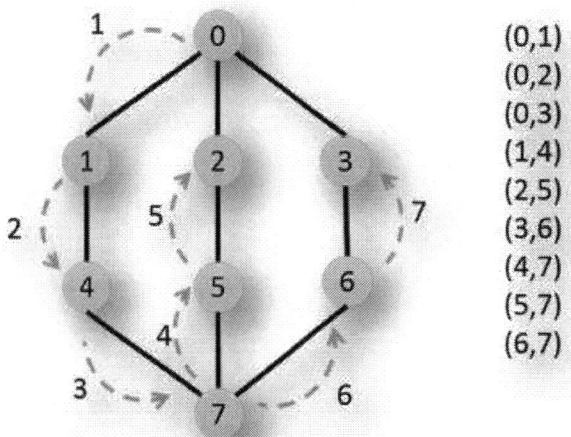

Depth First Traversal
0, 1, 4, 7 , 5, 2, 6, 3

Algorithm steps for DFS
1. Push the starting node in the stack.
2. Loop until the stack is empty.
3. Pop the node from the stack inside loop call this node current.
4. Process the current node. //Print, etc.
5. Traverse all the child nodes of the current node and push them into stack.
6. Repeat steps 3 to 5 until the stack is empty.

Stack based implementation of DFS

Example 13.2:

```java
public void DFSStack(Graph gph, int start) {
    int count = gph.count;
    int[] visited = new int[count];
    int curr;
    Stack<Integer> stk = new Stack<Integer>();
    for (int i = 0; i < count; i++)
        visited[i] = 0;

    visited[start] = 1;
    stk.push(start);

    while (stk.isEmpty() == false) {
        curr = stk.pop();
        AdjNode head = gph.array[curr].head;
        while(head != null) {
            if(visited[head.destination] == 0) {
                visited[head.destination] = 1;
                stk.push(head.destination);
            }
            head = head.next;
        }
    }
}
```

Recursion based implementation of DFS

Example 13.3:

```java
public static void DFSRec(Graph gph, int start, int[] visited)
{
    AdjNode head = gph.array[start].head;
    while(head != null)
    {
        if(visited[head.destination] == 0)
        {
            visited[head.destination] = 1;
            DFSRec( gph, head.destination, visited);
        }
        head = head.next;
    }
}
```

Breadth First Traversal

In BFS algorithm, a graph is traversed in layer-by-layer fashion. The graph is traversed closer to the starting point. The queue is used to implement BFS.

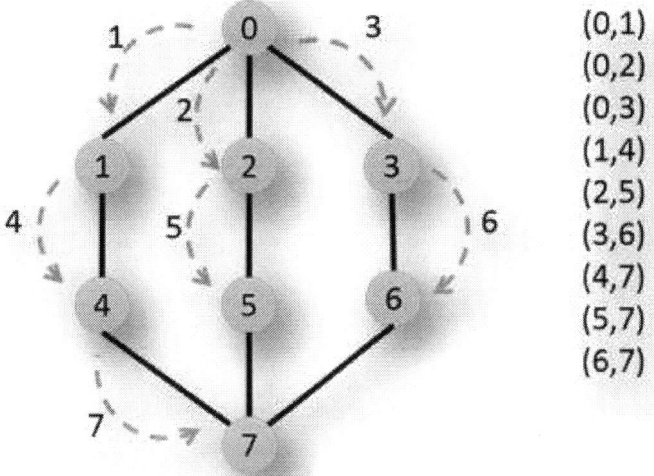

Breadth First Traversal
0, 1, 2, 3, 4, 5, 6, 7

Algorithm steps for BFS
1. Push the starting node into the Queue.
2. Loop until the Queue is empty.
3. Remove a node from the Queue inside loop, call this node current.
4. Process the current node.//print etc.
5. Traverse all the child nodes of the current node and push them into Queue.
6. Repeat steps 3 to 5 until Queue is empty.

Example 13.4:

```
Public void BFSQueue(Graph gph, int start, int[] visited)
{
    int curr;
    LinkedList<Integer> que= new LinkedList<Integer>();

    visited[start] = 1;
    que.add(start);

    while( que.isEmpty() == false)
    {
        curr=que.remove();
        AdjNode head = gph.array[curr].head;
        while(head != null)
        {
            if(visited[head.destination] == 0)
            {
                visited[head.destination] = 1;
                que.add(head.destination);
            }
            head = head.next;
        }
    }
}
```

A runtime analysis of DFS and BFS traversal is O(n+m) time, where n is the number of edges reachable from source node and m is the number of edges incident on s.

The following problems have O(m+n) time performance:
1. Determining a path from vertex u to vertex v, or report an error if there is no such path.
2. Given a starting vertex s, finding the minimum number of edges from vertex s to all the other vertices of the graph.
3. Testing of a graph G is connected.
4. Finding a spanning tree of a Graph.
5. Finding if there is some cycle in the graph.

Problems in Graph

Determining a path from vertex u to vertex v

IF there is a path from u to v and we are doing DFS from u then v must be visited. And if there is no path them report an error.

Example 13.5:

```
public static int PathExist(Graph gph, int source, int destination)
{
        int count = gph.count;
        int[] visited = new int[count];
        for (int i = 0; i < count; i++)
                visited[i] = 0;
        visited[source] = 1;
        DFSRec (gph, source, visited);
        return visited[destination];
}
```

Given a starting vertex s, finding the minimum number of edges from vertex s to all the other vertices of the graph

Look for single source shortest path algorithm for each edge cost as 1 unit.

Testing of a graph G is connected.

We simply start from any arbitrary vertex and do DFS search and should see if there is some vertex which is not visited. If all the vertices are visited then it is a connected graph.

Example 13.6:

```
public boolean isConnected(Graph gph)
{
        int count = gph.count;
        int[] visited = new int[count];
        for (int i = 0; i < count; i++)
                visited[i] = 0;
        visited[0] = 1;
        DFSRec(gph,0, visited);
        for (int i = 0; i < count; i++)
                if(visited[i] == 0)
                        return false;
        return true;
}
```

Finding if there is some cycle in the graph.

Modify DFS problem and get this done.

Directed Acyclic Graph

A Directed Acyclic Graph (DAG) is a directed graph with no cycle. A DAG represent relationship which is more general than a tree. Below is an example of DAG, this is how someone becomes ready for work. There are N other real life examples of DAG such as coerces selection to being graduated from college

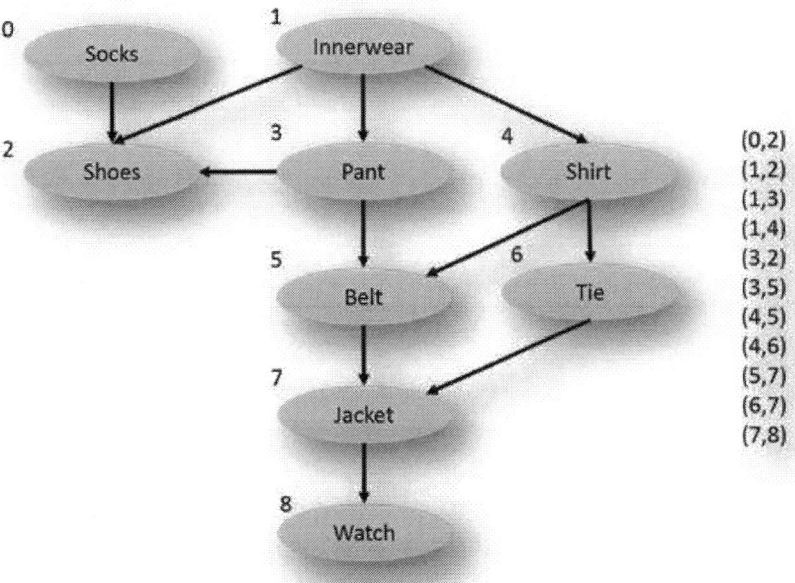

(0,2)
(1,2)
(1,3)
(1,4)
(3,2)
(3,5)
(4,5)
(4,6)
(5,7)
(6,7)
(7,8)

Topological Sort

A topological sort is a method of ordering the nodes of a directed graph in which nodes represent activities and the edges represent dependency among those tasks. For topological sorting to work it is required that the graph should be a DAG which means it should not have any cycle. Just use DFS to get topological sorting.

Example 13.7:

```java
public static void TopologicalSort(Graph gph)
{
    Stack<Integer> stk = new Stack<Integer>();
    int count = gph.count;
    int[] visited = new int[count];
    for (int i = 0; i < count; i++)
            visited[i] = 0;
    for (int i = 0; i < count; i++)
    {
            if (visited[i] == 0)
            {
                    visited[i] = 1;
                    TopologicalSortDFS(gph, i, visited, stk);
            }
    }
    while (stk.isEmpty() != true)
            System.out.print(" " + stk.pop());
}
```

```
private static void TopologicalSortDFS(Graph gph, int index, int[] visited,
Stack<Integer> stk)
{
        AdjNode head = gph.array[index].head;
        while(head != null)
        {
                if(visited[head.destination] == 0)
                {
                        visited[head.destination] = 1;
                        TopologicalSortDFS( gph, head.destination, visited, stk);
                }
                head = head.next;
        }
        stk.push(index);
}
```

Topology sort is DFS traversal of topology graph. First the children of node are added to the stack then only the current node is added. So the sorting order is maintained. Reader is requested to run some examples to fully understand this algo.

Minimum Spanning Trees (MST)

A Spanning Tree of a graph G is a tree which contains all the edges of the Graph G.
A Minimum Spanning Tree is a tree whose sum of length/weight of edges is minimum as possible. For example, if you want to setup communication between a set of cities, then you may want to use the least amount of wire as possible. MST can be used to find the network path and wire cost estimate.

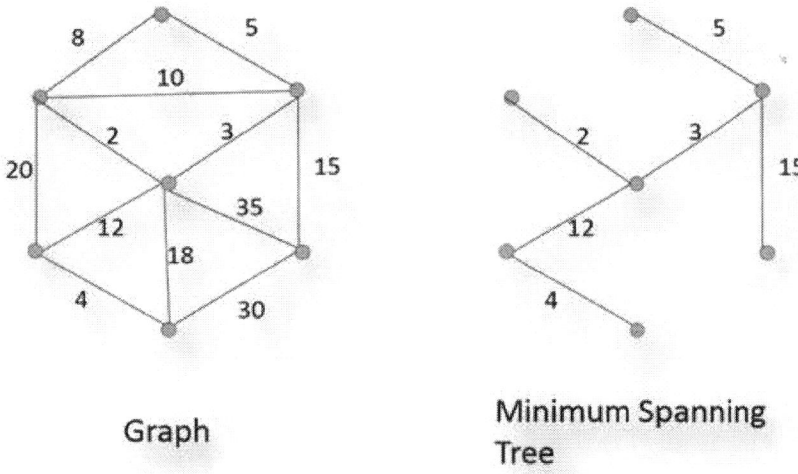

Graph   Minimum Spanning Tree

Prim's Algorithm for MST

Prim's algorithm grows a single tree T, one edge at a time, until it becomes a spanning tree.
We initialize T with zero edges. And U with single node. Where T is spanning tree edges set and U is spanning tree vertex set.
At each step, Prim's algorithm adds the smallest value edge with one endpoint in U and other not in us. Since each edge adds one new vertex to U, after n − 1 additions, U contain all the vertices of the spanning tree and T becomes a spanning tree.

Example 13.8:

```
// Returns the MST by Prim's Algorithm
// Input: A weighted connected graph G = (V, E)
// Output: Set of edges comprising a MST

Algorithm Prim(G)
      T = {}
      Let r be any vertex in G
      U = {r}
      for i = 1 to |V| - 1 do
            e = minimum-weight edge (u, v)
                  With u in U and v in V-U
            U = U + {v}
            T = T + {e}
      return T
```

Prim's Algorithm using a priority queue (min heap) to get the closest fringe vertex
Time Complexity will be O(m log n) where n vertices and m edges of the MST.

Example 13.9:

```java
public static void Prims(Graph gph) {
       int[] previous = new int[gph.count];
       int[] dist = new int[gph.count];
       int source = 1;

       for (int i = 0; i < gph.count; i++) {
              previous[i] = -1;
              dist[i] = Integer.MAX_VALUE;
       }

       dist[source] = 0;
       previous[source] = -1;

       AdjNodeComparator comp = new AdjNodeComparator();
       PriorityQueue<AdjNode> queue=new PriorityQueue<AdjNode>(100,comp);

       AdjNode node = new AdjNode(source, source, 0);
       queue.add(node);

       while (queue.isEmpty() != true) {
              node = queue.peek();
              queue.remove();

              if(dist[node.destination] < node.cost)
                     continue;

              dist[node.destination] = node.cost;
              previous[node.destination] = node.source;
```

```java
                AdjList adl = gph.array[node.destination];
                AdjNode adn = adl.head;

                while (adn != null) {
                        if (previous[adn.destination]==-1)
                        {
                                node = new AdjNode(adn.source, adn.destination, adn.cost);
                                queue.add(node);
                        }
                        adn = adn.next;
                }
        }
        AdjNodeComparator comp = new AdjNodeComparator();
        PriorityQueue<AdjNode> queue = new PriorityQueue<AdjNode>(100,comp);

        AdjNode node = new AdjNode(source, source, 0);
        queue.add(node);

        while (queue.isEmpty()!= true)
        {
                node = queue.peek();
                queue.remove();

                AdjList adl = gph.array[node.destination];
                AdjNode adn = adl.head;
                while(adn != null)
                {
                        int alt = adn.cost;
                        if(alt < dist[adn.destination]){
                                dist[adn.destination]= alt;
                                previous[adn.destination]=adn.source;
                                node = new AdjNode(adn.source, adn.destination, alt);
                                queue.add(node);
                        }
                        adn = adn.next;
                }
        }
        // Printing result.
        int count = gph.count;
        for (int i = 0; i < count; i++)
        {
                if(dist[i] == Integer.MAX_VALUE){
                        System.out.println(" node id " + i + "  prev " +  previous[i]
                                            + " distance : Unreachable");
                }
                else{
                        System.out.println(" node id " + i + "  prev " +  previous[i]
                                            + " distance : " + dist[i] );
                }
        }
}
```

```
public static void main(String[] args) {
    Graph gph = new Graph(9);
    gph.AddBiEdge( 0, 2, 1);
    gph.AddBiEdge( 1, 2, 5);
    gph.AddBiEdge( 1, 3, 7);
    gph.AddBiEdge( 1, 4, 9);
    gph.AddBiEdge( 3, 2, 2);
    gph.AddBiEdge( 3, 5, 4);
    gph.AddBiEdge( 4, 5, 6);
    gph.AddBiEdge( 4, 6, 3);
    gph.AddBiEdge( 5, 7, 5);
    gph.AddBiEdge( 6, 7, 7);
    gph.AddBiEdge( 7, 8, 17);

    //Dijkstra(gph,1);
    Prims(gph);
}
```

Kruskal's Algorithm

Kruskal's Algorithm repeatedly chooses the smallest-weight edge that does not form a cycle.
Sort the edges in non-decreasing order of cost: $c(e_1) \leq c(e_2) \leq \cdots \leq c(e_m)$.
Set T to be the empty tree. Add edges to tree one by one if it does not create a cycle.

Example 13.10:

```
// Returns the MST by Kruskal's Algorithm
// Input: A weighted connected graph G = (V, E)
// Output: Set of edges comprising a MST

Algorithm Kruskal(G)
    Sort the edges E by their weights
    T = {}
    while |T| + 1 < |V| do
        e = next edge in E
        if T + {e} does not have a cycle then
            T = T + {e}
    return T
```

Kruskal's Algorithm is $O(E \log V)$ using efficient cycle detection.

Shortest Path Algorithms in Graph

Single Source Shortest Path

For a graph G= (V, E), the single source shortest path problem is to find the shortest path from a given source vertex s to all the vertices of V.

Single Source Shortest Path for unweighted Graph.

Find single source shortest path for unweighted graph or a graph whose all the vertices have same weight.

Example 13.11:

```java
public void ShortestPath(Graph gph, int source)// unweighted graph
{
    int curr;
    int count = gph.count;
    int[] distance= new int[count];
    int[] path= new int[count];

    Queue<Integer> que= new LinkedList<Integer>();

    for (int i = 0; i < count; i++)
        distance[i] = -1;
    que.add(source);
    distance[source]=0;
    while(que.isEmpty() == false)
    {
        curr=que.remove();
        AdjNode head = gph.array[curr].head;
        while(head != null)
        {
            if(distance[head.destination] == -1)
            {
                distance[head.destination] = distance[curr] + 1;
                path[head.destination]=curr;
                que.add(head.destination);
            }
            head = head.next;
        }
    }
    for(int i=0;i<count;i++)
        System.out.println(path[i] + " to " + i +" weight " + distance[i]);
}
```

Dijkstra's algorithm

Dijkstra's algorithm for single-source shortest path problem for weighted edges with no negative weight. Given a weighted connected graph G, find shortest paths from the source vertex s to each of the other vertices. Dijkstra's algorithm is similar to prims algorithm. It maintains a set of nodes for which shortest path is known.

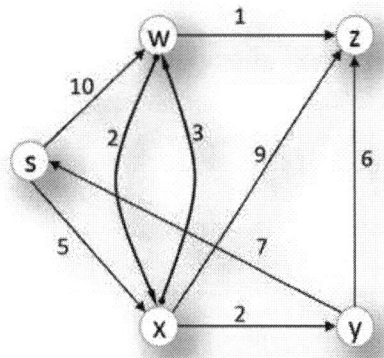
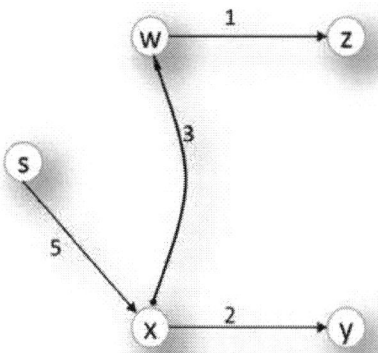

Single-Source shortest path

The algorithm starts by keeping track of the distance of each node and its parents. All the distance is set to infinite in the beginning as we don't know the actual path to the nodes and parents of all the vertices are set to null. All the vertices are added to a priority queue (min heap implementation)
At each step algorithm takes one vertex from the priority queue (which will be the source vertex in the beginning). Then update the distance array corresponding to all the adjacent vertices. When the queue is empty, then we will have the distance and parent array fully populated.

Example 13.12:

```
// Solves SSSP by Dijkstra's Algorithm
// Input: A weighted connected graph G = (V, E)
// with no negative weights, and source vertex v
// Output: The length and path from s to every v

Algorithm Dijkstra(G, s)
for each v in V do
            D[v] = infinite      // Unknown distance
            P[v] = null    //unknown previous node
            add v to PQ    //adding all nodes to priority queue

D[source] = 0 // Distance from source to source

while (PQ is not empty)
            u = vertex from PQ with smallest D[u]
            remove u from PQ
            for each v adjacent from u do
                        alt = D[u] + length ( u , v )
                        if  alt < D[v] then
                                    D[v] = alt
                                    P[v] = u
    Return D[] , P[]
```

Time Complexity will be $O(|E|\log|V|)$

Note: Dijkstra's algorithm does not work for graphs with negative edges weight.
Note: Dijkstra's algorithm is applicable to both undirected and directed graphs.

Example 13.13:

```java
public static void Dijkstra(Graph gph, int source) {
    class AdjNodeComparator implements Comparator<AdjNode>{
        public int compare(AdjNode x, AdjNode y){
            if(x.cost < y.cost)
                return -1;
            if (x.cost > y.cost)
                return 1;
            return 0;
        }
    };

    int[] previous=new int[gph.count];
    int[] dist=new int[gph.count];
    for(int i=0;i<gph.count;i++)
    {
        previous[i]=-1;
        dist[i]=Integer.MAX_VALUE; //infinite
    }

    dist[source] = 0;
    previous[source]= -1;

    AdjNodeComparator comp = new AdjNodeComparator();
    PriorityQueue<AdjNode> queue = new
                    PriorityQueue<AdjNode>(100,comp);

    AdjNode node = new AdjNode(source, source, 0);
    queue.add(node);
    while (queue.isEmpty()!= true)
    {
        node = queue.peek();
        queue.remove();

        AdjList adl = gph.array[node.destination];
        AdjNode adn = adl.head;
        while(adn != null)
        {
            int alt = adn.cost + dist[adn.source];
            if(alt < dist[adn.destination]){
                dist[adn.destination]= alt;
                previous[adn.destination]=adn.source;
                node = new AdjNode(adn.source, adn.destination, alt);
                queue.add(node);
            }
            adn = adn.next;
        }
    }
}
```

```java
        int count = gph.count;
        for (int i = 0; i < count; i++)
        {
                if(dist[i] == Integer.MAX_VALUE){
                        System.out.println(" node id " + i + "  prev " + previous[i]
                                + " distance : Unreachable" );
                }
                else{
                        System.out.println(" node id " + i + "  prev " + previous[i]
                                + " distance : " + dist[i] );
                }
        }
}
```

Bellman Ford Shortest Path

The bellman ford algorithm works even when there are negative weight edges in the graph. It does not work if there is some cycle in the graph whose total weight is negative.

Example 13.14:

```java
public void BellmanFordShortestPath(Graph gph, int source)
{
        int count = gph.count;
        int[] distance= new int[count];
        int[] path= new int[count];

        for (int i = 0; i < count; i++)
                distance[i] = Integer.MAX_VALUE;
        distance[source]=0;
        for (int i = 0; i < count -1; i++)
        {
                for (int j = 0; j < count; j++)
                {
                        AdjNode head = gph.array[j].head;
                        while(head!= null)
                        {
                                int newDistance = distance[j]+ head.cost;
                                if(distance[head.destination] > newDistance)
                                {
                                        distance[head.destination] = newDistance;
                                        path[head.destination]=j;
                                }
                                head = head.next;
                        }
                }
        }
        for(int i=0;i<count;i++)
                System.out.println(path[i] + " to " + i +" weight " + distance[i]);
}
```

All Pairs Shortest Paths

Given a weighted graph G(V, E), the all pair shortest path problem is to find the shortest path between all pairs of vertices u, v ∈ V.

Execute n instances of single source shortest path algorithm for each vertex of the graph.
The complexity of this algorithm will be $O(n^3)$

Exercise

1. In all the path finding algorithm we have created a path array which just store immediate parent of a node, print the complete path for it.

2. All the functions are implemented considering as if the graph is represented by adjacency list. To write all those functions for graph representation as adjacency matrix.

3. Given a start string, end string and a set of strings, find if there exists a path between the start string and end string via the set of strings.

   A path exists if we can get from start string to end the string by changing (no addition/removal) only one character at a time. The restriction is that the new string generated after changing one character has to be in the set.

   start: "cog"
   end: "bad"
   set: ["bag", "cag", "cat", "fag", "con", "rat", "sat", "fog"]
   one of the paths: "cog" -> "fog" -> "fag" -> "bag" -> "bad"

Chapter 14: String Algorithms

Introduction

String in JAVA programming is a sequence of character. String is so widely used in JAVA that they have their own class, which contains a number of methods to create and manipulate strings. A String is a object and if it is not initialized then it will point to null. We can initialize string by using a string literal, which is a text inside two double inverted commas. Or we can initialize string using one of many constructors provided by String class.

In this chapter, we will look into implementation of some of the constructors and methods of String class that are most frequently used and which are used in this chapter.

If string is just declared and not initialized then it will point to null. Just like any other object variable in JAVA.

```
String text;
```

The most easiest way to create a String is by string literal under double interted commas.

```
String text = "Hello, World!";
```

Another important method of creating String is by using String constructor that takes an array of characters as argument.

```
char[] arr = {'H','e','l','l','o',',',' ','W','o','r','l','d','!'};
String hello = new String(arr);
```

A length() function is there to find the length of string.

```
hello.length();
```

Concatination of string is done by concat() function provided in String class or just using '+' operator.

```
String first = "Hello, ";
String second = "World!";
String helloworld = first + second;
```

Below fragment of code also does the same job.

```
String helloworld = first.concat(second);
```

A single character at specific index in string can be extracted by calling charAt() method.

Below code will print 'W' to the output screen.

```
String text = "Hello, World!";
System.out.println(text.charAt(7));
```

Many times we need to manipulate char-by-char whole string and want to modify/ swap etc. So a string can be converted to character array. toCharArray() method returns an array of character for the string object on which this method is executed.

```
String text = "Hello, World!";
char[] array = text.toCharArray();
```

Remember that string is a object so if we call "==" to compare two string it will compare the memory location of the strings and if both the string objects are not pointing to same memory location / same string then it will fail. But many time we want to compare if the content of the two is same so string class had provided equals() method, which will compare if the given string content is same as the string object on whose equals() method is called.

Method equals() is case sensitive, it will consider upper case and lower case characters different. The below code will give true when str1 is compared with str2. But at the same time it will return false when str1 is compared with str3.

```
String str1 = "hello";
String str2 = "hello";
String str3 = "Hello";

System.out.println("str1 equals str2 :" + str1.equals(str2));
System.out.println("str1 equals str3 :" + str1.equals(str3));
```

Output:
```
str1 equals str2 :true
str1 equals str3 :false
```

Another method equalsIgnoreCase() ignore case.

```
String str1 = "hello";
String str2 = "Hello";

System.out.println("str1 equals str2 :" + str1.equalsIgnoreCase(str2));
```

Output:
```
str1 equals str2 :true
```

String Matching

Every word processing program has a search function in which you can search all occurrences of any particular word in a long text file. For this, we need string-matching algorithms.

Brute Force Search

We have a pattern that we want to search in the text. The pattern is of length m and the text is of length n. Where m < n.

The brute force search algorithm will check the pattern at all possible value of "i" in the text where the value of "i" range from 0 to n-m. The pattern is compared with the text, character by character from left to right. When a mismatch is detected, then pattern is compared by shifting the compare window by one character.

Example 14.1:

```
int BruteForceSearch(String text, String pattern){
        return BruteForceSearch(text.toCharArray(), pattern.toCharArray());
}

int BruteForceSearch(char[] text, char[] pattern)
{
        int i=0, j=0;
        final int n = text.length;
        final int m = pattern.length;
        while (i<=n-m)
        {
                j=0;
                while (j<m && pattern [j]== text [i+j])
                        j++;
                if (j==m)
                        return(i);
                i++;
        }
        return -1;
}
```

Worst case *Time Complexity* of the algorithm is O(m*n), we got the pattern at the end of the text or we didn't get the pattern at all.
Best case *Time Complexity* of this algorithm is O(m)
The average *Time Complexity* of this algorithm is ***O(n)***

Robin-Karp algorithm

Robin-Karp algorithm is somewhat similar to the brute force algorithm. Because the pattern is compared to each textbox. Instead of pattern at each position a hash code is compared, only one comparison is performed. The hash code of the pattern is compared with the hash code of the text window. We try to keep the hash code as unique as possible.

The two features of good hash code are:
- The collision should be excluded as much as possible.
- The hash code of text must be calculated in constant time.

A collision occurs when hash code matches, but the pattern does not.
Calculation in constant time, one member leaves the window and a new number enters a window.

Multiplication by 2 is same as left shift operation. Multiplication by 2^{m-1} is same as left shift m-1 times. We want this multiple times so just store it in variable pow(m) = 2^{m-1}

We do not want to do big multiplication operations so modular operation with a prime number is used.

Example 14.2:

```java
int RobinKarp(String text, String pattern) {
    return RobinKarp(text.toCharArray(), pattern.toCharArray());
}
```

```java
int RobinKarp(char[] text, char[] pattern) {
    int n=text.length;
    int m=pattern.length;
    int i, j;
    int prime = 101;
    int powm=1;
    int TextHash=0, PatternHash=0;
    if(m==0 || m>n)
        return -1;

    for (i = 0; i < m-1; i++)
        powm = (powm << 1) % prime;

    for (i = 0; i < m; i++)
    {
        PatternHash = ((PatternHash << 1) + pattern[i]) % prime;
        TextHash = ((TextHash << 1) + text[i]) % prime;
    }

    for( i=0;i<=(n-m);i++)
    {
        if(TextHash == PatternHash)
        {
            for (j = 0; j < m; j++)
            {
                if (text[i+j] != pattern[j])
                    break;
            }
            if (j == m)
                return i;
        }
        TextHash = (((TextHash-text[i]*powm)<< 1)+text[i+m])%prime;
        if (TextHash < 0)
            TextHash = (TextHash + prime);
    }
    return -1;
}
```

Knuth-Morris-Pratt algorithm

After a shift of the pattern, the brute force algorithm forgotten all the information about the previous matched symbols. This is because of which its worst case *Time Complexity* is O(mn).

The Knuth-Morris-Pratt algorithm make use of this information that is computed in the previous comparison. It never re compares the whole text.

It uses preprocessing of the pattern. The preprocessing takes O(m) time and whole algorithm is **O(n)**

Preprocessing step: we try to find the border of the pattern at a different prefix of the pattern.

A **prefix** is a string that comes at the start of a string.
A **proper prefix** is a prefix that is not the complete string. Its length is less than the length of the string.
A **suffix** is a string that comes at the end of a string.
A **proper suffix** is a suffix that is not the complete string. Its length is less than the length of the string.

A **border** is a string that is both proper prefix and a proper suffix.

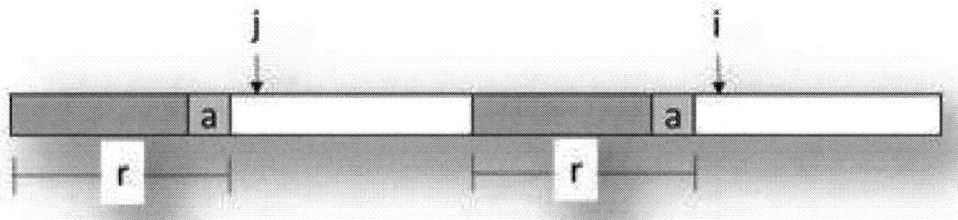

Example 14.3:

```
void KMPPreprocess(char[] pattern, int[] ShiftArr) {
    final int m = pattern.length;
    int i = 0, j = -1;
    ShiftArr[i] = -1;
    while(i < m) {
        while(j >= 0 && pattern[i] != pattern[j])
            j = ShiftArr[j];
        i++;
        j++;
        ShiftArr[i] = j;
    }
}
```

We have to loop outer loop for the text and inner loop for the pattern when we have matched the text and pattern mismatch, we shift the text such that the widest border is considered and then the rest of the pattern matching is resumed after this shift. If again a mismatch happens then the next mismatch is taken.

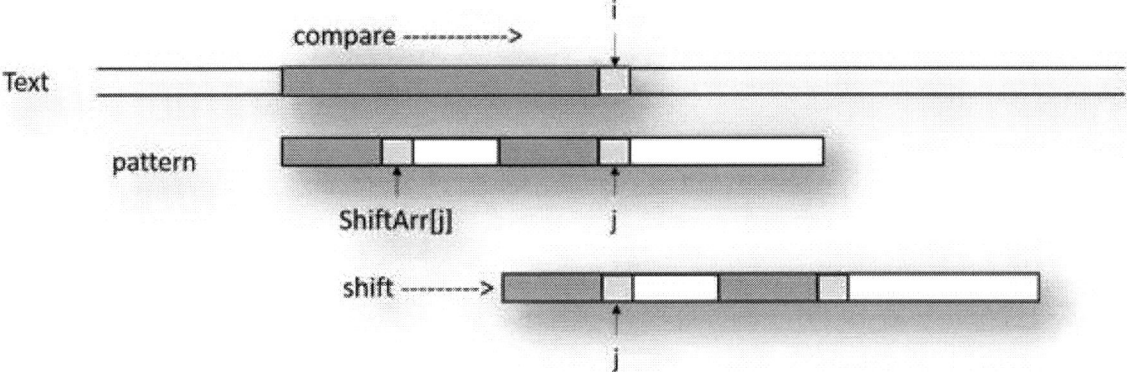

Example 14.4:

```
int KMP(String text, String pattern){
        return KMP(text.toCharArray(), pattern.toCharArray());
}
```

```
int KMP(char[] text, char[] pattern)
{
        int i=0, j=0;
        final int n = text.length;
        final int m = pattern.length;
        int[] ShiftArr = new int[m+1];
        KMPPreprocess(pattern,ShiftArr);
        while (i<n)
        {
                while (j>=0 && text[i]!=pattern[j])
                        j=ShiftArr[j];
                i++;
                j++;
                if (j==m)
                {
                        return (i - m);
                }
        }
        return -1;
}
```

Example 14.5: Use the same KMP algorithm to find the number of occurrences of the pattern in a text.

```
int KMPFindCount(char[] text, char[] pattern)
{
    int i=0, j=0, count = 0;
    final int n = text.length;
    final int m = pattern.length;
    int[] ShiftArr = new int[m+1];
    KMPPreprocess(pattern,ShiftArr);
    while (i<n)
    {
        while (j>=0 && text[i]!=pattern[j])
            j=ShiftArr[j];
        i++;
        j++;
        if (j==m)
        {
            count++;
            j=ShiftArr[j];
        }
    }
    return count;
}
```

Dictionary / Symbol Table

A symbol table is a mapping between a string (key) and a value that can be of any type. A value can be an integer such as occurrence count, dictionary meaning of a word and so on.

Binary Search Tree (BST) for Strings

Binary Search Tree (BST) is the simplest way to implement symbol table. Simple strcmp() function can be used to compare two strings. If all the keys are random, and the tree is balanced. Then on an average key lookup can be done in *O(logn)* time.

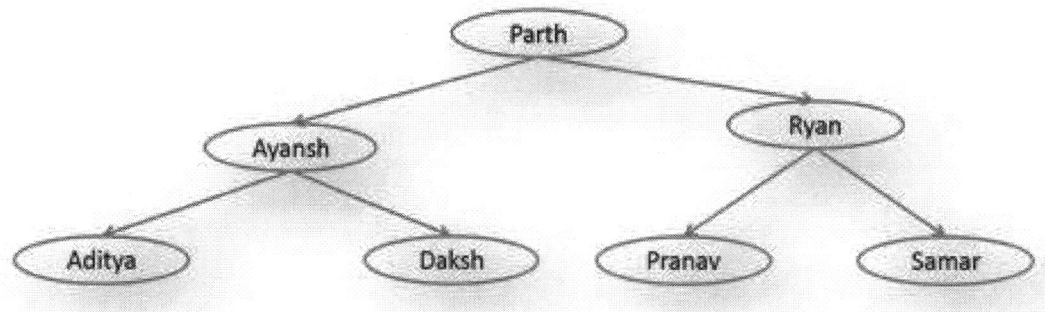

BINARY SEARCH TREE AS DICTIONARY

Below is an implementation of binary search tree to store string as key. This will keep track of the occurrence count of words in a text.

Example 14.6:

```java
public class StringTree {

    class Node{
        String value;
        int count;
        Node lChild;
        Node rChild;
    };
    Node root = null;

    //Other Methods.
}
```

```java
public void print(){
    print(root);
}

public void print(Node curr)/* pre order */
{
    if(curr != null)
    {
        System.out.print(" value is ::" + curr.value);
        System.out.println(" count is :: " + curr.count);
        print(curr.lChild);
        print(curr.rChild);
    }
}
```

```java
boolean find(String value) {
    boolean ret = find(root, value);
    System.out.println("Find " + value + " Return " + ret);
    return ret;
}
```

```java
boolean find(Node curr, String value)
{
    int compare;
    if(curr == null)
        return false;
    compare=curr.value.compareTo(value);
    if(compare==0)
        return true;
    else
    {
        if(compare==1)
            return find(curr.lChild,value);
        else
            return find(curr.rChild,value);
    }
}
```

```java
public void insert(String value) {
        root = insert(value, root);
}

Node insert(String value, Node curr) {
        int compare;
        if(curr==null)
        {
                curr=new Node();
                curr.value=value;
                curr.lChild=curr.rChild=null;
                curr.count=1;
        }
        else
        {
                compare = curr.value.compareTo(value);
                if(compare==0)
                        curr.count++;
                else if(compare==1)
                        curr.lChild=insert(value,curr.lChild);
                else
                        curr.rChild=insert(value,curr.rChild);
        }
        return curr;
}
```

```java
int frequency(String value)
{
        return frequency(root, value);

}
```

```java
int frequency(Node curr, String value)
{
        int compare;
        if(curr == null)
                return 0;

        compare=curr.value.compareTo(value);
        if(compare==0)
                return curr.count;
        else
        {
                if(compare>0)
                        return frequency(curr.lChild,value);
                else
                        return frequency(curr.rChild,value);
        }
}
```

```java
void freeTree() {
        root = null;
}
```

```java
public static void main(String[] args) {
    StringTree tt = new StringTree();
    tt.insert("banana");
    tt.insert("apple");
    tt.insert("mango");
    tt.insert("banana");
    tt.insert("apple");
    tt.insert("mango");
    System.out.println("Search results for apple, banana, grapes and mango :");
    tt.find("apple");
    tt.find("banana");
    tt.find("banan");
    tt.find("grapes");
    tt.find("mango");

    tt.print();
    System.out.println("frequency returned :: " + tt.frequency("apple"));
    System.out.println("frequency returned :: " + tt.frequency("banana"));
    System.out.println("frequency returned :: " + tt.frequency("mango"));
    System.out.println("frequency returned :: " + tt.frequency("hemant"));
}
```

Hash-Table

The Hash-Table is another data structure that can be used for symbol table implementation. Below Hash-Table diagram, we can see the name of that person is taken as the key, and their meaning is the value of the search. The first key is converted into a hash code by passing it to appropriate hash function. Inside hash function the size of Hash-Table is also passed, which is used to find the actual index where values will be stored. Finally, the value which is meaning of name is stored in the Hash-Table, or you can store a reference to the string which store meaning can be stored into the Hash-Table.

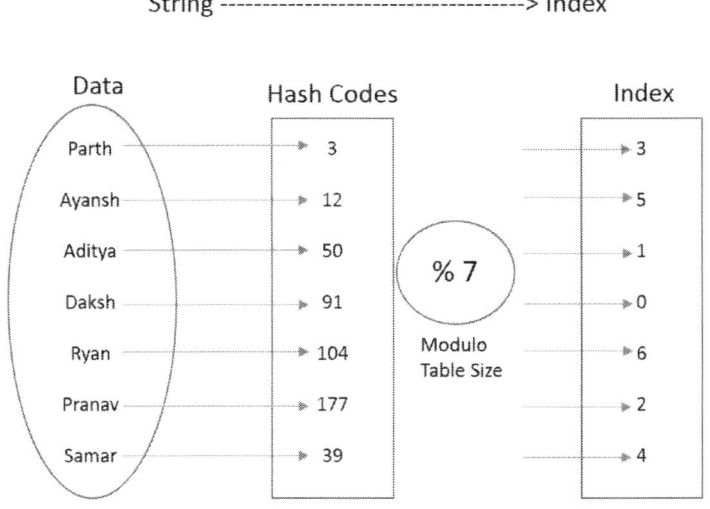
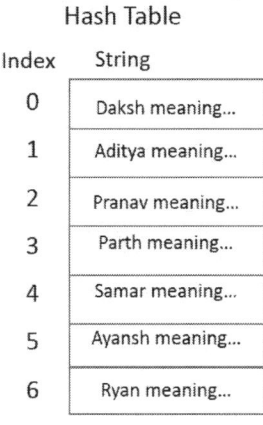

Hash-Table has an excellent lookup of *O(1)*.

Let us suppose we want to implement autocomplete the box feature of Google search. When you type some string to search in google search, it propose some complete string even before you have done typing. BST cannot solve this problem as related strings can be in both right and left subtree.

The Hash-Table is also not suited for this job. One cannot perform a partial match or range query on a Hash-Table. Hash function transforms string to a number. And a good hash function will give a fairly distributed hash bode even for partial string and there is no way to relate two strings in a Hash-Table.

Trie and Ternary Search tree are a special kind of tree that solves partial match and range query problem well.

Trie

Trie is a tree, in which we store only one character at each node. This final key value pair is stored in the leaves. Each node has R children, one for each possible character. For simplicity purpose, let's consider that the character set is 26, corresponds to different characters of English alphabets.

Trie is an efficient data structure. Using Trie we can search the key in O(M) time. Where M is the maximum string length. Trie is also suitable for solving partial match and range query problems.

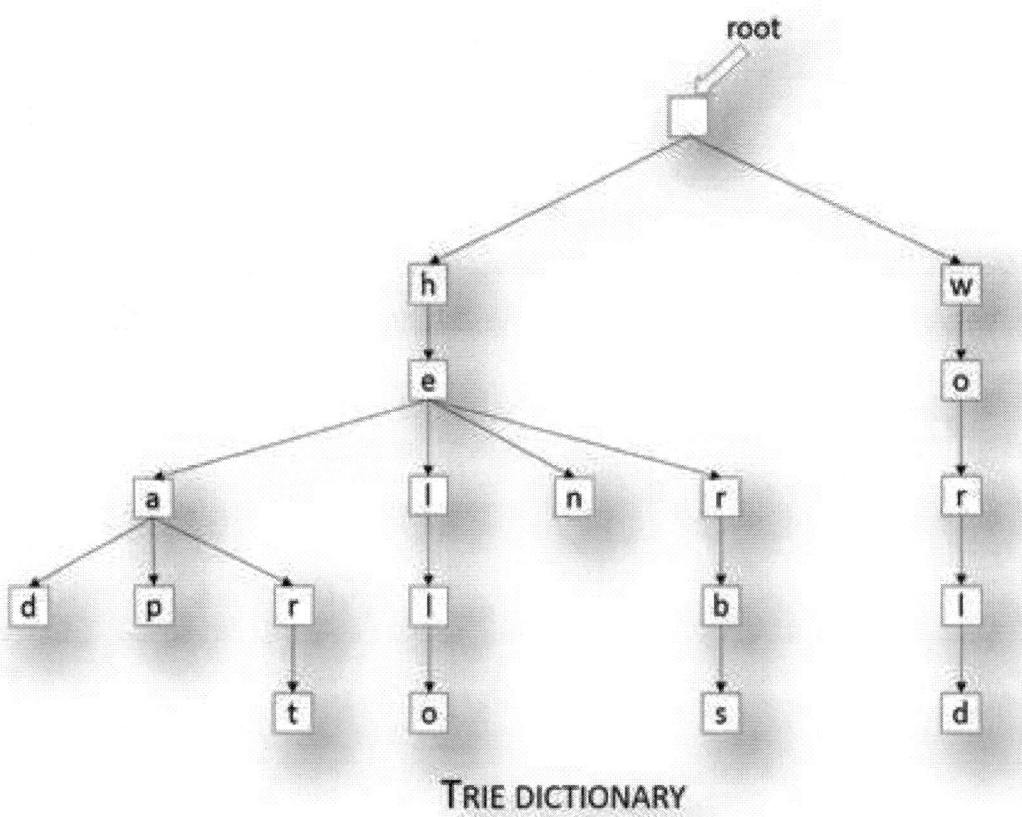

TRIE DICTIONARY

Example 14.7:

```java
public class Trie {
    private static final int CharCount = 26;
    Node root=null;

    private class Node{
        boolean isLastChar;
        char ch;
        Node[] child;
        public Node(char c){
            child = new Node[CharCount];
            for(int i=0;i<CharCount;i++)
                child[i]=null;
            isLastChar = false;
            ch = c;
        }
    };

    public Trie() {
        root = new Node(' ');        //first node with dummy value.
    }
}
```

```java
Node Insert(String str){
    if(str == null)
        return root;
    return Insert(root, str.toLowerCase(), 0);
}
```

```java
Node Insert(Node curr, String str, int index)
{
    if(curr==null)
        curr = new Node(str.charAt(index-1));

    if(str.length() == index)
        curr.isLastChar = true;
    else
        curr.child[str.charAt(index) - 'a'] =
            Insert(curr.child[str.charAt(index) - 'a'] , str, index+1);

    return curr;
}
```

```java
boolean Find(String str)
{
    if(str==null)
    {
        return false;
    }

    str = str.toLowerCase();
    return Find(root, str, 0);
}
```

```java
boolean Find(Node curr, String str, int index)
{
    if(curr == null)
        return false;

    if(str.length() == index)
        return curr.isLastChar;

    return Find(curr.child[str.charAt(index)- 'a'], str, index+1);
}
```

```java
void Remove(String str)
{
    if(str==null)
    {
        return;
    }

    str = str.toLowerCase();
    Remove(root, str, 0);
}
```

```java
void Remove(Node curr, String str, int index)
{
    if(curr == null)
        return;

    if(str.length() == index )
    {
        if(curr.isLastChar)
        {
            curr.isLastChar = false;
        }
        return;
    }

    Remove(curr.child[str.charAt(index)- 'a'], str, index+1);
}
```

```java
public static void main(String[] args) {
    Trie t = new Trie();
    String a="hemant";
    String b="heman";
    String c="hemantjain";
    t.Insert(a);
    t.Insert(c);
    System.out.println(t.Find(a));
    System.out.println(t.Find(b));
    System.out.println(t.Find(c));
}
```

Ternary Search Trie/ Ternary Search Tree

Tries have a very good search performance of O(M) where M is the maximum size of the search string. But tries have a very high space requirement. Every node Trie contains references to multiple nodes, each reference corresponds to possible characters of the key. To avoid this high space requirement Ternary Search Trie (TST) is used.

A TST avoid the heavy space requirement of the traditional Trie while still keeping many of its advantages. In a TST each node contains a character, an end of key indicator, and three references. The three references are corresponding to current char hold by the node (equal), characters less than and character greater than.

The *Time Complexity* of ternary search tree operation is proportional to the height of the ternary search tree. In the worst case, we need to traverse up to 3 times that many links. However, this case is rare.

Therefore, TST is a very good solution for implementing Symbol Table, Partial match and range query.

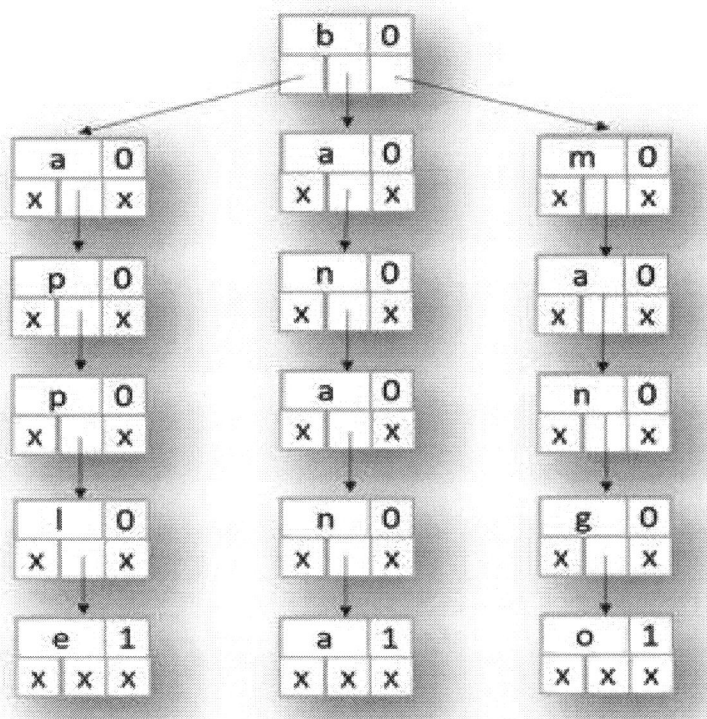

TERNARY Search Tree

Example 14.8:

```java
public class TST {
    Node root;

    private class Node
    {
        char data;
        boolean isLastChar;
        Node left, equal, right;

        private Node(char d)
        {
            data = d;
            isLastChar = false;
            left = equal = right = null;
        }
    };
}
```

```java
public void insert(String word)
{
    root = insert(root, word, 0);
}
```

```java
private Node insert(Node curr, String word, int wordIndex)
{
    if (curr == null)
        curr = new Node(word.charAt(wordIndex));
    if (word.charAt(wordIndex) < curr.data)
        curr.left = insert(curr.left, word, wordIndex);
    else if (word.charAt(wordIndex) > curr.data)
        curr.right = insert(curr.right, word, wordIndex);
    else
    {
        if (wordIndex < word.length() -1)
            curr.equal = insert(curr.equal, word, wordIndex + 1);
        else
            curr.isLastChar = true;
    }
    return curr;
}
```

```java
private boolean find(Node curr, String word, int wordIndex)
{
    if (curr==null)
        return false;
    if (word.charAt(wordIndex) < curr.data)
        return find(curr.left, word, wordIndex);
    else if (word.charAt(wordIndex) > curr.data)
        return find(curr.right, word, wordIndex);
    else
    {
        if ( wordIndex == word.length()-1)
            return curr.isLastChar;
        return find(curr.equal, word, wordIndex + 1);
    }
}
```

```java
public boolean find(String word)
{
    boolean ret = find(root,word,0);
    System.out.print(word + " :: ");
    if(ret)
        System.out.println(" Found ");
    else
        System.out.println("Not Found ");
    return ret;
}
```

```java
public static void main(String[] args) {
    TST tt = new TST();
    tt.insert("banana");
    tt.insert("apple");
    tt.insert("mango");
    tt.find("apple");
    tt.find("banana");
    tt.find("mango");
    tt.find("grapes");
}
```

Problems in String

Regular Expression Matching

Implement regular expression matching with the support of '?' and '*' special character.
'?' Matches any single character.
'*' Matches zero or more of the preceding element.

Example 14.9:

```
boolean matchExp(char[] exp, char[] str)
{
    return matchExpUtil(exp, str, 0, 0);
}
```

```
boolean matchExpUtil(char[] exp, char[] str, int i, int j)
{
    if (i == exp.length && j == str.length)
        return true;
    if ((i == exp.length && j != str.length)
            || (i != exp.length && j == str.length))
        return false;
    if (exp[i] == '?' || exp[i] == str[j])
        return matchExpUtil(exp, str, i + 1, j + 1);
    if (exp[i] == '*')
        return matchExpUtil(exp, str, i + 1, j) ||
                matchExpUtil(exp, str, i, j + 1)||
                matchExpUtil(exp, str, i + 1, j + 1);
    return false;
}
```

Order Matching

Given a long text string and a pattern string. Find if the characters of pattern string are in the same order in text string. Eg. Text String: ABCDEFGHIJKLMNOPQRSTUVWXYZ
Pattern string: JOST

Example 14.10:

```
boolean match(char[] source, char[] pattern) {
    int iSource = 0;
    int iPattern = 0;
    int sourceLen = source.length;
    int patternLen = pattern.length;
    for (iSource = 0; iSource < sourceLen; iSource++) {
        if (source[iSource] == pattern[iPattern]) {
            iPattern++;
        }
        if (iPattern == patternLen) {
            return true;
        }
    }
    return false;
}
```

ASCII to Integer Conversion

Write a function that take integer as a char array and convert it into an int.
Example 14.11:

```
int myAtoi(String str)
{
    int value=0;

    int size= str.length();
    for (int i = 0; i < size; i++)
    {
        char ch = str.charAt(i);
        value=(value<<3)+(value<<1)+(ch - '0');
    }
    return value;
}
```

To Upper Case

Write a function that will convert all lower case letters in a string to upper case.
Example 14.12: ToUpper

```
char ToUpper(char s)
{
    if(s >= 97 && s<=(97+25))
        s=(char) (s-32);
    return s;
}
```

To Lower Case

Write a function that will convert upper case letter in a string to lower case
Example 14.13: ToLower

```
char ToLower(char s)
{
    if(s>=65 && s<=(65+25))
        s=(char) (s+32);
    return s;
}
```

Unique Characters

Write a function that will take a string as input and return 1 if it contain all unique characters else return 0.
Example 14.14:

```
boolean isUniqueChar(String str)
{
    int[] bitarr = new int[26];
    for (int i = 0; i < 26; i++)
    {
        bitarr[i]=0;
    }
```

```java
        int size= str.length();
        for (int i = 0; i < size; i++)
        {
                char c = str.charAt(i);
                if ('A' <= c && 'Z' >= c)
                {
                        c = (char) (c - 'A');
                }
                else if ('a' <= c && 'z' >= c)
                {
                        c = (char) (c - 'a');
                }
                else
                {
                        System.out.println("Unknown Char!\n");
                        return false;
                }
                if (bitarr[c] != 0)
                {
                        System.out.println("Duplicate detected!\n");
                        return false;
                }
        }
        System.out.println("No duplicate detected!\n");
        return true;
}
```

Permutation Check

Write a function to check if two strings are permutation of each other.
Example 14.15:

```java
boolean isPermutation(String s1, String s2)
{
        int[] count = new int[256];
        int length = s1.length();
        if (s2.length() != length)
        {
                System.out.println("is permutation return false\n");
                return false;
        }
        for (int i = 0; i < 256; i++)
        {
                count[i] = 0;
        }
        for (int i = 0; i < length; i++)
        {
                char ch = s1.charAt(i);
                count[ch]++;
                ch = s2.charAt(i);
                count[ch]--;
        }
```

```java
        for (int i = 0; i < length; i++)
        {
                if (count[i] != 0)
                {
                        System.out.println("is permutation return false\n");
                        return false;
                }
        }
        System.out.println("is permutation return true\n");
        return true;
}
```

Palindrome Check

Given a string as an array of characters find if the string is a palindrome or not?

Example 14.16:

```java
boolean isPalindrome(String str) {
        int i=0,j=str.length()-1;

        while(i<j && str.charAt(i) == str.charAt(j)) {
                i++;
                j--;
        }

        if(i<j) {
                System.out.println("String is not a Palindrome");
                return false;
        }
        else {
                System.out.println("String is a Palindrome");
                return true;
        }
}
```

Time Complexity is **O(n)** and Space Complexity is **O(1)**

Reverse Case function

Write a function that will convert Lower case letter in a string to upper case and upper case letter to lower case.

Example 14.17:

```java
char LowerUpper(char s) {
        if(s>=97 && s<=(97+25))
                s=(char)(s-32);
        else if(s>=65 && s<=(65+25))
                s=(char)(s+32);
        return s;
}
```

Power function

Write a function which will calculate x^n, Taking x and n as argument.
Example 14.18: Power function

```
int pow(int x, int n)
{
        int value;
        if(n==0)
                return(1);
        else if(n%2==0)
        {
                value=pow(x,n/2);
                return(value*value);
        }
        else
        {
                value=pow(x,n/2);
                return(x*value*value);
        }
}
```

String Compare function

Write a function strcmp() to compare two strings. The function return values should be:
The return value is 0 indicates that both first and second strings are equal.
The return value is negative indicates the first string is less than the second string.
The return value is positive indicates that the first string is greater than the second string.
Example 14.19:

```
int myStrcmp(String a, String b)
{
        int index =0;
        int len1 = a.length();
        int len2 = b.length();
        int minlen = len1;
        if(len1 > len2)
                minlen=len2;

        while(index < minlen && a.charAt(index) == b.charAt(index))
        {
                index++;
        }

        if (index == len1 && index == len2)
                return 0;
        else if(len1 == index)
                return -1;
        else if(len2 == index)
                return 1;
        else
                return a.charAt(index) - b.charAt(index);
}
```

String duplicate function

Write a function that will return a reference to a new string which is a duplicate of the input string. Memory for the new string is obtained with malloc and freed with free.

Example 14.20:

```
char[] myStrdup(char[] src)
{
        int index = 0;
        char[] dst = new char[src.length];
        for(char ch : src){
                dst[index] = ch;
        }
        return dst;
}
```

Reverse String

Example 14.21: Reverse all the characters of a string.

```
void reverseString(char[] a)
{
        int lower=0;
        int upper= a.length -1;
        char tempChar;
        while(lower<upper)
        {
                tempChar=a[lower];
                a[lower]=a[upper];
                a[upper]=tempChar;
                lower++;
                upper--;
        }
}
```

```
void reverseString(char[] a, int lower, int upper)
{
        char tempChar;
        while(lower<upper)
        {
                tempChar=a[lower];
                a[lower]=a[upper];
                a[upper]=tempChar;
                lower++;
                upper--;
        }
}
```

Reverse Words

Example 14.22: Reverse order of words in a string sentence.

```
void reverseWords(char[] a)
{
    int length= a.length;
    int lower,upper=-1;
    lower=0;
    for(int i=0;i<=length;i++)
    {
        if(a[i]==' '||a[i]=='\0')
        {
            reverseString(a,lower,upper);
            lower=i+1;
            upper=i;
        }
        else
        {
            upper++;
        }
    }
    reverseString(a,0,length-1); //-1 because we do not want to reverse '\0'
}
```

Print Anagram

Example 14.23: Given a string as character array, print all the anagram of the string.

```
void printAnagram (char[] a)
{
    int n = a.length;
    printAnagram (a,n,n);
}
```

```
void printAnagram (char[] a,int max, int n)
{
    if(max==1)
        System.out.println(a.toString());
    for(int i=-1;i<max-1;i++)
    {
        if(i!=-1)
            a[i]^=a[max-1]^=a[i]^=a[max-1];
        printAnagram(a,max-1,n);
        if(i!=-1)
            a[i]^=a[max-1]^=a[i]^=a[max-1];
    }
}
```

Shuffle String

Example 14.24: Write a program to convert array ABCDE12345 to A1B2C3D4E5

```
void shuffle(char[] ar,int n)
{
        int count=0;
        int k=1;
        char temp='\0';
        for(int i=1;i<n;i=i+2)
        {
                temp=ar[i];
                k=i;
                do{
                        k=(2*k)%(2*n-1);
                        temp^=ar[k]^=temp^=ar[k];
                        count++;
                }while(i!=k);
                if(count == (2*n-2))
                {
                        break;
                }
        }
}
```

Binary Addition

Example 14.25: Given two binary string, find the sum of these two binary strings.

```
char[] addBinary(char[] first, char[] second)
{
        int size1 = first.length;
        int size2 = second.length;
        int totalIndex;
        char[] total;
        if(size1>size2)
        {
                total = new char[size1+2];
                totalIndex = size1;
        }
        else
        {
                total = new char[size2+2];
                totalIndex = size2;
        }
        total[totalIndex + 1] = '\0';
        int carry =0 ;
        size1--;
        size2--;
        while(size1 >=0 || size2 >=0)
        {
                int firstValue = (size1 < 0) ? 0 : first[size1]-'0';
                int secondValue = (size2 < 0)? 0: second[size2]-'0';
```

```
            int sum = firstValue + secondValue + carry;
            carry = sum >> 1;
            sum = sum & 1;
            total[totalIndex] = (sum==0) ? '0' : '1';
            totalIndex--;
            size1--;
            size2--;
        }
        total[totalIndex] = (carry==0) ? '0' : '1';
        return total;
}
```

Exercise

1. Given a string, find the longest substring without reputed characters.

2. The function memset() copies ch into the first 'n' characters of the string

3. Serialize a collection of string into a single string and de serializes the string into that collection of strings.

4. Write a smart input function, which take 20 characters as input from the user. Without cutting some word.
   User input: "Harry Potter must not go"
   First 20 chars: "Harry Potter must no"
   Smart input: "Harry Potter must"

5. Write a code that returns if a string is palindrome and it should return true for below inputs too.
   Stella won no wallets.
   No, it is open on one position.
   Rise to vote, Sir.
   Won't lovers revolt now?

6. Write an ASCII to *integer* function which ignore the non-integral character and give the *integer*. For example, if the input is "12AS5" it should return 125.

7. Write code that would parse a Bash brace expansion.
   Example: the expression "(a, b, c) d, e" and would output all the possible strings: ad, bd, cd, e

8. Given a string write a function to return the length of the longest substring with only unique characters

9. Replace all occurrences of "a" with "the"

10. Replace all occurrences of %20 with ' '.

E.g. Input: www.Hello%20World.com
Output: www.Hello World. com

11. Write an expansion function that will take an input string like "1..5,8,11..14,18,20,26..30" and will print "1,2,3,4,5,8,11,12,13,14,18,20,26,27,28,29,30"

12. Suppose you have a string like "Thisisasentence". Write a function that would separate these words. And will print whole sentence with spaces.

13. Given three string str1, str2 and str3. Write a complement function to find the smallest sub-sequence in str1 which contains all the characters in str2 and but not those in str3.

14. Given two strings A and B, find whether any anagram of string A is a sub string of string B. For eg: If A = xyz and B = afdgzyxksldfm then the program should return true.

15. Given a string, find whether it contains any permutation of another string. For example, given "abcdefgh" and "ba", the function should return true, because "abcdefgh" has substring "ab", which is a permutation of the given string "ba".

16. Give an algorithm which removes the occurrence of "a" by "bc" from a string? The algorithm must be in-place.

17. Given a string "1010101010" in base2 convert it into string with base4. Do not use an extra space.

18. In Binary Search tree to store strings, delete() function is not implemented implement the same.

19. If you implement delete() function, then you need to make changes in find() function. Do the needful.

Chapter 15: Algorithm Design Techniques

Introduction

In real life when we are asked to do some work, we try to correlate it with our experience and then try to solve it. Similarly, when we get a new problem to solve. We first try to find the similarity of the current problem with some problems for which we already know the solution. Then solve the current problem and get our desired result.

This method provides following benefits:
1) It provides a template for solving a wide range of problems.
2) It provides us the idea of the suitable data structure for the problem.
3) It helps us in analyzing, space and *Time Complexity* of algorithms.

In the previous chapters, we have used various algorithms to solve different kind of problems. In this chapter, we will read about various techniques of solving algorithmic problems.

Various Algorithm design techniques are:
1) Brute Force
2) Greedy Algorithms
3) Divide-and-Conquer, Decrease-and-Conquer
4) Dynamic Programming
5) Reduction / Transform-and-Conquer
6) Backtracking and Branch-and-Bound

Brute Force Algorithm

Brute Force is a straightforward approach of solving a problem based on the problem statement. It is one of the easiest approaches to solve a particular problem. It is useful for solving small size dataset problem.

Some examples of brute force algorithms are:
- Bubble-Sort
- Selection-Sort
- Sequential search in an array
- Computing pow(a, n) by multiplying a, n times.
- Convex hull problem
- String matching
- Exhaustive search: Traveling salesman, Knapsack, and Assignment problems

Greedy Algorithm

In greedy algorithm, solution is constructed through a sequence of steps. At each step, choice is made which is locally optimal. Greedy algorithms are generally used to solve optimization problems. We always take the next data to be processed depending upon the dataset which we have already processed and then choose the next optimum data to be processed. Greedy algorithms does not always give optimum solution.

Some examples of brute force algorithms are:
- Minimal spanning tree: Prim's algorithm, Kruskal's algorithm
- Dijkstra's algorithm for single-source shortest path problem
- Greedy algorithm for the Knapsack problem
- The coin exchange problem
- Huffman trees for optimal encoding

Divide-and-Conquer, Decrease-and-Conquer

Divide-and-Conquer algorithms involve basic three steps, first split the problem into several smaller sub-problems, second solve each sub problem and then finally combine the sub problems results so as to produce the final result.

In divide-and-conquer the size of the problem is reduced by a factor (half, one-third, etc.), While in decrease-and-conquer the size of the problem is reduced by a constant.

Examples of divide-and-conquer algorithms:
- Merge-Sort algorithm (using recursion)
- Quicksort algorithm (using recursion)
- Computing the length of the longest path in a binary tree (using recursion)
- Computing Fibonacci numbers (using recursion)
- Quick-hull

Examples of decrease-and-conquer algorithms:
- Computing pow(a, n) by calculating pow(a, n/2) using recursion.
- Binary search in a sorted array (using recursion)
- Searching in BST
- Insertion-Sort
- Graph traversal algorithms (DFS and BFS)
- Topological sort
- Warshall's algorithm (using recursion)
- Permutations (Minimal change approach, Johnson-Trotter algorithm)
- Computing a median, Topological sorting, Fake-coin problem (Ternary search)

Consider the problem of exponentiation Compute x^n

Brute Force:	n-1 multiplications
Divide and conquer:	T(n) = 2*T(n/2) + 1 = n-1
Decrease by one:	T(n) = T(n-1) + 1 = n-1
Decrease by constant factor:	T(n) = T(n/a) + a-1 = (a-1) n = n when a = 2

Dynamic Programming

While solving problems using Divide-and-Conquer method, there may be a case when recursively sub-problems can result in the same computation being performed multiple times. This problem arises when there are identical sub-problems arise repeatedly in a recursion.

Dynamic programming is used to avoid the requirement of repeated calculation of same sub-problem. In this method, we usually store the result of sub - problems in a table and refer that table to find if we have already calculated the solution of sub - problems before calculating it again.

Dynamic programming is a bottom up technique in which the smaller sub-problems are solved first and the result of these are sued to find the solution of the larger sub-problems.

Examples:
- Fibonacci numbers computed by iteration.
- Warshall's algorithm for transitive closure implemented by iterations
- Floyd's algorithms for all-pairs shortest paths

```
int fibonacci(int n)
{
        if (n <= 1)
                return n;
        return fibonacci(n - 1) + fibonacci(n - 2);
}
```

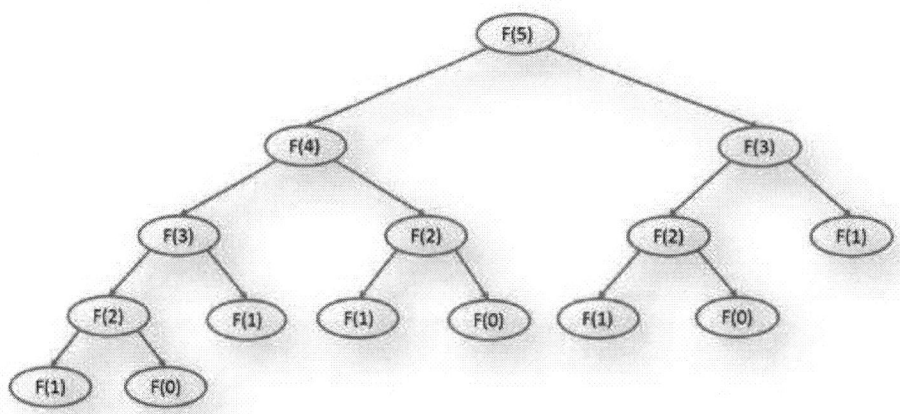

Using divide and conquer the same sub problem is solved again and again, which reduce the performance of the algorithm. This algorithm has an exponential *Time Complexity*. And linear *Space Complexity*.

```
int fibo(int n)
{
        int first = 0, second = 1;
        int temp, i;

        if (n == 0)
                return first;
        else if (n == 1)
                return second;

        for (i = 2; i <= n; i++)
        {
                temp = first + second;
                first = second;
                second = temp;
        }
        return temp;
}
```

Using this algorithm, we will get Fibonacci in linear *Time Complexity* and constant *Space Complexity*.

Reduction / Transform-and-Conquer

These methods works as two-stage procedure. First, the problem is transformed into a known problem for which we know optimal solution. In the second stage, the problem is solved.

The most common types of transformation are sort of an array.

For example: Given an array of numbers finds the two closest number.

The brute force solution for this problem will take distance between each element in the array and will try to keep the minimum distance pair; this approach will have a *Time Complexity* of $O(n^2)$

Transform and conquer solution, will be first sort the array in *O(nlogn)* time and then find the closest number by scanning the array in another *O(n)*. Which will give the total *Time Complexity* of *O(nlogn)*.

Examples:
- Gaussian elimination
- Heaps and Heapsort

Backtracking

In real life, let us suppose someone gave you a lock with a number (three digit lock, number range from 1 to 9). And you don't have the exact password key for the lock. You need to test each and every combination till you got the right one. Obviously, you need to test starting from something like "111", then "112" and so on. And you will get your key before you reach "999". So what you are doing is backtracking.

Suppose the lock produce some sound "click" correct digit is selected for any level. If we can listen to this sound such intelligence/ heuristics will help you to reach your goal much faster. These functions are called Pruning function or bounding functions.

Backtracking is a method by which solution is found by exhaustively searching through large but finite number of states, with some pruning or bounding function which will narrow down our search. For all the problems like (NP hard problems) for which there does not exist any other method we use backtracking.

Backtracking problems have the following components:
1. Initial state
2. Target / Goal state
3. Intermediate states
4. Path from the initial state to the target / goal state
5. Operators to get from one state to another
6. Pruning function (optional)

The solving process of backtracking algorithm starts with the construction of state's tree, whose nodes represents the states. The root node is the initial state and one or more leaf node will be our target state. Each edge of the tree represents some operation. The solution is obtained by searching the tree until a Target state is found.

Backtracking uses depth-first search:
1) Store the initial state in a stack
2) While the stack is not empty, repeat:
3) Read a node from the stack.
4) While there are available operators, do:
    a. Apply an operator to generate a child
    b. If the child is a goal state – return solution
    c. If it is a new state, and pruning function does not discard it push the child into the stack.

There are three monks and three demons at one side of a river. We want to move all of them to the other side using a small boat. The boat can carry only two persons at a time. Given if on any shore the number of demons will be more than monks then they will eat the monks. How can we move all of these people to the other side of the river safely?

Same as the above problem there is a farmer who has a goat, a cabbage and a wolf. If the farmer leaves, goat with cabbage, goat will eat the cabbage. If the farmer leaves wolf alone with goat, wolf will kill the goat. How can the farmer move all his belongings to the other side of the river?

You are given two jugs, a 4-gallon one and a 3-gallon one. There are no measuring markers on jugs. There is a tap that can be used to fill the jugs with water. How can you get 2 gallons of water in the 4-gallon jug?

Branch-and-bound

Branch and bound method is used when we can evaluate cost of visiting each node by a utility functions. At each step we choose the node with lowest cost to proceed further. Branch-and bound algorithms are implemented using a priority queue. In branch and bound we traverse the nodes in breadth-first manner.

A* Algorithm

A* is sort of an elaboration on branch-and-bound. In branch-and-bound, at each iteration we expand the shortest path that we have found so far. In A*, instead of just picking the path with the shortest length so far, we pick the path with the shortest estimated total length from start to goal, where the total length is estimated as length traversed so far plus a heuristic estimate of the remaining distance from the goal.

Branch-and-bound will always find an optimal solution which is shortest path. A* will always find an optimal solution if the heuristic is correct. Choosing a good heuristic is the the most important part of A* algorithm.

Conclusion

Usually a given problem can be solved using a number of methods, however it is not wise to settle for the first method that comes to our mind. Some methods result in a much more efficient solutions than others.

For example, the Fibonacci numbers calculated recursively (decrease-and-conquer approach), and computed by iterations (dynamic programming). In the first case the complexity is **$O(2^n)$,** and in the other case the complexity is **$O(n)$**.

Another example, consider sorting based on the Insertion-Sort and basic bubble sort. For almost sorted files Insertion-Sort will give almost linear complexity, while bubble sort sorting algorithms have quadratic complexity.

So the most important question is, how to choose the best method?
First, you should understand the problem statement.
Second by knowing various problems and there solutions.

CHAPTER 16: BRUTE FORCE ALGORITHM

Introduction

Brute Force is a straightforward approach of solving a problem based on the problem statement. It is one of the easiest approaches to solve a particular problem. It is useful for solving small size dataset problem.

Many times, there are other algorithm techniques that can be used to get a better solution of the same problem.

Some examples of brute force algorithms are:
- Bubble-Sort
- Selection-Sort
- Sequential search in an array
- Computing pow (a, n) by multiplying a, n times.
- Convex hull problem
- String matching
- Exhaustive search
- Traveling salesman
- Knapsack
- Assignment problems

Problems in Brute Force Algorithm

Bubble-Sort

In Bubble-Sort, adjacent elements of the list are compared and are exchanged if they are out of order.

```
// Sorts a given array by Bubble Sort
// Input: An array A of orderable elements
// Output: Array A[0..n - 1] sorted in ascending order

Algorithm BubbleSort(A[0..n - 1])
sorted = false
while !sorted do
        sorted = true
        for j = 0 to n - 2 do
                if A[j] > A[j + 1] then
                        swap A[j] and A[j + 1]
                        sorted = false
```

The *Time Complexity* of the algorithm is $\theta(n^2)$

Selection-Sort

The entire given list of N elements is traversed to find its smallest element and exchange it with the first element. Then, the list is traversed again to find the second element and exchanged it with the second element. After N-1 passes, the list will be fully sorted.

```
//Sorts a given array by selection sort
//Input: An array A[0..n-1] of orderable elements
//Output: Array A[0..n-1] sorted in ascending order

Algorithm SelectionSort (A[0..n-1])
for i = 0 to n - 2 do
        min = i
        for j = i + 1 to n - 1 do
                if A[j] < A[min]
                        min = j
        swap A[i] and A[min]
```

The *Time Complexity* of the algorithm is $\theta(n^2)$

Sequential Search

The algorithm compares consecutive elements of a given list with a given search keyword until either a match is found or the list is exhausted.

```
Algorithm SequentialSearch (A[0..n], K)
i = 0
While A [i] ≠ K do
        i = i + 1
        if i < n
                return i
        else
                return -1
```

Worst case *Time Complexity* is Θ (n).

Computing pow (a, n)

Computing a^n (a > 0, and n is a nonnegative *integer*) based on the definition of exponentiation. N-1 multiplications are required in brute force method.

```
// Input: A real number a and an integer n = 0
// Output: a power n

Algorithm Power(a, n)
result = 1
for i = 1 to n do
        result = result * a
return result
```

The algorithm requires Θ (n)

String matching

A brute force string matching algorithm takes two inputs, first text consists of n characters and a pattern consist of m character (m<=n). The algorithm starts by comparing the pattern with the beginning of the text. Each character of the patters is compared to the corresponding character of the text. Comparison starts from left to right until all the characters are matched or a mismatch is found. The same process is repeated until a match is found. Each time the comparison starts one position to the right.

```
//Input: An array T[0..n - 1] of n characters representing a text
// an array P[0..m - 1] of m characters representing a pattern
//Output: The position of the first character in the text that starts the first
// matching substring if the search is successful and -1 otherwise.

Algorithm BruteForceStringMatch (T[0..n - 1], P[0..m - 1])
for i = 0 to n - m do
        j = 0
        while j < m and P[j] = T[i + j] do
                j = j + 1
        if j = m then
                return i
return -1
```

In the worst case, the algorithm is O(mn).

Closest-Pair Brute-Force Algorithm

The closest-pair problem is to find the two closest points in a set of n points in a 2-dimensional space. A brute force implementation of this problem computes the distance between each pair of distinct points and find the smallest distance pair.

```
// Finds two closest points by brute force
// Input: A list P of n >= 2 points
// Output: The closest pair
Algorithm BruteForceClosestPair(P)
dmin = infinite
for i = 1 to n - 1 do
        for j = i + 1 to n do
                d = (xi − xj )² + (yi − yj )²
                if d < dmin then
                        dmin = d
                        imin = i
                        jmin = j
return imin, jmin
```

In the *Time Complexity* of the algorithm is $\theta(n^2)$

Convex-Hull Problem

Convex-hull of a set of points is the smallest convex polygon containing all the points. All the points of the set will lie on the convex hull or inside the convex hull. Illustrate the rubber-band interpretation of the convex hull. The convex-hull of a set of points is a subset of points in the given sets.

How to find this subset?
Answer: The rest of the points of the set are all on one side.

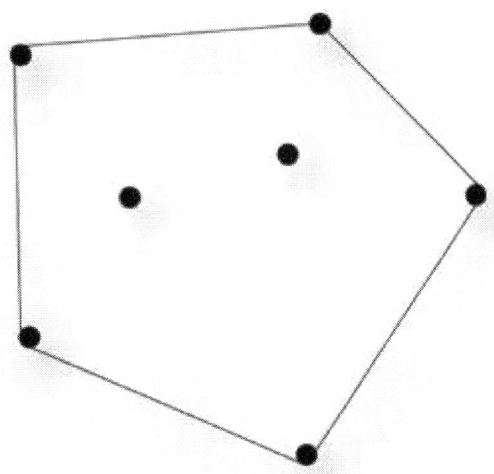

Two points (x_1, y_1), (x_2, y_2) make the line $ax + by = c$
Where $a = y_2-y_1$, $b = x_1-x_2$, and $c = x_1y_2 - y_1x_2$

And divides the plane by $ax + by - c < 0$ and $ax + by - c > 0$
So we need to only check $ax + by - c$ for the rest of the points

If we find all the points in the set lies one side of the line with either all have $ax + by - c < 0$ or all the points have $ax + by - c > 0$ then we will add these points to the desired convex hull point set.

For each of $n(n-1)/2$ pairs of distinct points, one needs to find the sign of $ax + by - c$ in each of the other $n - 2$ points.
What is the worst case cost of the algorithm? $O(n^3)$

```
Algorithm ConvexHull
for i=0 to n-1
   for j=0 to n-1
     if (xi,yi) !=(xj,yj)
         draw a line from (xi,yi) to (xj,yj)
         for k=0 to n-1
           if(i!=k and j!=k)
             if ( all other points lie on the same side of the line (xi,yi) and (xj,yj))
                add  (xi,yi) to (xj,yj) to the convex hull set
```

Exhaustive Search

Exhaustive search is a brute force approach applies to combinatorial problems.
In exhaustive search we generate all the possible combinations. See if the combinations satisfy the problem constraints and then finding the desired solution.
Examples of exhaustive search are:
- Traveling salesman problem
- Knapsack problem
- Assignment problem

Traveling Salesman Problem (TSP)

In the traveling salesman problem we need to find the shortest tour through a given set of N cities that salesman visits each city exactly once before returning to the city where he started.

Alternatively: Finding the shortest Hamiltonian circuit in a weighted connected graph. A cycle that passes through all the vertices of the graph exactly once.

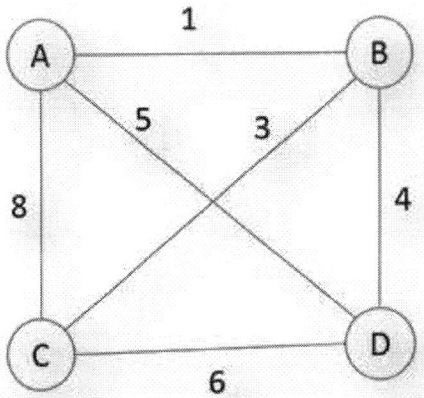

Tours where A is starting city:

Tour	Cost
A→B→C→D→A	1+3+6+5 = 15
A→B→D→C→A	1+4+6+8 = 19
A→C→B→D→A	8+3+4+5 = 20
A→C→D→B→A	8+6+4+1 = 19
A→D→B→C→A	5+4+3+8 = 20
A→D→C→B→A	5+6+3+1 = 15

```
Algorithm TSP
Select a city
MinTourCost = infinite
For ( All permutations of cities ) do
        If( LengthOfPathSinglePermutation < MinTourCost )
             MinTourCost = LengthOfPath
```

Total number of possible combinations = (n-1)!

Cost for calculating the path? Θ(n)
So the total cost for finding the shortest path? Θ(n!)

Knapsack Problem

Given an item with cost C1, C2,..., Cn, and volume V1, V2,..., Vn and knapsack of capacity Vmax, find the most valuable (max $\sum Cj$) that fit in the knapsack ($\sum Vj$ ≤ Vmax).

The solution is one of all the subset of the set of object taking 1 to n objects at a time, so the *Time Complexity* will be O(2^n)

```
Algorithm KnapsackBruteForce
MaxProfit = 0
For ( All permutations of objects ) do
        CurrProfit = sum of objects selected
        If( MaxProfit < CurrProfit )
                MaxProfit = CurrProfit
                Store the current set of objects selected
```

Conclusion

Brute force is the first algorithm that comes into mind when we see some problem. They are the simplest algorithms which are very easy to understand. But these algorithms rarely provide an optimum solution. Many cases we will find other effective algorithm which is more efficient than the brute force method.
This is the most simple to understand the kind of problem solving technique.

Chapter 17: Greedy Algorithm

Introduction

Greedy algorithms are generally used to solve optimization problems. To find the solution that minimizes or maximizes some value (cost/profit/count etc.).

In greedy algorithm solution is constructed through a sequence of steps. At each step choice is made which is locally optimal. We always take the next data to be processed depending upon the dataset which we have already processed and then choose the next optimum data to be processed.

Greedy algorithms does not always give optimum solution. For some problems, greedy algorithm gives an optimal solution. For most, they don't, but can be useful for fast approximations.

Greedy is a strategy that works well on optimization problems with the following characteristics:
1. Greedy choice: A global optimum can be arrived at by selecting a local optimum.
2. Optimal substructure: An optimal solution to the problem is made from optimal solutions of sub problems.

Some examples of brute force algorithms are:
Optimal solutions:
- Minimal spanning tree:
    - Prim's algorithm,
    - Kruskal's algorithm
- Dijkstra's algorithm for single-source shortest path
- Huffman trees for optimal encoding
- Scheduling problems

Approximate solutions:
- Greedy algorithm for the Knapsack problem
- Coin exchange problem

Problems on Greedy Algorithm

Coin exchange problem

How can a given amount of money N be made with the least number of coins of given denominations D= {d1... dn}?

The Indian coin system {5, 10, 20, 25, 50,100}

Suppose we want to give change of a certain amount of 40 paisa.

We can make a solution by repeatedly choosing a coin ≤ to the current amount, resulting in a new amount. The greedy solution is to always choose the largest coin value possible without exceeding the total amount.

For 40 paisa: {25, 10, and 5}
The optimal solution will be {20, 20}
The greedy algorithm did not give us optimal solution, but it gave a fair approximation.

```
Algorithm  MAKE-CHANGE (N)
C = {5, 20, 25, 50, 100}    // constant.
S = {}                       // set that will hold the solution set.
Value = N
WHILE Value != 0
        x = largest item in set C such that x < Value
        IF no such item THEN
                RETURN    "No Solution"
        S = S + x
        Value = Value - x
RETURN S
```

Minimum Spanning Tree

A spanning tree of a connected graph is a tree containing all the vertices.
A minimum spanning tree of a weighted graph is a spanning tree with the smallest sum of the edge weights.

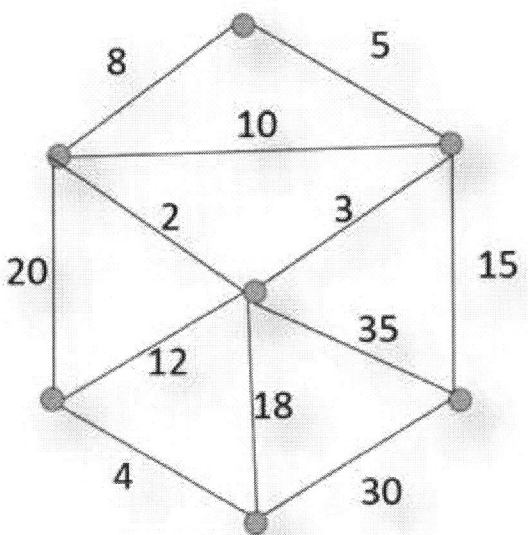
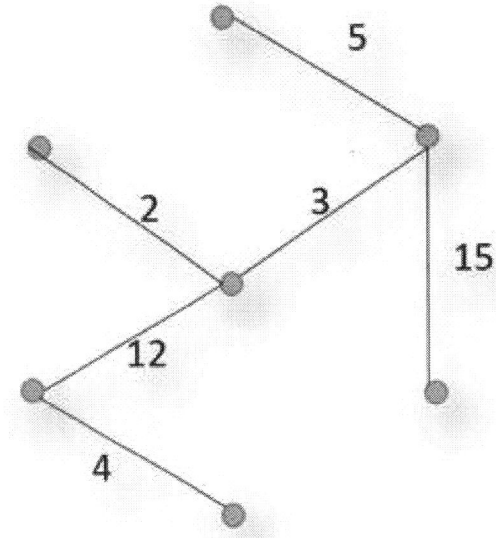

Prim's Algorithm

Prim's algorithm grows a single tree T, one edge at a time, until it becomes a spanning tree.
We initialize T with zero edges. And U with single node. Where T is spanning tree edges set and U is spanning tree vertex set.

At each step, Prim's algorithm adds the smallest value edge with one endpoint in U and other not in us.

Since each edge adds one new vertex to U, after n − 1 additions, U contain all the vertices of the spanning tree and T becomes a spanning tree.

```
// Returns the MST by Prim's Algorithm
// Input: A weighted connected graph G = (V, E)
// Output: Set of edges comprising a MST
Algorithm Prim(G)
T = {}
Let r be any vertex in G
U = {r}
for i = 1 to |V| - 1 do
        e = minimum-weight edge (u, v)
               With u in U and v in V-U
        U = U + {v}
        T = T + {e}
return T
```

Prim's Algorithm using a priority queue (min heap) to get the closest fringe vertex
Time Complexity will be O(m log n) where n vertices and m edges of the MST.

Kruskal's Algorithm

Kruskal's Algorithm is used to create minimum spanning tree. Spanning tree is created by choosing smallest weight edge that does not form a cycle. And repeating this process till all the edges from the original set is exhausted.

Sort the edges in non-decreasing order of cost: $c(e_1) \leq c(e_2) \leq \cdots \leq c(e_m)$.
Set T to be the empty tree. Add edges to tree one by one if it does not create a cycle. (If the new edge form cycle then ignore that edge.)

```
// Returns the MST by Kruskal's Algorithm
// Input: A weighted connected graph G = (V, E)
// Output: Set of edges comprising a MST

Algorithm Kruskal(G)
Sort the edges E by their weights
T = {}
while |T| + 1 < |V| do
        e = next edge in E
        if T + {e} does not have a cycle then
               T = T + {e}
return T
```

Kruskal's Algorithm is O(E log V) using efficient cycle detection.

Dijkstra's algorithm for single-source shortest path problem

Dijkstra's algorithm for single-source shortest path problem for weighted edges with no negative weight. It
determine the length of the shortest path from the source to each of the other nodes of the graph. Given a weighted graph G, we need to find shortest paths from the source vertex s to each of the other vertices.

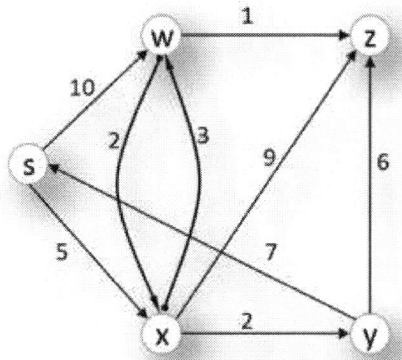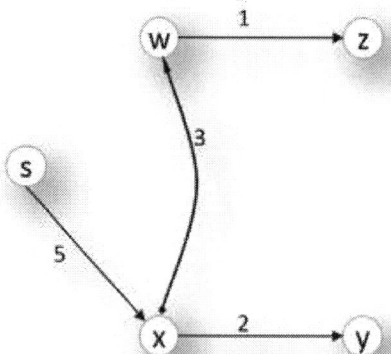

Single-Source shortest path

The algorithm starts by keeping track of the distance of each node and its parents. All the distance is set to infinite in the beginning as we don't know the actual path to the nodes and parents of all the vertices are set to null. All the vertices are added to a priority queue (min heap implementation)
At each step algorithm takes one vertex from the priority queue (which will be the source vertex in the beginning). Then update the distance array corresponding to all the adjacent vertices. When the queue is empty, then we will have the distance and parent array fully populated.

```
// Solves SSSP by Dijkstra's Algorithm
// Input: A weighted connected graph G = (V, E)
// with no negative weights, and source vertex v
// Output: The length and path from s to every v

Algorithm Dijkstra(G, s)
for each v in V do
        D[v] = infinite     // Unknown distance
        P[v] = null    //unknown previous node
        add v to PQ    //adding all nodes to priority queue

D[source] = 0 // Distance from source to source

while (Q is not empty)
        u = vertex from PQ with smallest D[u]
        remove u from PQ
                for each v adjacent from u do
                        alt = D[u] + length ( u , v)
                        if  alt < D[v] then
                                D[v] = alt
                                P[v] = u
Return D[] , P[]
```

Time Complexity will be O(|E|log|V|).

Note: Dijkstra's algorithm does not work for graphs with negative edges weight.
Note: Dijkstra's algorithm is applicable to both undirected and directed graphs.

Huffman trees for optimal encoding

Coding is basically an assignment of bit strings of alphabet characters.
There are two types of encoding:
- Fixed-length encoding (eg., ASCII)
- Variable-length encoding (eg., Huffman code)

Variable length encoding can only work on prefix free encoding. Which means that no code word is a prefix of another code word.

Huffman codes are the best prefix free code. Any binary tree with edges labeled as 0 and 1 will produce a prefix free code of characters assigned to its leaf nodes.

Huffman's algorithm is used to construct a binary tree whose leaf value is assigned a code which is optimal for the compression of the whole text need to be processed. For example, the most frequently occurring words will get the smallest code so that the final encoded text is compressed.

Initialize n one-node trees with words and the tree weights with their frequencies. Join the two binary tree with smallest weight into one and the weight of the new formed tree as the sum of weight of the two small trees. Repeat the above process N-1 times and when there is just one big tree left you are done.

Mark edges leading to left and right subtrees with 0's and 1's, respectively.

Word	Frequency
Apple	30
Banana	25
Mango	21
Orange	14
Pineapple	10

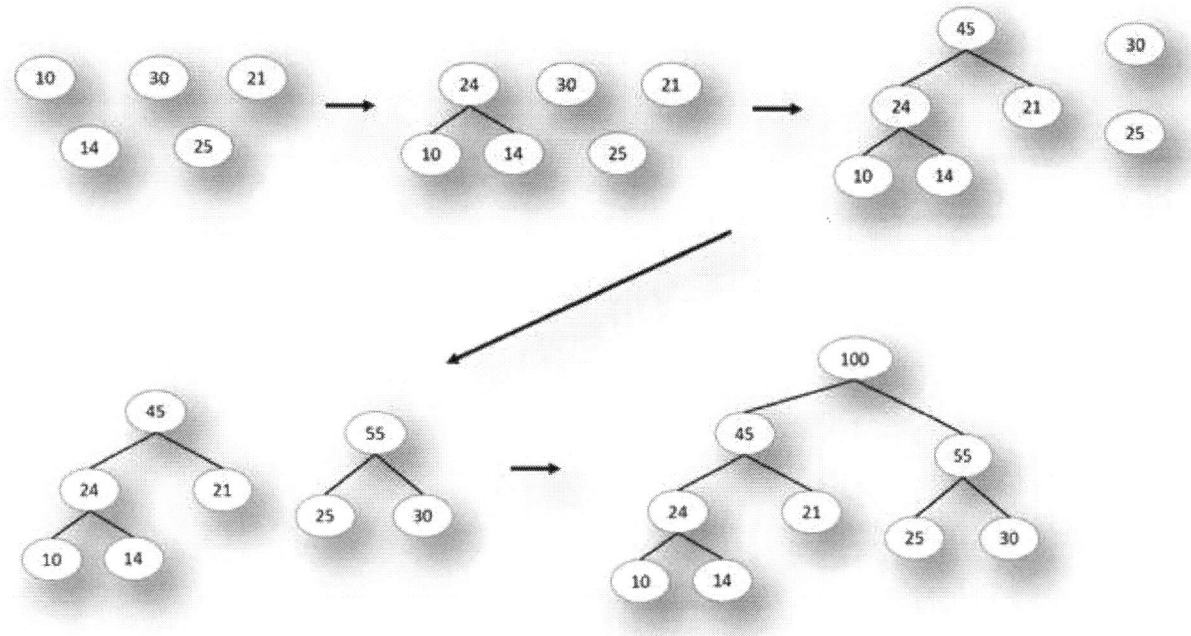

Word	Value	Code
Apple	30	11
Banana	25	10
Mango	21	01
Orange	14	001
Pineapple	10	000

It is clear that more frequency words gets smaller Huffman's code.

```
// Computes optimal prefix code.
// Input: Array W of character probabilities
// Output: The Huffman tree.

Algorithm Huffman(C[0..n - 1], W[0..n - 1])
PQ = {} // priority queue
for i = 0 to n - 1 do
        T.char = C[i]
        T.weight = W[i]
        add T to priority queue PQ

for i = 0 to n - 2 do
        L = remove min from PQ
        R = remove min from PQ
        T = node with children L and R
        T.weight = L.weight + R.weight
        add T to priority queue PQ
return T
```

The Time Complexity is **O(nlogn)**.

Activity Selection Problem

Suppose that activities require exclusive use of common resources, and you want to schedule as many activities as possible.
Let S = {a1,..., an} be a set of n activities.

Each activity ai needs the resource during a time period starting at si and finishing before fi, i.e., during [si, fi).
The optimization problem is to select the non-overlapping largest set of activities from S.

We assume that activities S = {a1,..., an} are sorted in finish time $f_1 \leq f_2 \leq ... f_{n-1} \leq f_n$ (this can be done in $\Theta(n \lg n)$).

Example
Consider these activities:

i	1	2	3	4	5	6	7	8	9	10	11
S[i]	1	3	0	5	3	5	6	8	8	2	11
F[i]	4	5	6	7	8	9	10	11	12	13	14

Here is a graphic representation:

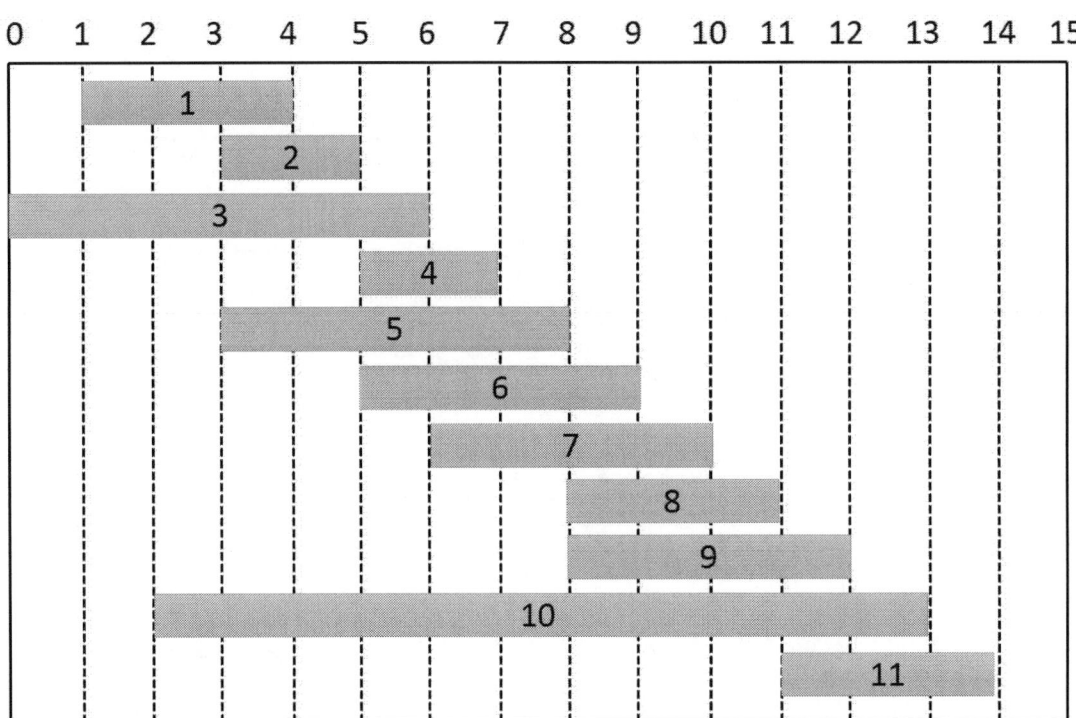

We chose an activities that start first, and then look for the next activity that starts after it is finished. This could result in {a4, a7, a8}, but this solution is not optimal.
An optimal solution is {a1, a3, a6, a8}. (It maximizes the objective function of a number of activities scheduled.)
Another one is {a2, a5, a7, a9}. (Optimal solutions are not necessarily unique.)
How do we find (one of) these optimal solutions? Let's consider it as a dynamic programming problem...

We are trying to optimize the number of activities. Let's be greedy!
- The more time left after running an activity, the more subsequent activities we can fit in.
- If we choose the first activity to finish, the more time will be left.
- Since activities are sorted by finish time, we will always start with a_1.
- Then we can solve the single sub problem of activity scheduling in this remaining time.

```
Algorithm ActivitySelection(S[], F[], N)
Sort S[] and F [] in increasing order of finishing time
A = {a1}
K = 1
For m = 2 to N do
        If S[m] >= F[k]
            A = A + {am}
            K = m
Return A
```

Knapsack Problem

A thief enters a store and sees a number of items with their cost and weight mentioned. His Knapsack can hold a max weight. What should he steal to maximize profit?

Fractional Knapsack problem

A thief can take a fraction of an item (they are divisible substances, like gold powder).

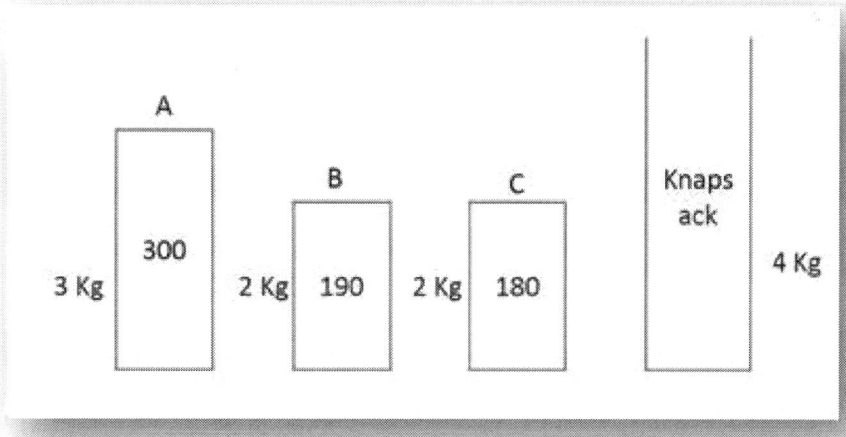

The fractional knapsack problem has a greedy solution one should first sort the items in term of cost density against weight. Then fill up as much of the most valuable substance by weight as one can hold, then as much of the next most valuable substance, etc. Until W is reached.

Item	A	B	C
Cost	300	190	180
Weight	3	2	2
Cost/weight	100	95	90

For a knapsack of capacity of 4 kg.
The optimum solution of the above will take 3kg of A and 1 kg of B.

```
Algorithm FractionalKnapsack(W[], C[], Wk)
For i = 1 to n do
        X[i] = 0
Weight = 0
//Use Max heap
H = BuildMaxHeap(C/W)
While Weight < Wk do
        i = H.GetMax()
        If(Weight + W[i] <= Wk) do
                X[i] = 1
                Weight = Weight + W[i]
        Else
                X[i] = (Wk - Weight)/W[i]
                Weight = Wk
Return X
```

0/1 Knapsack Problem

A thief can only take or leave the item. He can't take a fraction.
A greedy strategy same as above could result in empty space, reducing the overall cost density of the knapsack.

In the above example, after choosing object A there is no place for B or C so there leaves empty space of 1kg. And the result of the greedy solution is not optimal.
The optimal solution will be when we take object B and C. This problem can be solved by dynamic programming which we will see in the coming chapter.

CHAPTER 18: DIVIDE-AND-CONQUER, DECREASE-AND-CONQUER

Introduction

Divide-and-Conquer algorithms works by recursively breaking down a problem into two or more sub-problems (divide), until these sub problems become simple enough so that can be solved directly (conquer). The solution of these sub problems is then combined to give a solution of the original problem.

Divide-and-Conquer algorithms involve basic three steps
1. Divide the problem into smaller problems.
2. Conquer by solving these problems.
3. Combine these results together.

In divide-and-conquer the size of the problem is reduced by a factor (half, one-third etc.), While in decrease-and-conquer the size of the problem is reduced by a constant.

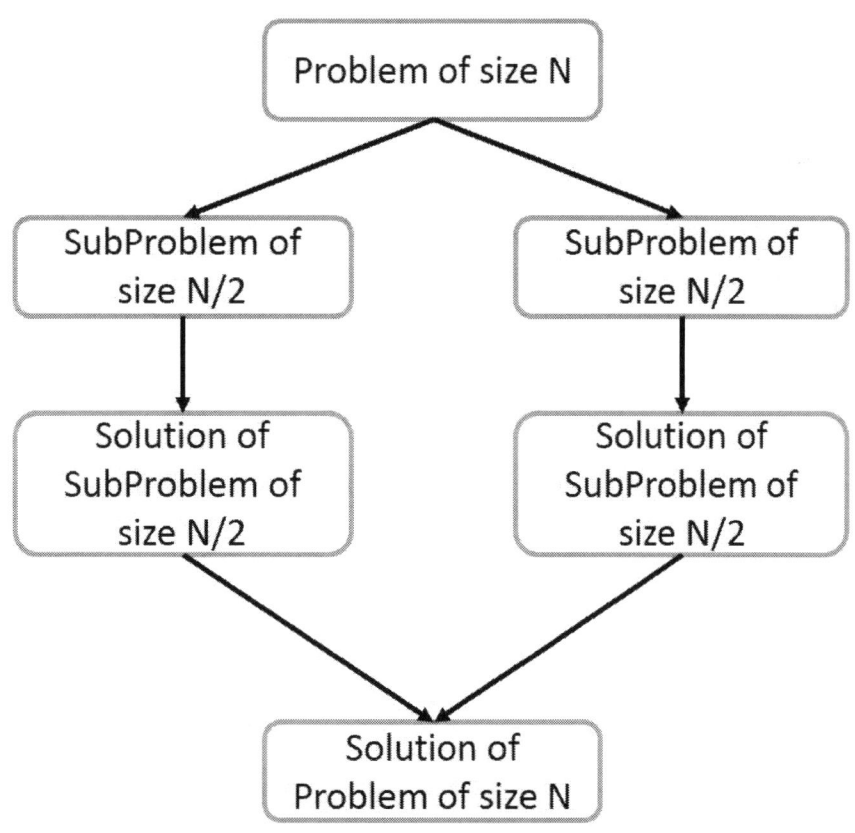

Divide-and-Conquer algorithms

Examples of divide-and-conquer algorithms:
- Merge-Sort algorithm (recursion)
- Quicksort algorithm (recursion)
- Computing the length of the longest path in a binary tree (recursion)
- Computing Fibonacci numbers (recursion)
- Convex Hull

Examples of decrease-and-conquer algorithms:
- Computing POW (a, n) by calculating POW (a, n/2) using recursion
- Binary search in a sorted array (recursion)
- Searching in BST
- Insertion-Sort
- Graph traversal algorithms (DFS and BFS)
- Topological sort
- Warshall's algorithm (recursion)
- Permutations (Minimal change approach, Johnson-Trotter algorithm)
- Fake-coin problem (Ternary search)
- Computing a median

General Divide-and-Conquer Recurrence

T(n) = aT(n/b) + f (n)
- Where a ≥ 1 and b > 1.
- "n" is the size of a problem.
- "a" is a number of sub-problem in the recursion.
- "n/b" is the size of each sub-problem.
- "f(n)" is the cost of the division of the problem into sub problem or merge of the results of sub-problem to get the final result.

Master Theorem

The master theorem solves recurrence relations of the form:
$T(n) = a\ T(n/b) + f(n)$

It is possible to determine an asymptotic tight bound in these three cases:
Case 1: when f(n) = $O(n^{\log_b a - \epsilon})$ and constant $\epsilon > 1$, than the final *Time Complexity* will be:
T(n) = $\Theta(n^{\log_b a})$

Case 2: when f(n) = $\Theta(n^{\log_b a} \log^k n)$ and constant k ≥ 0, than the final *Time Complexity* will be:
T(n) = $\Theta(n^{\log_b a} \log^{k+1} n)$

Case 3: when f(n) = $\Omega(n^{\log_b a + \epsilon})$ and constant $\epsilon > 1$, Then the final *Time Complexity* will be:
T(n) = $\Theta(f(n))$

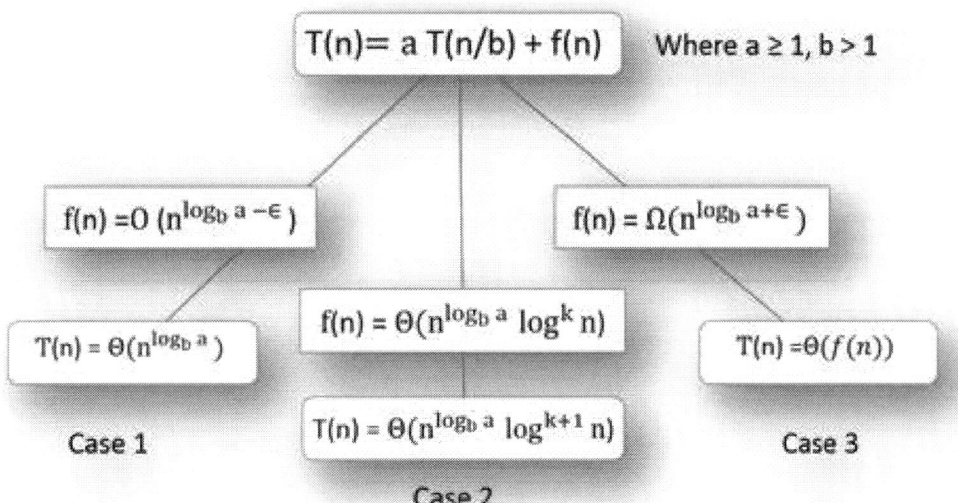

Modified Master theorem: This is a shortcut to solving the same problem easily and fast. If the recurrence relation is in the form of $T(n) = a\,T(n/b) + dx^s$

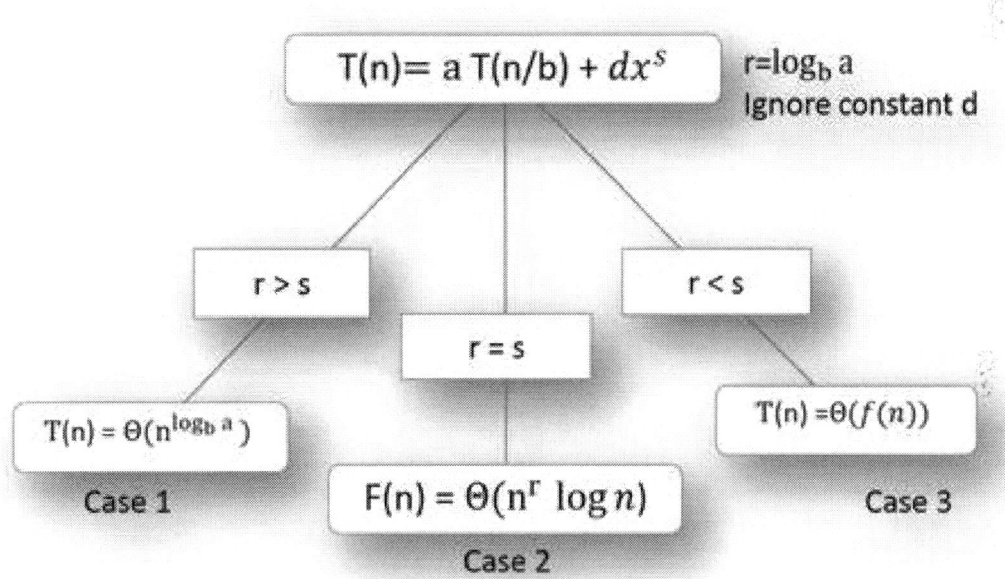

Example 1: Take an example of Merge-Sort, $T(n) = 2\,T(n/2) + n$
Sol:-
$\log_b a = \log_2 2 = 1$
$f(n) = n = \Theta(n^{\log_2 2} \log^0 n)$
Case 2 applies and $T(n) = \Theta(n^{\log_2 2} \log^{0+1} n)$
$T(n) = \Theta(n \log(n))$

Example 2: Binary Search $T(n) = T(n/2) + O(1)$
$log_b a = log_2 1 = 0$
$f(n) = 1 = \Theta(n^{log_2 1} log^0 n)$
Case 2 applies and $T(n) = \Theta(n^{log_2 1} log^{0+1} n)$
$T(n) = \Theta(log(n))$

Example 3: Binary tree traversal $T(n) = 2T(n/2) + O(1)$
$log_b a = log_2 2 = 1$
$f(n) = 1 = O(n^{log_2 2 - 1})$
Case 1 applies and $T(n) = \Theta(n^{log_2 2})$
$T(n) = \Theta(n)$

Problems on Divide-and-Conquer Algorithm

Merge-Sort algorithm

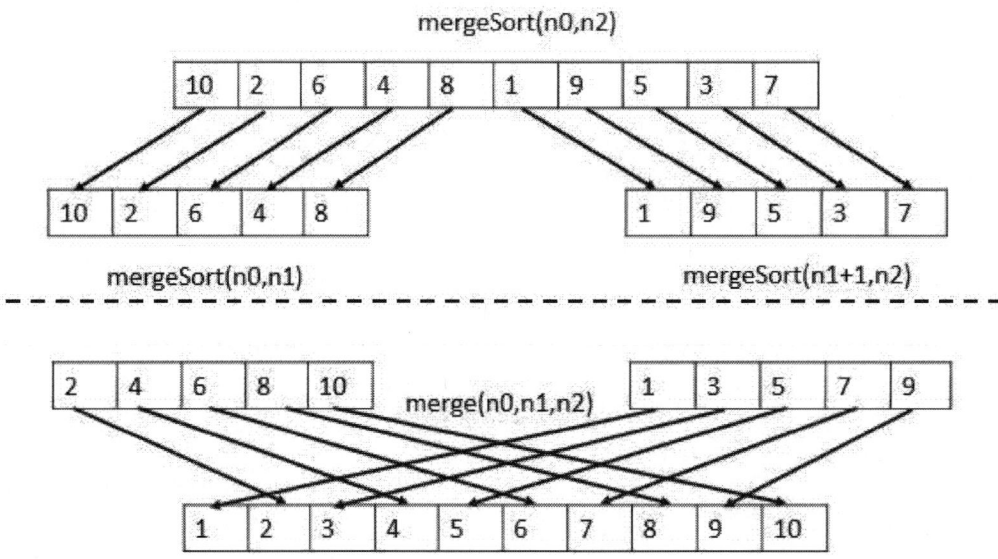

```
// Sorts a given array by mergesort
// Input: An array A of orderable elements
// Output: Array A[0..n − 1] in ascending order

Algorithm Mergesort(A[0..n − 1])
if n ≤ 1 then
        return;
copy A[0..⌊n/2⌋ − 1] to B[0..⌊n/2⌋ − 1]
copy A[⌊n/2⌋..n − 1] to C[0..⌊n/2⌋ − 1]
Mergesort(B)
Mergesort(C)
Merge(B, C, A)
```

```
// Merges two sorted arrays into one array
// Input: Sorted arrays B and C
// Output: Sorted array A
Algorithm Merge(B[0..p − 1], C[0..q − 1], A[0..p + q − 1])
i = 0
j = 0
for k = 0 to p + q − 1 do
        if i < p and (j = q or B[i] ≤ C[j]) then
                A[k] = B[i]
                i = i + 1
        else
                A[k] = C[j]
                j = j + 1
```

Time Complexity: **O(nlogn)**
Space Complexity: **O(n)**
The *Time Complexity* of Merge-Sort is *O(nlogn)* in all 3 cases (worst, average and best) as Merge-Sort always divides the array into two halves and take linear time to merge two halves.

It requires the equal amount of additional space as the unsorted list. Hence, it's not at all recommended for searching large unsorted lists.

Quick-Sort

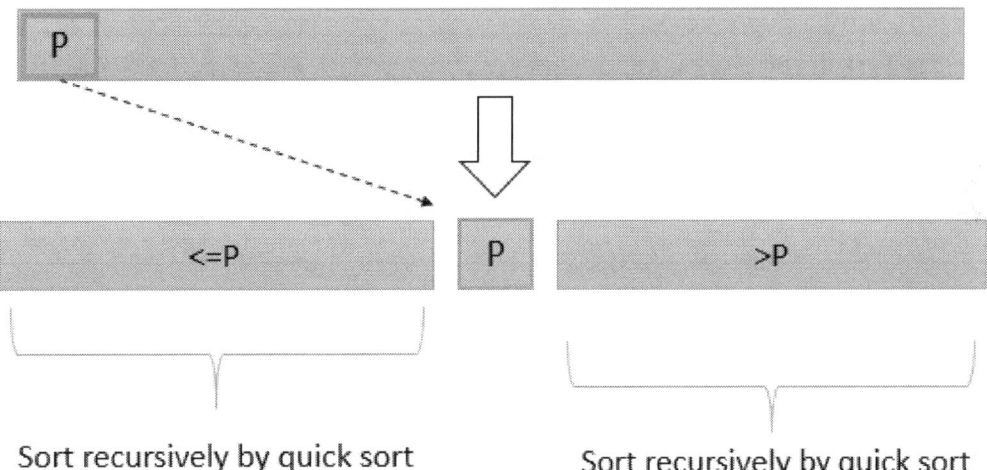

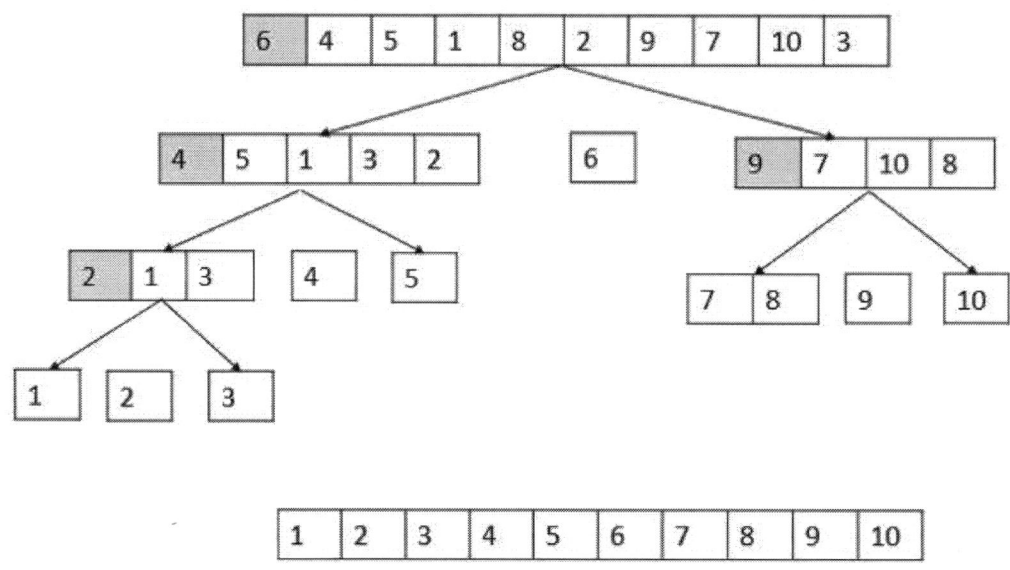

```
// Sorts a subarray by quicksort
// Input: An subarray of A
// Output: Array A[l..r] in ascending order

Algorithm Quicksort(A[l..r])
if l < r then
        p ← Partition(A[l..r]) // p is index of pivot
        Quicksort(A[l..p − 1])
        Quicksort(A[p + 1..r])
```

```
// Partitions a subarray using A[..] as pivot
// Input: Subarray of A
// Output: Final position of pivot

Algorithm Partition(A[], left, right)
pivot = A[left]
lower = left
upper= right
while lower < upper
        while A[lower] <= pivot
                lower = lower + 1
        while A[upper] > pivot
                upper = upper − 1
        if lower < upper then
                swap A[lower] and A[upper]
swap A[lower] and A[upper] //upper is the pivot position
return upper
```

Worst Case *Time Complexity:* **O(n²)**
Best Case Time Complexity: **O(nlogn)**
Average Time Complexity: **O(nlogn)**
Space Complexity: **O(nlogn)**

The space required by Quick-Sort is very less, only *O(nlogn)* additional space is required.

Quicksort is not a stable sorting technique, so it might change the occurrence of two similar elements in the list while sorting.

External Sorting

External sorting is also done using divide and conquer algorithm.

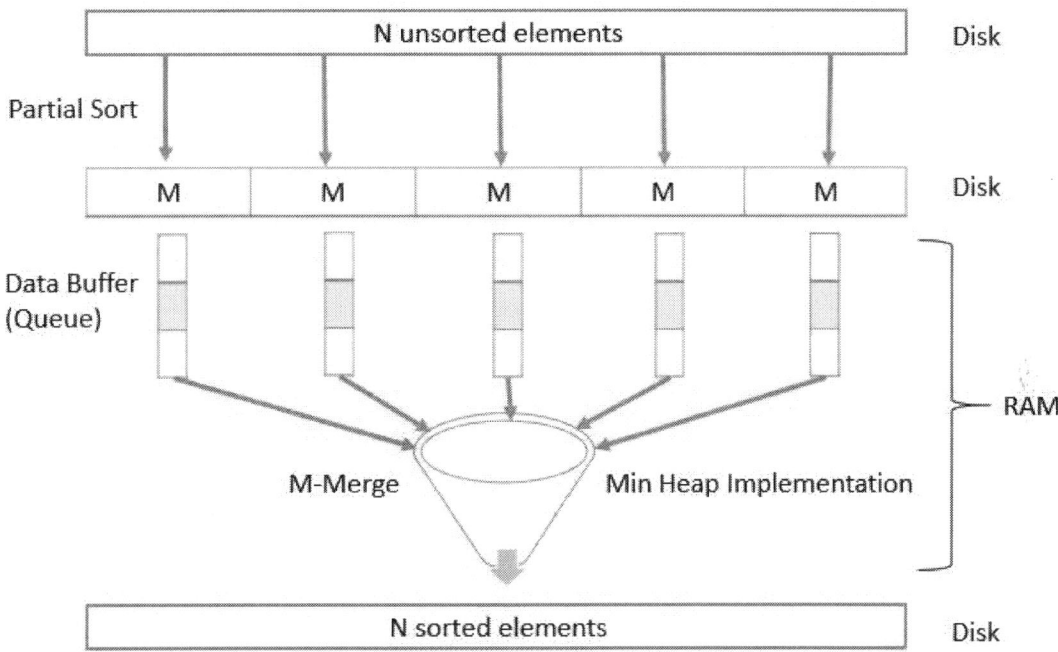

Binary Search

We get the middle point from the sorted array and start comparing with the desired value.
Note: Binary search requires the array to be sorted otherwise binary search cannot be applied.

```
// Searches a value in a sorted array using binary search
// Input: An sorted array A and a key K
// Output: The index of K or −1

Algorithm BinarySearch(A[0..N − 1], N, K) // iterative solution
low = 0
high = N-1
while low <= high do
        mid = ⌊ (low + high)/2⌋
                if K = A[mid] then
                        return mid
                else if A[mid] < K
                        low = mid + 1
                else
                        high = mid - 1
return −1
```

```
// Searches a value in a sorted array using binary search
// Input: An sorted array A and a key K
// Output: The index of K or −1

Algorithm BinarySearch(A[], low, high, K)  //Recursive solution
      If low > high
            return -1
      mid = ⌊(low + high)/2⌋
      if K = A[mid] then
            return mid
      else if A[mid] < K
            return BinarySearch(A[],mid + 1, high, K)
      else
            return BinarySearch(A[],low, mid - 1, K)
```

Time Complexity: **O(logn)**. If you notice the above programs, you see that we always take half input and throwing out the other half. So the recurrence relation for binary search is $T(n) = T(n/2) + c$. Using a divide and conquer master theorem, we get $T(n)$ = **O(logn)**.
Space Complexity: **O(1)**

Power function

```
// Compute Nth power of X using divode and conquer using recursion
// Input: Value X and power N
// Output: Power( X, N)

Algorithm Power( X, N)
      If N = 0
            Return 1
      Else if N % 2 == 0
            Value = Power(X, N/2)
            Return Value * Value
      Else
            Value = Power(X, N/2)
            Return Value * Value * X
```

Convex Hull

Sort points by X-coordinates
Divide points into equal halves A and B
Recursively compute HA and HB
Merge HA and HB to obtain CH

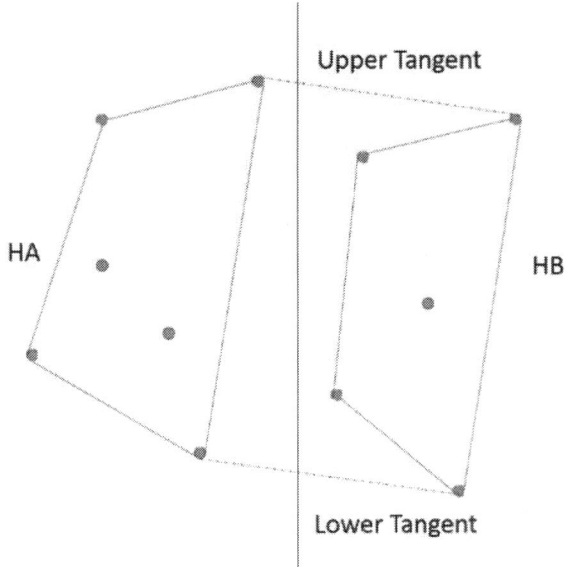

```
LowerTangent(HA, HB)
A = rightmost point of HA
B = leftmost point of HB
While ab is not a lower tangent for HA and HB do
        While ab is not a lower tangent to HA do
                a = a − 1 (move a clockwise)
        While ab is not a lower tangent to HB do
                b = b + 1 (move b counterclockwise)
Return ab
```

Similarly find upper tangent and combine the two hulls.

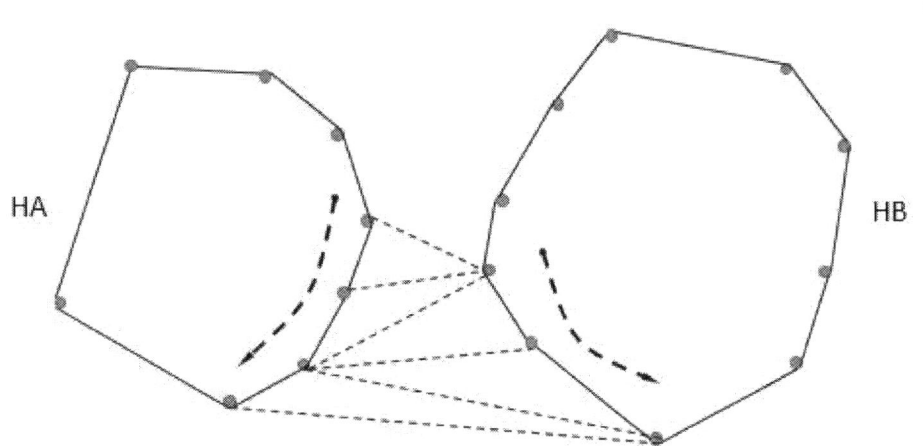

Initial sorting takes **O(nlogn)** time
Recurrence relation T (N) = 2T (N/2) + **O(N)**
Where, **O(N)** time for tangent computation inside merging
Final *Time Complexity* will be T (N) = **O(nlogn)**.

Closest Pair

Given N points in 2-dimensional plane, find two points whose mutual distance is smallest.

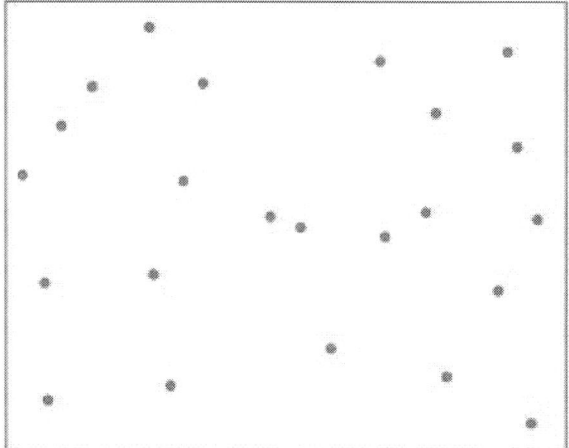

A brute force algorithm takes each and every point and find its distance with all the other points in the plane. And keep track of the minimum distance points and minimum distance. The closest pair will be found in O(n^2) time.

Let us suppose there is a vertical line which divide the graph into two separate parts (let's call it left and right part). The brute force algorithm, we will notice that we are comparing all the points in the left half with the points in the right half. This is the point where we are doing some extra work.

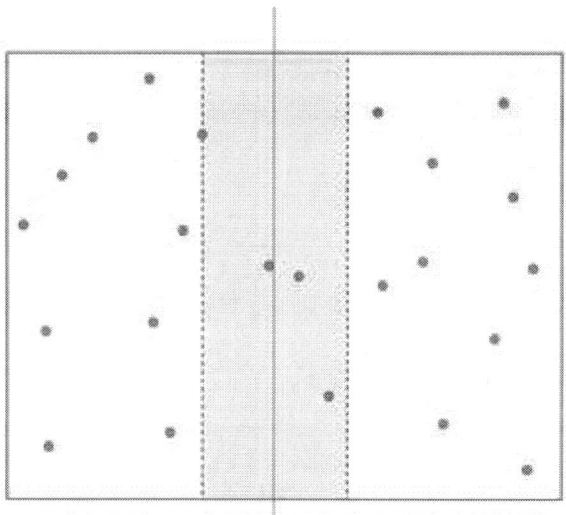

To find the minimum we need to consider only three cases:
  1) Closest pair in the right half
  2) Closest pair in the left half.
  3) Closest pair in the boundary region of the two halves. (Gray)

Every time we will divide the space S into two parts S1 and S2 by a vertical line. Recursively we will compute the closest pair in both S1 and S2. Let's call minimum distance in space S1 as δ1 and minimum distance in space S2 as δ2.

We will find δ = min (δ1, δ2)

Now we will find the closest pair in the boundary region. By taking one point each from S1 and S2 in the boundary range of δ width on both sides.

The candidate pair of point (p, q) where p ∈ S1 and q ∈ S2.

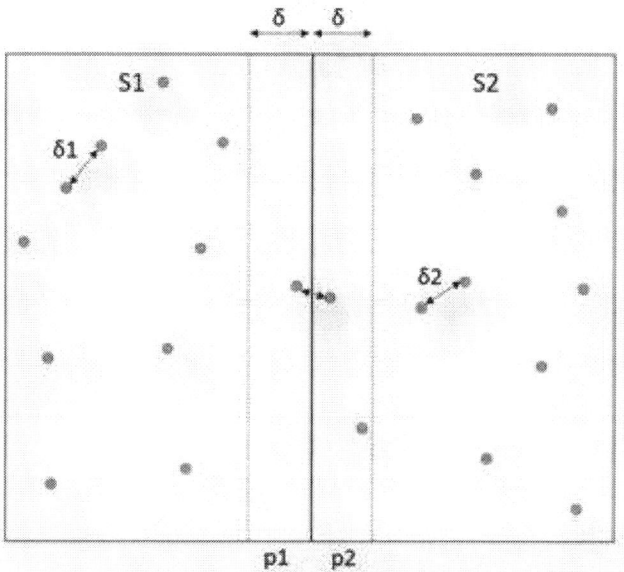

We can find the points which lie in this region in linear time *O(N)* by just scanning through all the points and finding which all points lie in this region.

Now we can sort them in increasing order in Y axis in just *O(nlogn)* time. And then scan through them and get the minimum in just one more linear pass. Closest pair can't be far apart from each other.

Let's look into the next figure.

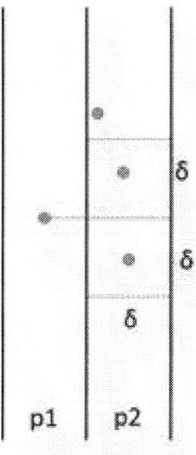

Then the question is how many points we need to compare. We need to compare the points sorted in Y axis only in the range of δ. So the number of points will come down to only 6 points.

By doing this we are getting equation.
T(N) = 2T(N/2) + N + NlogN + 6N = O($n(logn)^2$)

Can we optimize this further?
Yes

Initially, when we are sorting the points in X coordinate we are sorting them in Y coordinate too. When we divide the problem, then we traverse through the Y coordinate list too, and construct the corresponding Y coordinate list for both S1 and S2. And pass that list to them.

Since we have the Y coordinate list passed to a function the δ region points can be found sorted in the Y coordinates in just one single pass in just **O(N)** time.

T(N) = 2T(N/2) + N + N + 6N = ***O(nlogn)***

```
// Finds closest pair of points
// Input: A set of n points sorted by coordinates
// Output: Distance between closest pair

Algorithm ClosestPair(P)
if n < 2 then
        return ∞
else if n = 2 then
        return distance between pair
else
        m = median value for x coordinate
        δ 1 = ClosestPair(points with x < m)
        δ 2 = ClosestPair(points with x > m)
        δ = min(δ 1, δ 2)
        δ 3 = process points with m − δ < x < m + δ
return min(δ, δ 3)
```

First pre-process the points by sorting them in X and Y coordinates. Use two separate lists to keep this sorted points.

Before recursively solving sub-problem pass the sorted list for that sub-problem.

CHAPTER 19: DYNAMIC PROGRAMMING

Introduction

While solving problems using Divide-and-Conquer method, there may be a case when recursively sub-problems can result in the same computation being performed multiple times. This problem arises when there are identical sub-problems arise repeatedly in a recursion.

Dynamic programming is used to avoid the requirement of repeated calculation of same sub-problem. In this method we usually store the result of sub - problems in some data structure (like a table) and refer it to find if we have already calculated the solution of sub - problems before calculating it again.

Dynamic programming is applied to solve problems with the following properties:
1. Optimal Substructure: An optimal solution constructed from the optimal solutions of its sub-problems.
2. Overlapping Sub problems: While calculating the optimal solution of sub problems same computation is repeated again and again.

Examples:
- Fibonacci numbers computed by iteration.
- Assembly-line Scheduling
- Matrix-chain Multiplication
- 0/1 Knapsack Problem
- Longest Common Subsequence
- Optimal Binary Tree
- Warshall's algorithm for transitive closure implemented by iterations
- Floyd's algorithms for all-pairs shortest paths
- Optimal Polygon Triangulation
- Floyd-Warshall's Algorithm

Steps for solving / recognizing if DP applies.
1) Optimal Substructure: Try to find if there is a recursive relation between problem and sub-problem.
2) Write recursive relation of the problem. (Observe Overlapping Sub problems at this step.)
3) Compute the value of sub problems in a bottom up fashion and store this value in some table.
4) Construct the optimal solution from the value stored in step 3.
5) Repeat step 3 and 4 till you get your solution.

Problems on Dynamic programming Algorithm

Fibonacci numbers

```
int fibonacci(int n)
{
    if (n <= 1)
        return n;
    return fibonacci(n - 1) + fibonacci(n - 2);
}
```

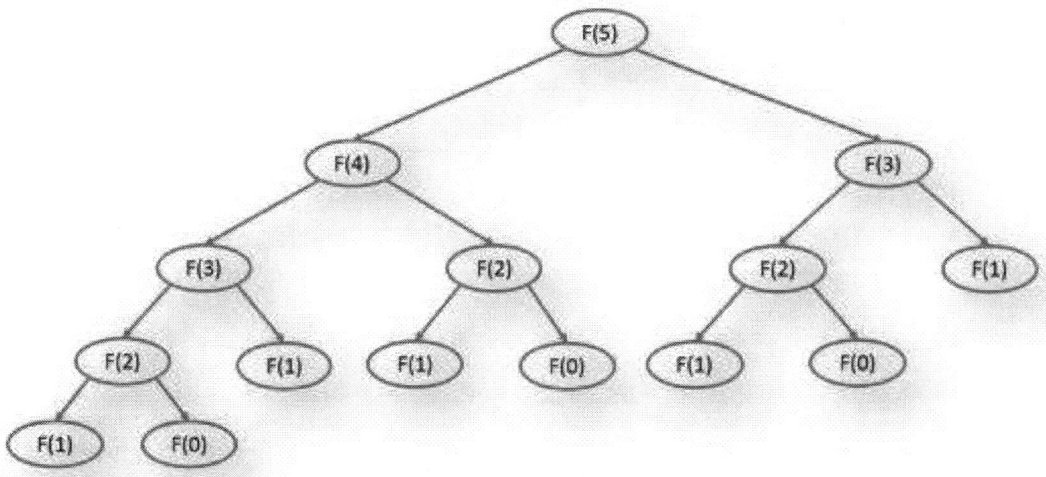

Using divide and conquer same sub-problem is solved again and again, which reduce the performance of the algorithm. This algorithm has an exponential *Time Complexity*.

Same problem of Fibonacci can be solved in linear time if we sort the results of sub problems.

```
int fibonacci (int n)
{
    int first = 0, second = 1;
    int temp, i;

    if (n == 0)
        return first;
    else if (n == 1)
        return second;

    for (i = 2; i <= n; i++)
    {
        temp = first + second;
        first = second;
        second = temp;
    }
    return temp;
}
```

Using this algorithm we will get Fibonacci in linear *Time Complexity* and constant *Space Complexity*.

Assembly-line Scheduling

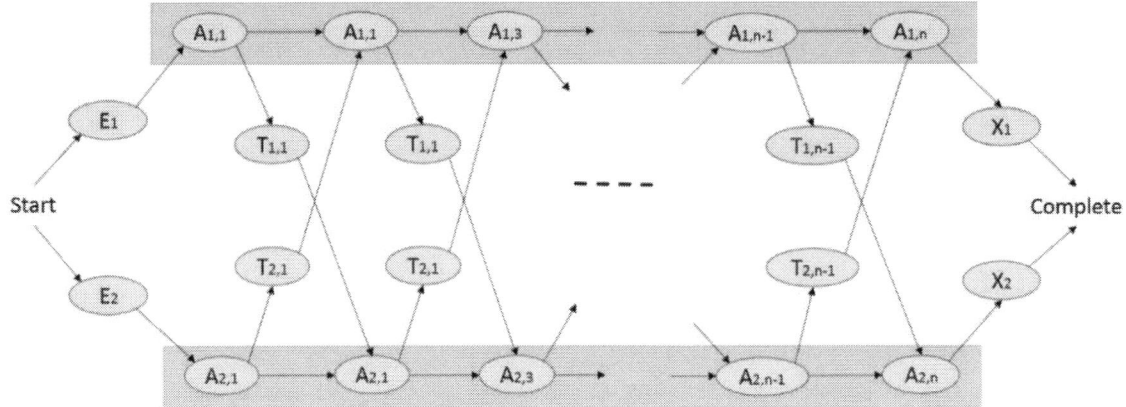

We consider the problem of calculating the least amount of time necessary to build a car when using a manufacturing chain with two assembling lines, as shown in the figure
The problem variables:
- e[i]: entry time in assembly line i
- x[i]: exit time from assembly line i
- a[i,j]: Time required at station S[i,j] (assembly line i, stage j)
- t[i,j]: Time required to transit from station S[i,j] to the other assembly line

Your program must calculate:
- The least amount of time needed to build a car
- The list of stations to traverse in order to assemble a car as fast as possible.

The manufacturing chain will have no more than 50 stations.

If we want to solve this problem in the brute force approach, there will be in total 2^n Different combinations so the *Time Complexity* will be $O(2^n)$

Step 1: Characterizing the structure of the optimal solution
To calculate the fastest assembly time, we only need to know the fastest time to S1;n and the fastest time to S2;n, including the assembly time for the nth part. Then we choose between the two exiting points by taking into consideration the extra time required, x1 and x2. To compute the fastest time to S1;n we only need to know the fastest time to S1;n1 and to S2;n1. Then there are only two choices...

Step 2: A recursive definition of the values to be computed

$$f1[j] = \begin{cases} e1 + a1,1 & if\ j = 1 \\ \min(f1[j-1] + a1,j, \quad f2[j-1] + t2, j-1 + a1, j) & if j \geq 2 \end{cases}$$

$$f2[j] = \begin{cases} e2 + a2,1 & if\ j = 1 \\ \min(f2[j-1] + a2,j, \quad f1[j-1] + t1, j-1 + a2, j) & if j \geq 2 \end{cases}$$

Step 3: Computing the fastest time finally, compute f* as

Step 4: Computing the fastest path compute as li[j] as the choice made for fi[j] (whether the first or the second term gives the minimum). Also, compute the choice for f* as l*.

```
FASTEST-WAY(a, t, e, x, n)
f1[1] ← e1 + a1,1
f2[1] ←e2 + a2,1
for j ← 2 to n
        do if f1[j - 1] + a1,j ≤ f2[j - 1] + t2,j-1 + a1,j
                then f1[j] ← f1[j - 1] + a1, j
                      l1[j] ← 1
                else f1[j] ← f2[j - 1] + t2,j-1 + a1,j
                      l1[j] ← 2
           if f2[j - 1] + a2,j ≤ f1[j - 1] + t1,j-1 + a2,j
                then f2[j] ← f2[j - 1] + a2,j
                      l2[j] ← 2
                else f2[j] ∞ f1[j - 1] + t1,j-1 + a2,j
                      l2[j] ← 1
        if f1[n] + x1 ≤ f2[n] + x2
                then f* = f1[n] + x1
                      l* = 1
                else f* = f2[n] + x2
                      l* = 2
```

Matrix chain multiplication

Same problem is also known as Matrix Chain Ordering Problem or Optimal-parenthesization of matrix problem.

Given a sequence of matrices, M = M1,..., Mn. The goal of this problem is to find the most efficient way to multiply these matrices. The guild is not to perform the actual multiplication, but to decide the sequence of the matrix multiplications, so that the result will be calculated in minimal operations.

To compute the product of two matrices of dimensions pXq and qXr, pqr number of operations will be required. Matrix multiplication operations are associative in nature. So matrix multiplication can be done in many ways.
For example, M1, M2, M3 and M4, can be fully parenthesized as:
(M1· (M2· (M3·M4)))
(M1· ((M2·M3)· M4))
((M1·M2)· (M3·M4))
(((M1·M2)· M3)· M4)
((M1· (M2·M3))· M4)

For example,
Let M1 dimensions are 10 × 100, M2 dimensions are 100 × 10, and M3 dimensions are 10 × 50.
((M1·M2)· M3) = (10*100*10) + (10*10*50) = 15000
(M1· (M2·M3) = (100*10*50) + (10*100*50) = 100000

So in this problem we need to parenthesize the matrix chain so that total multiplication cost is minimized.

Given a sequence of n matrices M1, M2,... Mn. And their dimensions are p0, p1, p2,..., pn. Where matrix Ai has dimension pi − 1 × pi for 1 ≤ i ≤ n. Determine the order of multiplication that minimizes the total number of multiplications.

If you try to solve this problem using the brute - force method, then you will find all possible parenthesization. Then will compute the cost of multiplication. Then will pick the best solution. This approach will be exponential in nature.

There is an insufficiency in the brute force approach. Take an example of M1, M2,..., Mn. When you have calculated that ((M1·M2) · M3) is better than (M1· (M2·M3) so there is no point of calculating then combinations of (M1· (M2·M3) with (M4, M5.... Mn).

Optimal substructure:
Assume that M (1, N) is the optimum cost of production of the M1,..., Mn.

An array p [] to record the dimensions of the matrices.
P [0] = row of the M1
p[i] = col of Mi   1<=i<=N

For some k
M(1,N) = M(1,K) + M(K+1,N) + p0*pk*pn

If M (1, N) is minimal then both M (1, K) & M (K+1, N) are minimal.

Otherwise, if there is some M'(1, K) is there whose cost is less than M (1.. K), then M (1.. N) can't be minimal and there is a more optimal solution possible.

For some general i and j.
M(i,j) = M(i,K) + M(K+1,j) + pi-1*pk*pj

Recurrence relation:

$$M(i,j) = \begin{cases} 0 & \text{if } i = j \\ \min \{M(i,k) + M(k,j) + pi-1 * pk * pj\} & i \leq k < j \end{cases}$$

Overlapping Sub problems:

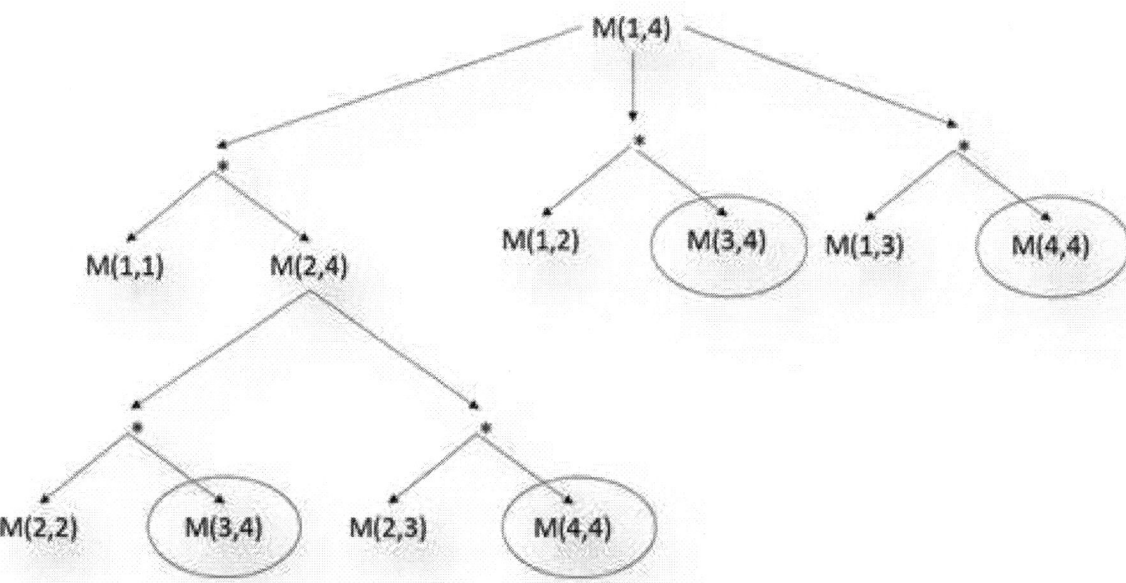

Directly calling recursive function will lead to calculation of same sub-problem multiple times. This will lead to exponential solution.

```
Algorithm MatrixChainMultiplication(p[])
   for i := 1 to n
      M[i, i] := 0;
      for l = 2 to n  // l is the moving line
         for i = 1 to n − l +1
            j = i + l − 1;
            M[i, j] =  min {M(i,k) + M(k,j) + pi − 1 ∗ pk ∗ pj }
                      i ≤ k < j
```

Time Complexity will $O(n^3)$
Constructing optimal parenthesis Solution

Use another table s[1..n, 1..n]. Each entry s[i, j] records the value of k such that the optimal parenthesization of Mi Mi+1...Mj splits the product between Mk and Mk+1.

```
Algorithm MatrixChainMultiplication(p[])
for i := 1 to n
      M[i, i] := 0;
      for l = 2 to n  // l is the moving line
         for i = 1 to n − l +1
            j = i + l − 1;
               M[i, j] =  min {M(i,k) + M(k,j) + pi − 1 ∗ pk ∗ pj }
                         i ≤ k < j
               S[i, j] = k for  min {M(i,k) + M(k,j) + pi − 1 ∗ pk ∗ pj }
                                i ≤ k < j
```

```
Algorithm MatrixChainMultiplication(p[])
    for i := 1 to n
    M[i, i] := 0;
    for l = 2 to n  // l is the moving line
        for i = 1 to n − l +1
            j = i + l − 1;
            for k = i to j
                if( (M(i,k) + M(k,j) + pi − 1 * pk * pj ) < M[i,j]
                    M[i, j] =  (M(i,k) + M(k,j) + pi − 1 * pk * pj )
                    S[i, j] =  k
```

```
Algorithm PrintOptimalParenthesis(s[], i, j)
    If i = j
        Print Ai
    Else
        Print "("
        PrintOptimalParenthesis(s[], i,s[i, j])
        PrintOptimalParenthesis(s[], s[i, j],j)
        Print ")"
```

Longest Common Subsequence

Let X = {x1, x2,...., xm} is a sequence of characters. And Y = {y1, y2,..., yn} is another sequence. Z is a subsequence of X if it can be driven by deleting some elements of X. Z is a subsequence of Y if it can be driven by deleting some elements form Y. Z is LCS of it is subsequence to both X and Y, and there is no subsequence whose length is greater than Z.

Optimal Substructure:

Let X = < x1, x2, ..., xm > and Y = < y1, y2, ..., yn > be two sequences, and let Z = < z1, z2, ..., zk > be a LCS of X and Y.

- If xm = yn, then zk = xm = yn ⇒ Zk−1 is a LCS of Xm−1 and Yn−1
- If xm != yn, then:
    - zk != xm ⇒ Z is an LCS of Xm−1 and Y.
    - zk != yn ⇒ Z is an LCS of X and Yn−1.

Recurrence relation

Let c[i, j] be the length of the longest common subsequence between X = {x1, x2,...., xi} and Y = {y1, y2,..., yj}.
Then c[n, m] contains the length of an LCS of X and Y

$$c[i,j] = \begin{cases} 0 & if\ i = 0\ or\ j = 0 \\ c[i-1, j-1] + 1 & if\ i, j > 0\ and\ x_i = y_j \\ \max(c[i-1, j], c[i, j-1]) & otherwise \end{cases}$$

```
Algorithm LCS(X[], m, Y[], n)
for i = 1 to m
        c[i,0] = 0
for j = 1 to n
        c[0,j] = 0;
for i = 1 to m
        for j = 1 to n
                if X[i] == Y[j]
                        c[i,j] = c[i-1,j-1] + 1
                        b[i,j] = ↖
                else
                        if c[i-1,j] ≥ c[i,j-1]
                                c[i,j] = c[i-1,j]
                                b[i,j] = ↑
                        else
                                c[i,j] = c[i,j-1]
                                b[i,j] = ←
```

```
Algorithm PrintLCS(b[],X[], i, j)
if i = 0
        return
if j = 0
        return
if b[i, j] = ↖
        PrintLCS (b[],X[], i − 1, j − 1)
        print X[i]
else if b[i, j] = ↑
        PrintLCS (b[],X[], i − 1, j)
else
        PrintLCS (b[],X[], i, j − 1)
```

Coin Exchanging problem

How can a given amount of money N be made with the least number of coins of given denominations D= {d1... dn}?

For example, Indian coin system {5, 10, 20, 25, 50,100}. Suppose we want to give change of a certain amount of 40 paisa.

We can make a solution by repeatedly choosing a coin ≤ to the current amount, resulting in a new amount. The greedy solution is to always choose the largest coin value possible.
For 40 paisa: {25, 10, and 5}

This is how billions of people around the globe do change every day. That is an approximate solution of the problem. But this is not the optimal way, the optimal solution for the above problem is {20, 20}

Step (I): Characterize the structure of a coin-change solution.
Define C [j] to be the minimum number of coins we need to make a change for j cents.

If we knew that an optimal solution for the problem of making change for j cents used a coin of denomination di, we would have:
C[j] = 1+C[j − di]

Strep (II): Recursively defines the value of an optimal solution.

$$c[j] = \begin{cases} \text{infinite} & \text{if } j < 0 \\ 0 & \text{if } j = 0 \\ 1 + \min(c[j - d_i]) \ 1 \leq i \leq k & \text{if } j \geq 1 \end{cases}$$

Step (III): Compute values in a bottom-up fashion.

```
Algorithm CoinExchange(n, d[], k)
C[0] = 0
for j = 1 to n do
        C[j] = infinite
for i = 1 to k do
        if j < di and 1+C[j − di] < C[j] then
                C[j] = 1+C[j − di]
return C
```

Complexity: O(nk)

Step (iv): Construct an optimal solution
We use an additional array Deno[1.. n], where Deno[j] is the denomination of a coin used in an optimal solution.

```
Algorithm CoinExchange(n, d[], k)
C[0] = 0
for j = 1 to n do
        C[j] = infinite
        for i = 1 to k do
                if j < di and 1+C[j − di] < C[j] then
                        C[j] = 1+C[j − di]
                        Deno[j] = di
return C
```

```
Algorithm PrintCoins( Deno[], j)
if j > 0
        PrintCoins (Deno, j −Deno[j])
        print Deno[j]
```

CHAPTER 20: BACKTRACKING AND BRANCH-AND-BOUND

Introduction

Suppose the lock produce some sound "click" correct digit is selected for any level. You just will find the first digit, then find the second digit, then find the third digit and done. This will be a greedy algorithm and you will find the solution very quickly.

But let us suppose the lock is some old one and it creates same sound not only at the correct digit but at some other digits also. So when you are trying to find the digit of the first ring, then it may product sound at multiple instances. So at this point you are not directly going straight to the solution, but you need to test various states and in case those states are not the solution you are looking for, then you need to backtrack one step at a time and find the next solution. But sure this intelligence/ heuristics of click sound will help you to reach your goal much faster. These functions are called Pruning function or bounding functions.

Problems on Backtracking Algorithm

N Queens Problem

There are N queens given, you need to arrange them in a chess board on NxN such that no queen should attach each other.

```
public static void NQueens(int[] Q, int k, int n) {
    if( k == n) {
        print( Q, n);
        return;
    }
    for ( int i = 0; i< n; i++) {
        Q[k]=i;
        if(Feasible(Q, k))
            NQueens(Q, k+1, n);
    }
}
```

```
public static boolean Feasible(int[] Q, int k) {
    for ( int i = 0; i< k; i++) {
        if(Q[k] == Q[i] || Math.abs(Q[i] - Q[k]) == Math.abs(i-k))
            return false;
    }
    return true;
}
```

```
public static void print(int[] Q, int n)
{
        for ( int i = 0; i< n; i++)
                System.out.print(" "+ Q[i]);
        System.out.println(" ");
}
```

```
public static void main(String[] args) {
        int[] Q= new int[8];;
        NQueens(Q,0,8);
}
```

Tower of Hanoi

The Tower of Hanoi puzzle, disks need to be moved from one pillar to another such that any large disk cannot rest above any small disk.

This is a famous puzzle in the programming world, its origins can be tracked back to India.
"There is a story about an Indian temple in Kashi Viswanathan which contains a large room with three timeworn posts in it surrounded by 64 golden disks. Brahmin priests, acting out the command of an ancient Hindu prophecy, have been moving these disks, in accordance with the immutable rules of the Brahma the creator of universe, since the beginning of time. The puzzle is therefore also known as the Tower of Brahma puzzle. According to the prophecy, when the last move of the puzzle will be completed, the world will end."   ;) ;) ;)

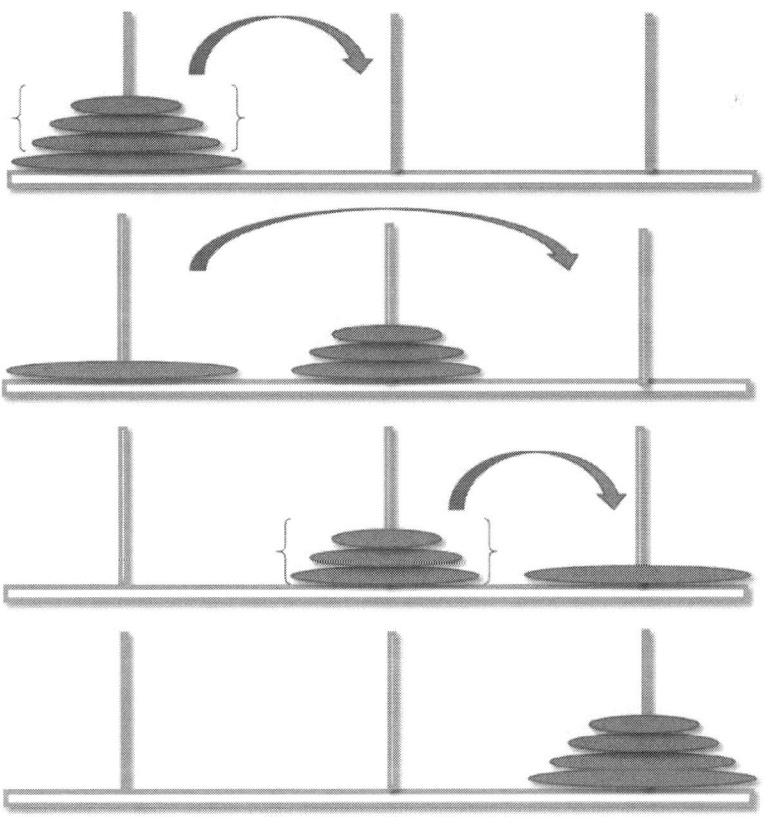

```java
public static void TOHUtil(int num, char from, char to, char temp)
{
    if (num < 1)
        return;

    TOHUtil(num - 1, from, temp, to);
    System.out.println("Move disk "+num+" from peg "+from+" to peg "+to);
    TOHUtil(num - 1, temp, to, from);
}
```

```java
public static void TowersOfHanoi(int num)
{
    System.out.println("The moves involved in the Tower of Hanoi are :");
    TOHUtil(num, 'A', 'C', 'B');
}
```

Chapter 21: Complexity Theory and NP Completeness

Introduction

Computational complexity is the measurement of how much resources are required to solve some problem.

There are two types of resources:
1. Time: how many steps it takes to solve a problem
2. Space: how much memory it takes to solve a problem.

Decision problem

Much of Complexity theory deals with decision problems. A decision problem always has a yes or no answer.

Many problems can be converted to a decision problem which have answered as yes or no. For example:
1. Searching: The problem of searching element can be a decision problem if we ask to find if a particular number is there in the list?

2. Sorting of list and to find if the list is sorted you can make a decision problem is the list is sorted in increasing order or not?

3. Graph coloring algorithms: this is also can be converted to a decision problem. Can we do the graph coloring by using X number of colors?

4. Hamiltonian cycle: Is there is a path from all the nodes, each node is visited exactly once and come back to the starting node without breaking?

Complexity Classes

Problems are divided into many classes such that how difficult to solve them or how difficult to find if the given solution is correct or not.

Class P problems

The class P consists of a set of problems that can be solved in polynomial time. The complexity of a P problem is $O(n^k)$ Where n is input size and k is some constant (it can't depend on n).

Class P Definition: The class P contains all decision problems for which there exists a Turing machine algorithm that leads to the "yes/no" answer in a definite number of steps bounded by a polynomial function.

For example:
Given a sequence a1, a2, a3…. an. Find if a number X is in this array.
We can search, the number X in this array in linear time (polynomial time)

Another example:
Given a sequence a1, a2, a3…. an. If we are asked to sort the sequence.
We can sort and array in polynomial time using Bubble-Sort, this is also linear time.

Note: **O(logn)** is also polynomial. Any algorithm which has complexity less than some O(n^k) is also polynomial.

Some problem of P class is:
1. Shortest path
2. Minimum spanning tree
3. Maximum problem.
4. Max flow graph problem.
5. Convex hull

Class NP problems

Set of problems for which there is a polynomial time checking algorithm. Given a solution if we can check in a polynomial time if that solution is correct or not then, the problem is NP problem.

Class NP Definition: The class NP contains all decision problems for which, given a solution, there exists a polynomial time "proof" or "certificate" that can verify if the solution is the right "yes/no" answer

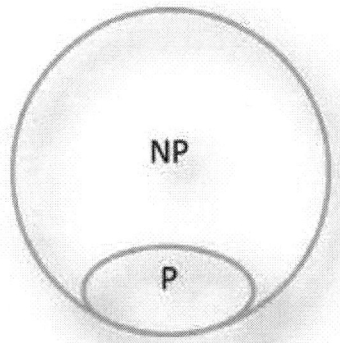

Note: There is no guarantee that you will be able to solve this problem in polynomial time. But if a problem is an NP problem, then you can verify an answer in polynomial time.

NP does not mean non polynomial. Actually, it is Non-Deterministic polynomial type of problem. They are the kind of problems which can be solved in polynomial time by a Non-Deterministic Turing machine. At each point all the possibilities are executed in parallel. If there are n possible choices, then all n cases will be executed in parallel. We don't have non deterministic computers. Don't confuse it with parallel computing because the number of CPU is limited in parallel computing it may be 16 core or 32 core but it can't be N-Core.

In short NP problems are those problems for which, if a solution is given. We can verify that solution (if it is correct or not) in polynomial time.

Boolean Satisfiability problem

A Boolean formula is satisfied if there exist some assignment of the values 0 and 1 to its variables that causes it to evaluate to 1.

$(A1 \lor A2 \ldots) \land (A2 \lor A4..) \ldots \land (..\lor AN)$

There are in total N Different Boolean Variables A1, A2… AN. There are an M number of brackets. Each bracket has K variables.

There is N variable so the number of solutions will be 2^n
And to verify if the solutions really evaluate the equation to 1 will take total $2^n * km$ steps
Given solution of this problem you can find if the formula satisfies or not in KM steps.

Hamiltonian cycle

Hamiltonian cycle is a path from all the nodes of a graph, each node is visited exactly once and come back to the starting node without breaking.
Is an NP problem, if you have a solution to it, then you just need to see if all the nodes are there in the path and you came back to where you started and you are done? The checking is done in linear time and you are done.

Determining whether a directed graph has a Hamiltonian cycle doesn't have a polynomial time algorithm. O(n!)

However, if someone have given you a sequence of vertices, determining whether or not that sequence forms a Hamiltonian cycle can be done in polynomial time (Linear time).
Hamiltonian cycles are in NP

Clique Problem

In a graph given is there is a clique of size K or more. A clique is a subset of nodes which are fully connected to each other.
This problem is NP problem. Given a set of nodes you can very easily find out whether it is a clique or not.
For example:

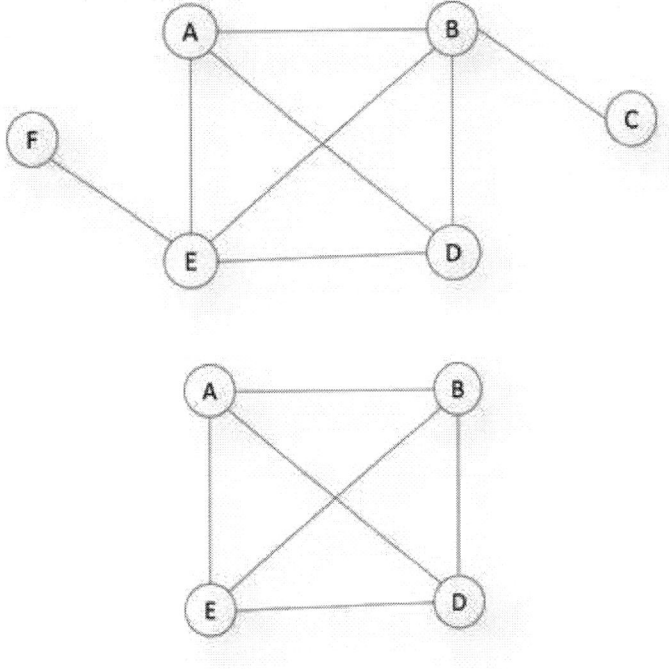

Clique of size 4

Prime Number

Finding Prime number is NP. Given a solution, it is easy to find if it is a Prime or not in polynomial time. Finding prime numbers is important as cryptography heavily uses prime numbers.

```
boolean isPrime(int n){
    boolean answer = (n>1)? true: false;

    for(int i = 2; i*i <= n; ++i)
    {
        if(n%i == 0)
        {
            answer = true;
            break;
        }
    }
    return answer;
}
```

Checking will happen till the square root of number so the *Time Complexity* will be $O(\sqrt{n})$. Hence prime number finding is an NP problem as we can verify the solution in polynomial time.

Graph theory have wonderful set of problems
- Shortest path algorithms?
- Longest path is NP complete.
- Eulerian tours is a polynomial time problem.
- Hamiltonian tours is a NP complete

Class co-NP

Set of problems for which there is a polynomial time checking algorithm. Given a solution if we can check in a polynomial time if that solution is incorrect the problem is co-NP problem.

Class co-NP Definition: The class co-NP contains all decision problems such that there exists a polynomial time proof that can verify if the problem does not have the right "yes/no" answer.

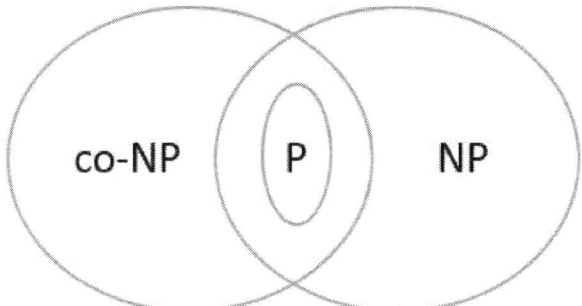

Relationship between P, NP and co-NP

Class P is Subset of Class NP

All problems which are P also are NP ($P \subseteq NP$). Problem set P is a subset of problem set NP.

Searching

If we have some number sequence a1, a2, a3…. an. We already know that searching a number X inside this array is of type P.

If it is given that number X is inside this sequence, then we can verify by looking into each and every entry again and find if the answer is correct in polynomial time (linear time.)

Sorting

Another example of sorting a number sequence, if it is given that the array b1, b2, b3.. bn is a sorted then we can loop through this given array and find if the list is really sorted in polynomial time (linear time again.)

NP–Hard:

A problem is NP-Hard if all the problems in NP can be reduced to it in polynomial time.

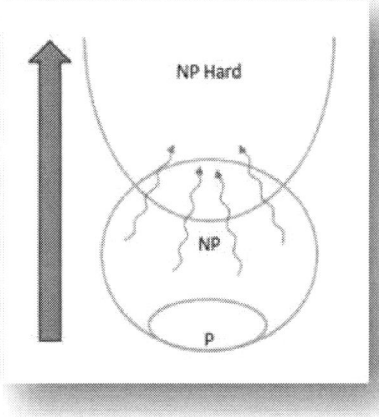

NP–Complete Problems

Set of problem is NP-Complete if it is an NP problem and also an NP-Hard problem.
It should follow both the properties:
1) Its solutions can be verified in a polynomial time.
2) All problems of NP are reduced to NP complete problems in polynomial time.

You can always reduce any NP problem into an NP-Complete in polynomial time. And when you get the answer to the problem, then you can verify this solution in polynomial time.

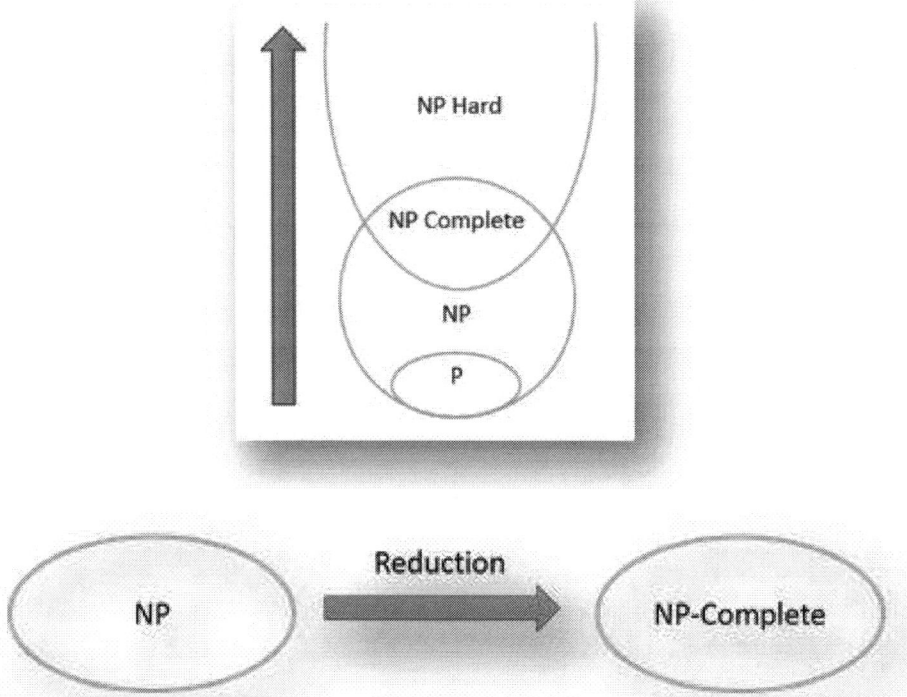

Any NP problem is polynomial reduced to NP-Complete problem, if we can find a solution to a single NP-Complete problem in polynomial time, then we can solve all the NP problems in polynomial time. But so far no one is able to find any solution of NP-Complete problem in polynomial time.

P ≠ NP

Reduction

It is a process of transformation of one problem into another problem. The transformation time should be polynomial. If a problem A is transformed into B and we know the solution of B in polynomial time, then A can also be solved in polynomial time.

For example,

Quadratic Equation Solver: We have a Quadratic Equation Solver, which solves equation of the form $ax^2 + bx + c = 0$. It takes Input a, b, c and generate output r1, r2.

Now try to solve a linear equation 2x+4=0. Using reduction second equation can be transformed to the first equation.

$2x+4 = 0x^2 + 2x + 4 = 0$

ATLAS: We have an atlas and we need to color maps so that no two countries have the same color. Let us suppose below is the various countries. And different pattern represents different color.

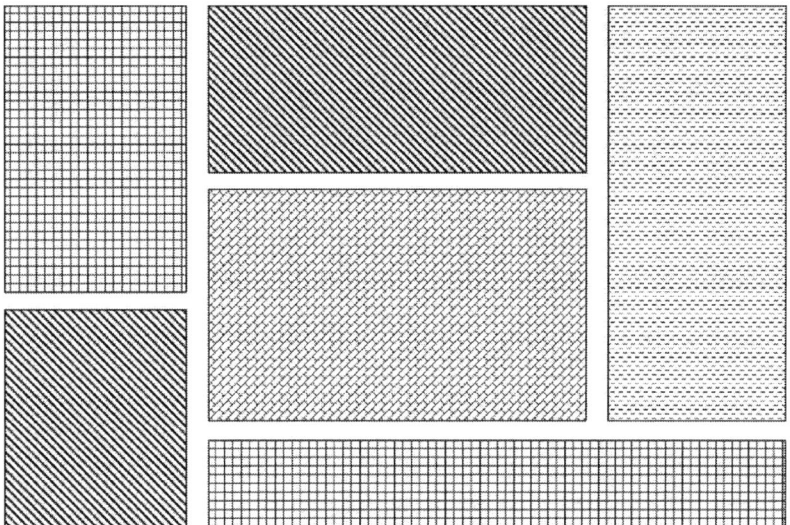

We can see that same problem of atlas coloring can be reduced to graph coloring and if we know the solution of graph coloring then same solution can work for atlas coloring too. Where each node of the graph represents one country and the adjacent country relation is represented by the edges between nodes.

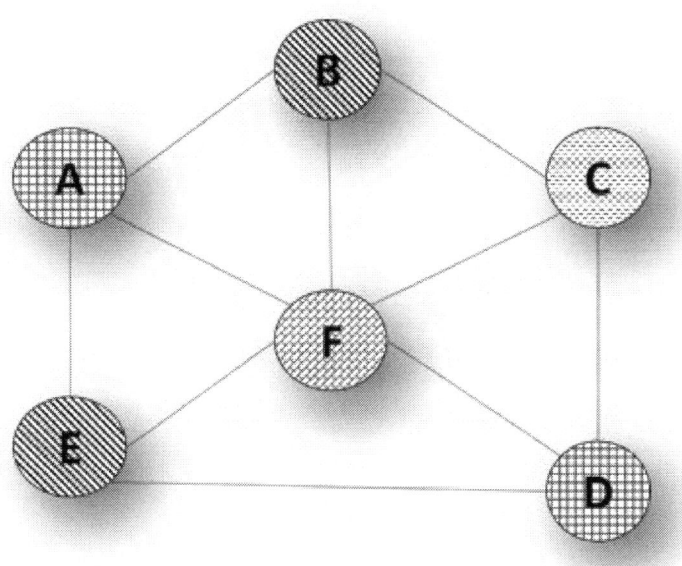

The sorting problem reduces (≤) to Convex Hull problem.
SAT reduces (≤) to 3SAT

Traveling Salesman Problem (TSP)

The traveling salesman problem tries to find the shortest tour through a given set of n cities that visits each city exactly once before returning to the city where it started.
Alternatively: Finding the shortest Hamiltonian circuit in a weighted connected graph. A cycle that passes through all the vertices of the graph exactly once.

```
Algorithm TSP
Select a city
MinTourCost = infinite
For ( All permutations of cities ) do
        If( LengthOfPathSinglePermutation < MinTourCost )
            MinTourCost = LengthOfPath
```

Total number of possible combinations = (n-1)!
Cost for calculating the path? $\Theta(n)$
So the total cost for finding the shortest path? $\Theta(n!)$

It is an NP-Hard problem there is no efficient algorithm to find its solution. Even if some solution is given, it is equally hard to verify that this is a correct solution or not. But there are some approximate algorithms which can be used to find a fairly good solution. We will not always get the best solution but will get a fairly good solution.

Our approximate algorithm is based on the minimum spanning tree problem. In which we have to construct a tree from a graph such that every node is connected by edges of the graph and the total sum of the cost of all the edges it minimum.

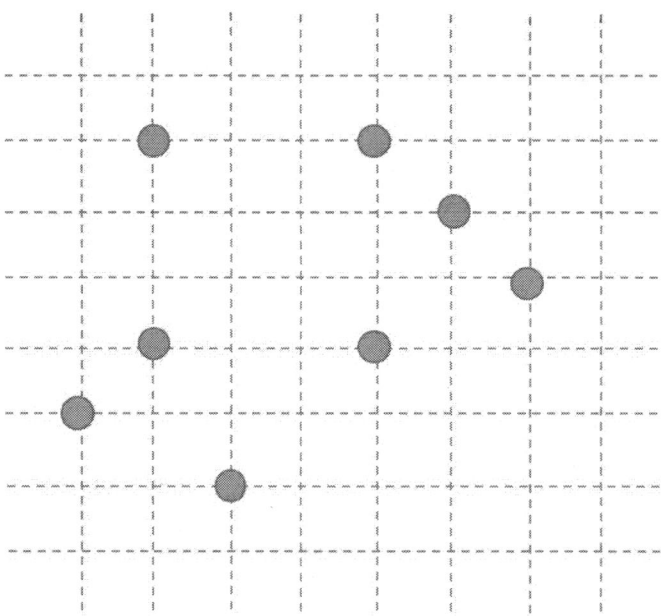

In the above diagram, we have a group of cities (each city is represented by a circle.) Which are located in the grid and the distance between the cities is same as per the actual distance. And there is a path from each city to another city which is a straight path from one to another.

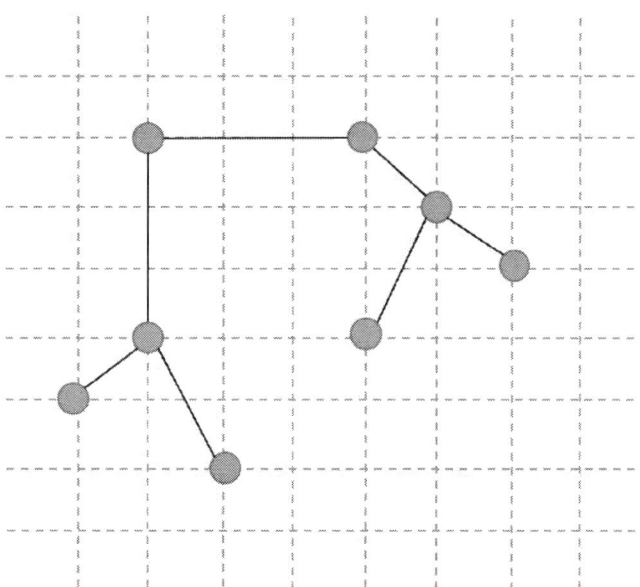

We have made a minimum spanning tree for the above city graph.

What we want to prove that the shortest path in a TSP will always be greater than the length of MST. Since all nodes are connected to the next node which is the minimum distance from the group of node so some node is removed and new nodes will be added to make it a path so TSP path will always be greater than MST.

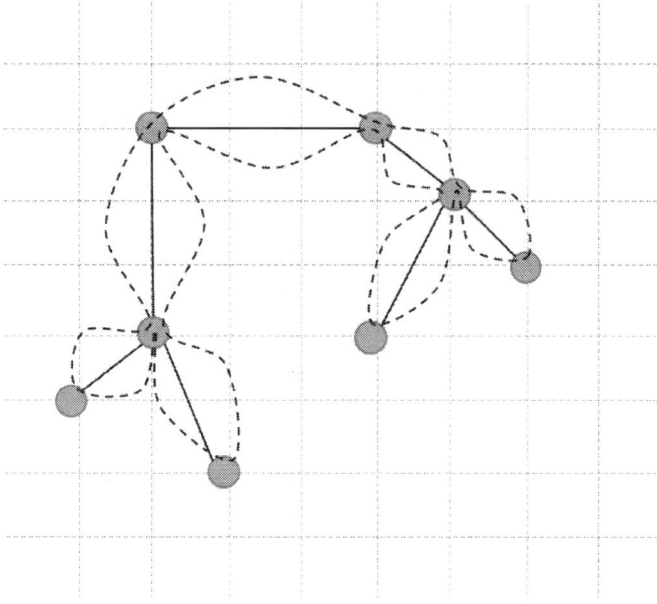

Now let's take a path from starting node and traverse each node on the way given above and then come back to the starting node. The total cost of the path is 2MST. The only difference is that we are visiting many nodes multiple times.

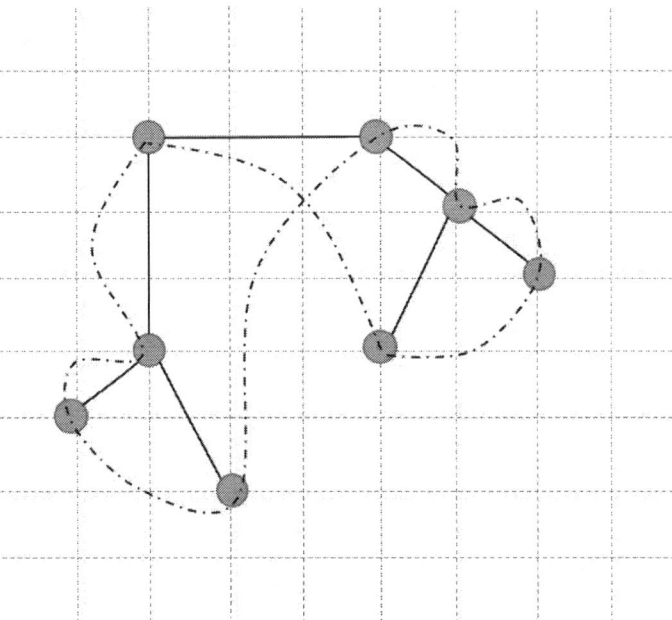

Now let's change our traversal algorithm so that it will become TSP in our traversal, we didn't visit an already visited node we will skip them and will visit the next unvisited node. In this way we will reach the next node by as shorter path. (The sum of the length of all the edges of a polygon is always greater than a single edge.) Ultimately we will get the TSP and its path length is no more than twice the optimal solution. So the proposed algorithm gives fairly good results.

End Note

Nobody has come up with such a polynomial-time algorithm to solve a NP-Complete problem. Many important algorithms depends upon it. But at the same time nobody has proven that no polynomial time algorithm is possible. There's a million US dollars for anyone who can do either solve any NP Complete problem in polynomial time. The whole economy of the world will fall as most of the banks depends on public key encryption will be easy to break if P=NP solution is found.

CHAPTER 22: INTERVIEW STRATEGY

Introduction

Success in tech interview depends on so many factors, your non-technical skills, your technical skills, etc. But above all the interviewers should be convinced that they would enjoy working with you.

Resume

The best resumes are those that communicate your skills and accomplishments in a clear and effective way.

A good resume format has the following attributes:
1. Multiple Columns: Multiple columns make it easier for someone to quickly skim your company name, positions, collage, and other key facts.
2. Short and Sweet: Interviewer is going to spend about 30 Sec reading your resume. You should just focus on the highlights. One page is all you need, but if you are 10+ years of experience, then you can justify two pages.
3. No Junk: No objective, No oath, Summary section/Key skills section may be fine, if your resume is short and concise then you don't need a summary section.
4. Use Tables: You can use tables, but it should not waste space.
5. Highlights: highlights should be short. Keep your highlights to one liner.
6. Neat: Keep your resume neat and clean. Use appropriate Fonts and Formatting. Bold to represent highlights and maybe italics in some places.

Nontechnical questions

Prepare for various non-technical questions. The first thing to do is to prepare answers of any question that is related to your resume. The interviewer is going to look into it and ask a few questions to get an idea about you. So go through all the past/current job and projects and make sure you know what they were about and your role.

These questions may be like:
1. What was the most challenging activity you have done in project ABC?
2. What did you learn from project ABC?
3. What are your responsibilities in the current job?
4. What was the most interesting thing you have done in your current job?
5. Which course in university did you like most and why?

Technical questions

Solving a technical question is not just about knowing the algorithms and designing a good software system. The interviewer wants to know you approach towards any given problem.

Many people make mistakes like they don't ask clarifying questions about a given problem? They assume a lot of things and begin working with that. Well the truth is the interviewer to actually expect you to ask constraints questions. There are a lot of data that is missing that you need to collect from your interviewer before beginning to solve a problem.

For example: Let us suppose the interviewer ask you to give a best sorting algorithm.
Some interviewee will directly jump to Quick-Sort **O(nlogn)**. Oops, mistake you need to ask many questions before beginning to solve this problem.

Questions:
1. What kind of data we are talking about? Are they *integer*s?
2. How much data are we going to sort?
3. What exactly is this data about?
4. What kind of data-structure used to hold this data?
5. Can we modify the given data-structure? And many, many more...?

Answer:
1. Yes, they are *integer*s.
2. May be thousands.
3. They store a person's age.
4. Data are given in the form of some array.
5. No you can't modify the data structure provided.

Ok from the first answer we will deduce that the data is *integer*. The data is not so big it just contains a few thousand entries. The third answer is interesting from this we deduce that the range of data is 1-150. Data is provided in an array. From fifths answer we deduce that we have to create our own data structure and we cannot modify the array provided. So finally we conclude, we can just use bucket sort to sort the data. The range is just 1-150 so we need just 151 capacity integral array. Data is under thousands so we don't have to worry about data overflow and we get the solution in linear time **O(N)**.

Chapter 23: System Design

System Design

The section we will look into questions in which interviewer asks to design a high-level architecture of any software system.

Note: - This is an advance chapter It may be that the user is not able to understand it completely. I would suggest that give it some time read the chapter and try to read online. The more time you give to this chapter the better understanding you will get. It may also help if you give multiple rounds of reading.

There are two kinds of questions in this and which will be asked depends on the type of companies. The first kind of questions is to design some kind of elevator system, valet parking system, etc. In this, the interviewer just wants to test how well you are able to design a system, especially how well your classes are interacting.

The Second kind of system design problems is more interesting, in which the interviewer asks you to design some kind of website or some kind of service or some API interface. For example, design google search engine or design some feature of Facebook like how friends mapping is done on Facebook, design a web-based game that allows 4 people play poker etc. They are interesting one and in this, the interviewer can ask about scalability aspect.

Now comes a question to our mind, how would you design google search engine in 10-15 minutes? Well, the answer is you can't. It took many days if not years by a group of a smart engineer to design google search engine. The interviewer is expecting a Higher-level architecture of the system that can address the given Use-Cases and Constraints of the problem in hand. There is no single right solution. The same problem can be solved in a number of ways. The most important thing is that you should be able to justify your solution.

System Design Process

Let's look into a 5 Steps approach for solving system design problems:
1. Use Cases Generation
2. Constraints and Analysis
3. Basic Design
4. Bottlenecks
5. Scalability

Use Cases

Just like algorithm design problems, the system design questions are also most likely weakly defined. There is so much information that is missing and without them the design is impossible. So first thing in the design process is that you should gather all the possible use cases. You should ask questions to the interviewer to find the use case of the system. The interviewer wants to see your **requirement gathering capability**. Same as algorithm questions never assume things, which are not stated.

Constraints and Analysis

This is the step in which you will define various constraints of the system and then analyse them. Your system design will depend on the analysis that you do in this step. In this step, you need to find answers to questions like. How many users will be using the system? What kind of data that we are going to store? Etc.

Basic Design

In this step, you will design the most basic design of the system. Draw your main components and make connections between them. In this step, you need to design a system with the supposition that there is no memory limitation and all data can fit in one single machine. You should be able to justify your idea. In this step, you need to handle all the use-cases.

Bottlenecks Analysis

In this step, you will find the one or more bottlenecks on the basic design you had proposed. The "Scalability Theory" given below will help to identify the bottlenecks. You need to know the below theory which experts had developed over time. In this step, you will consider how much data your proposed system can handle, memory limitations etc.

Scalability

In this step, you will remove all the bottlenecks of the system and you are done. There may be multiple iterations between "Bottlenecks analysis" and "Scalability" until we reach our final solution. We will be reading various concepts like Vertical scaling, Horizontal scaling, Load-Balancer, Redundancy and Caching in this chapter. "Scalability Theory" given below will help you to understand these concepts.

Scalability Theory

In this section, we will be designing a generic web server, which will be handling a large number of requests. You can imagine it as some sort of website like Facebook in which large number of users are accessing it.

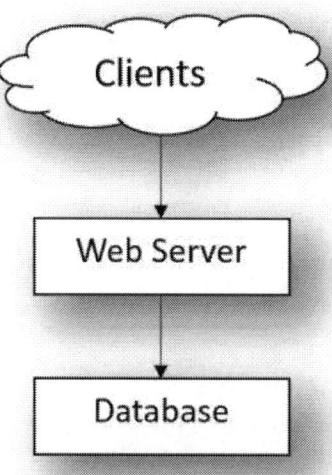

Vertical scaling

Vertical scaling means that you scale by adding more resources (Higher speed CPU, More RAM etc.) To your existing machine.

Vertical scaling has its own limit it can help you to handle more load, but until its limit is reached, then we have to go for horizontal scaling.

Horizontal scaling

Horizontal scaling means that you scale by adding more machines to your pool of resources.

Distribute the request by distributing the request among more than one web server. In doing this we need to have a load balancer, which will distribute the request among the servers.

Load Balancer (Application layer)

Load balancer has to decide which server should serve the next request. So distributing the load can be made using different strategies:

1) **Round Robbin**: Round robin is the way of distributing requests in a sequential fashion. The request is sent to the server 1 then the next request is sent to server 2 and so on till we reach the end of the server list. Then when we reach the end, it is sent again to server 1. Round robin has a problem that a server, which is already busy, may get another request. Round robin also has a problem with sticky sessions. We want that a request to be sent to the same server the next time.

2) Another approach is to select server corresponds to the hash value of the data. Find the hash value of the data, mod the hash value by the number of servers. Assign the job to a machine whose value we got after mod. Stick session problem is already solved in hash value approach. However, the problem of uneven load distribution is there, there is possible to have a more load sent to a server, which is already busy.

3) May be the load balancer know, how much load each server has or how busy each the server is. Moreover, will send the next request to the least busy server.

4) The server can be a specialized one serving image some serving video and some serving other data.

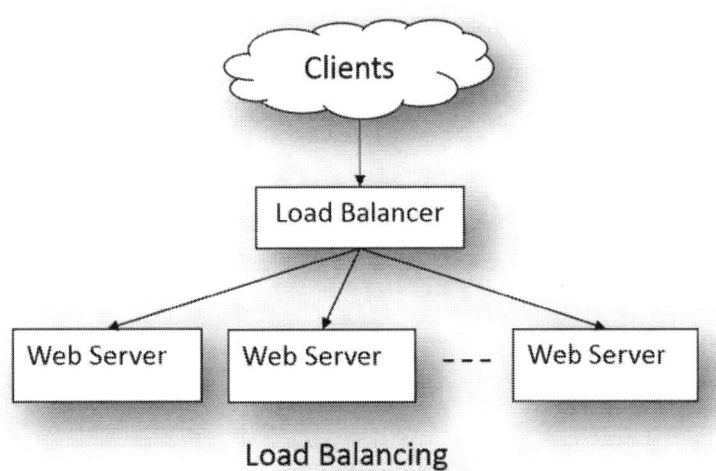

Load Balancing

Problems of load Balancing

Consider a customer who had selected some items in his buy cart on Amazon. When he selects another item then it should be added to the same cart so it should be sent to the same server. Also, the user profile which is saved to one server and if the user request reaches the other server his profile will be empty this is also not a good idea.

This problem can be solved by making the load balancer decide that a particular user request would always go to the same server. The user profile and cart details should be saved in some database.

Stick session: same sessions should lend to the same server. How to get this done. The first approach is that we store the IP address returned by load balancing into a cookie and then use this IP address in the subsequent requests. However, this reveals the IP address of the server to the world that we do not want. Therefore, another solution is that we use some session id that is a number that the load balancer knows that belongs to which server. By this, we are preventing our servers being exposed to the outer world and prevent it from being attacked.

Load Balancer (Database layer)

1. The most basic approach is **Round Robin**. Data is distributed in a circular fashion. First, data go to the first database, the second will go to the second database and so one. Each database server had an equal load. However, it has a disadvantage that the data lookup is complex. And need a large lookup table.

2. Another approach is to divide the data in such a way that all the data will go to the first machine until it reaches its maximum capacity. When maximum capacity is reached, then data goes to the second machine and so on. This approach has an advantage that only the required number of machines is used. However, it has a disadvantage that the data lookup is complex. And need a large lookup table.

3. Another approach is to select database corresponds to the hash value of the data. Find the hash value of the data. Mod the hash value by the number of databases. The data are then stored in the database value we got after modulus. For has a value approach we do not require any lookup table. We can find the database, which is storing the data, by finding the hash value. However, the problem of uneven distribution of data is there, there is possible to have a more data sent to a database, which has already reached its maximum capacity. In this case, we need to find a better load-balancing key or split the data from the database into a number of databases.

4. In the hash value, based distribution of data there is no relation between the data that is stored in a particular database. Information about the data can be used to make the database accessible faster. For example, in social networking like Facebook, if someone who lives in India is more likely to have friends from India. And someone who lives in the USA is more likely to have friends in the USA.

5. Perhaps location aware (approach 4) and the hash value based (approach 3) distribution of data may be the best approach to keep the data so that it can take advantage of both the approaches. Country code and user ID can be used to get the location of the database.

Redundancy

There is one problem in our system, there is a redundancy in the servers but our load balancer is now our single point of failure. We add a secondary load balancer in case the primary load balancer dies, then secondary load balancer becomes primary and then all the requests will be handled by it.

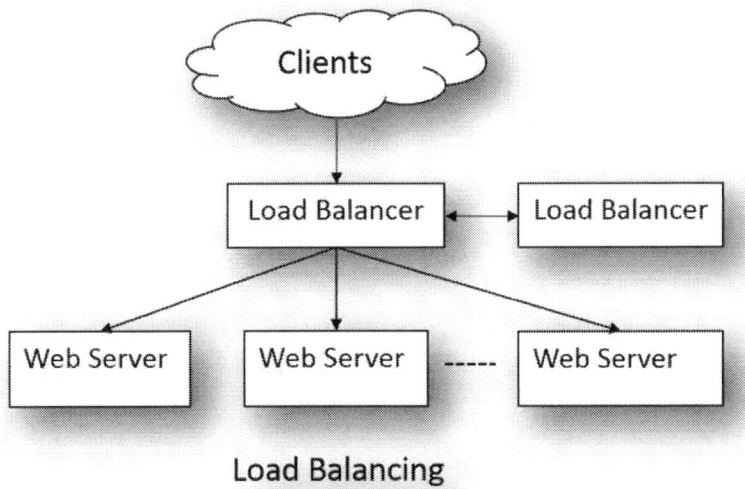

Load Balancing

Raid (Redundant Array of Inexpensive Disk): Raid is a technology to create redundancy in the databases. Multiple hard drives are used to replicate data, thereby proving redundancy.

Caching

A cache is a simple key-value store and it should reside as a buffering layer between your application and your data storage. Whenever your application wants to read, it first tries to retrieve the data from your cache. Only if data is not present in the cache, then only it tries to get the data from the main database.

Caching improves application performance by storing portion of data in memory for low-latency access. We need databases and access to the database is slow, so we use multiple types of caching to make our system faster. Database servers itself does caching so do the other entities in between the user and the database.

Memcached: It is a server software, what it does is it kept whatever you access in memory. It can run on the same server as the webserver or it can run on a separate machine all together.

Redis: It is a data structure server based on **"NoSQL"** which is a key-value data store. Data is stored as the value with respect to corresponding key. This data is later retrieved by the use of the key. Redis is used for caching it is best to store the whole object as one instance so that the data can be accessed in parallel and data expiration will flush out the whole object.

There is a problem since ram is finite, then the cache will get full. The expired object will be removed so everything that is accessed then its expiry will be reset and if there is an object that is not used for some time then it was deleted. Cache is more important when the website that we are designing is more read heavier than a write.

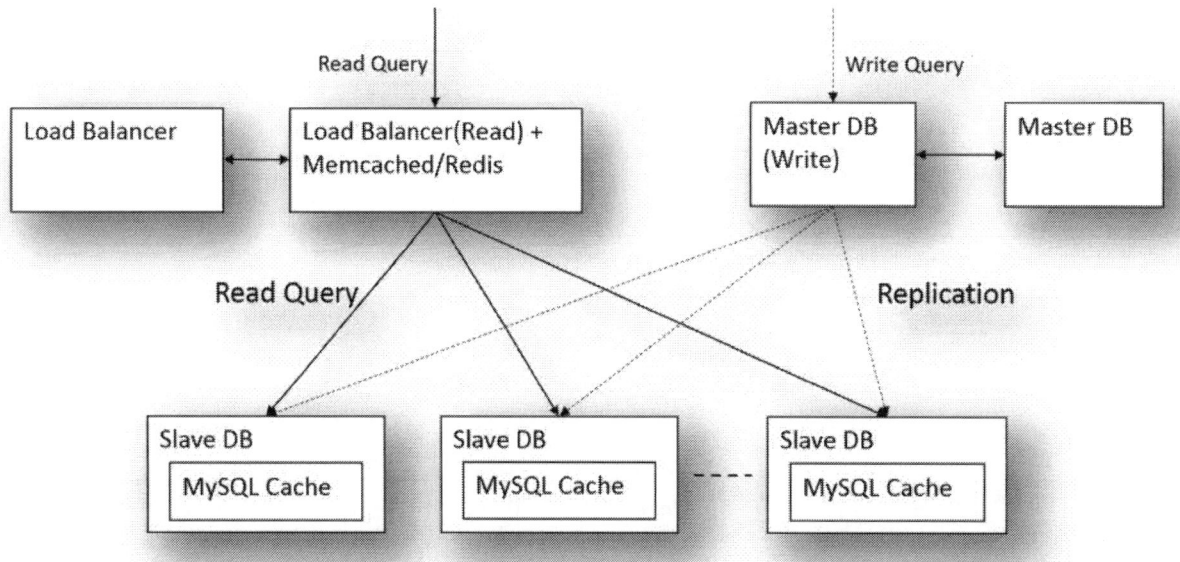

A complete web server implementation

The summary of the above system.
1. The Web-Servers of scalable web service is hidden behind a load balancer. The load balancer evenly distributes load across all the servers.
2. The user should get the same result from web-server regardless which server is actually serving the request. Therefore, every server should be identical to each other. Servers should not contain any data like session information or user profile.
3. Session need to be stored in a centralized data store (DB) which is accessible to all the servers. Data can be stored in some external database. Redundancy in the database is provided by raid technology.
4. The database is slow, so we need a cache. In-memory based cache like Redis or Memcached.
5. However, the cache has a problem of expiring. When a table changes, then the cache is outdated.
6. For Memcached there are two options:
a. We can save queries to the DB
b. We can save the whole object that will keep us close to web-server.
7. CDN (Content delivery networks) can be used to provide a pre-processed web page.
Below diagram will give you a complete picture of the whole system.

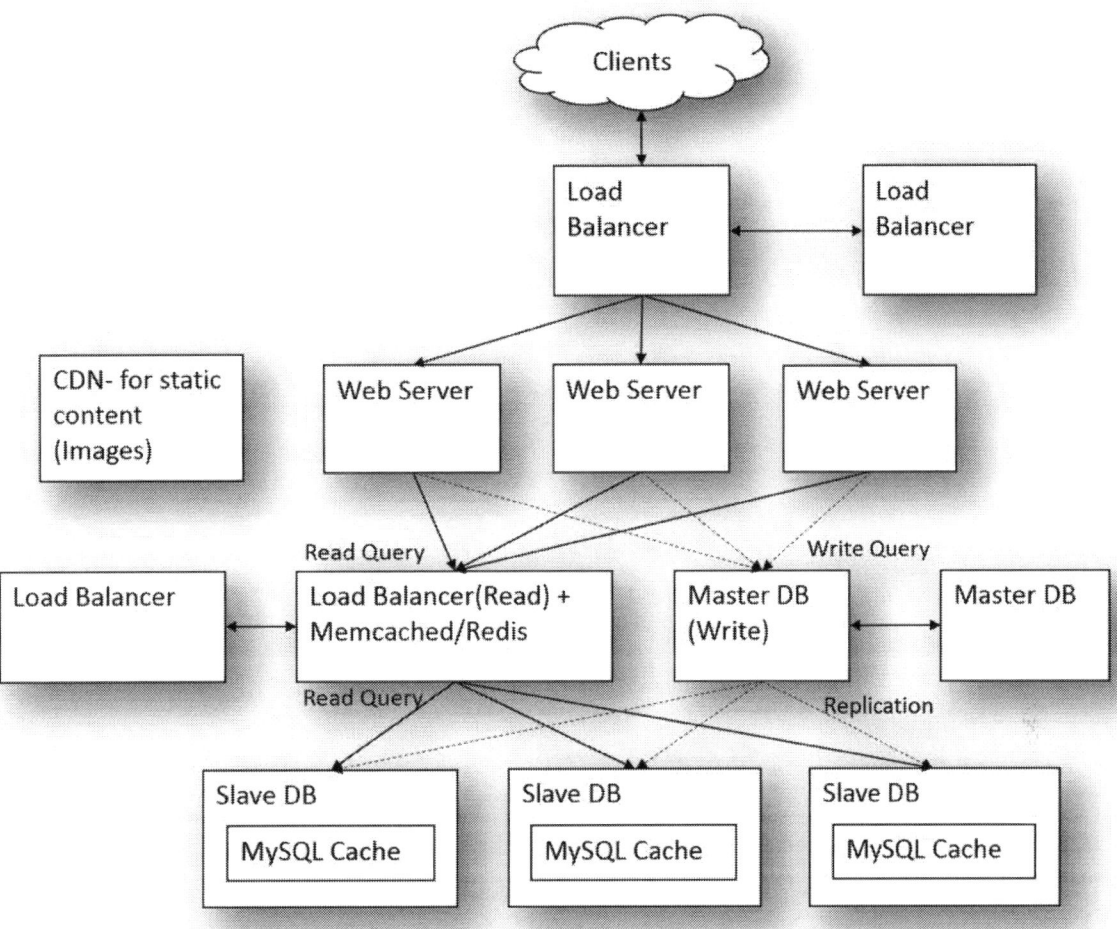

Design simplified Facebook

Design simplified Facebook where people can add other people as friends. In addition, where people can post messages and that messages are visible on their friend's page. The design should be such that it can handle 10 million of people. There may be, on an average 100 friends each person has. Every day each person posts some 10 messages on an average.

Use Case

1. A user can create their own profile.
2. A user can add other users to his friend list.
3. Users can post messages to their timeline.
4. The system should display posts of friends to the display board/timeline.
5. People can like a post.
6. People can share their friends post to their own display board/timeline.

Constraints

1. Consider a whole network of people as represented by a graph. Each person is a node and each friend relationship is an edge of the graph.
2. Total number of distinct users / nodes: 10 million

3. Total number of distinct friend's relationship / edges in the graph: 100 * 10 million
4. Number of messages posted by a single user per day: 10
5. Total number of messages posted by the whole network per day: 10 * 10 million

Basic Design

Our system architecture is divided into two parts:
1. First, the web server which will handle all the incoming requests.
2. The second database, which will store the entire person's profile, their friend relations and posts.

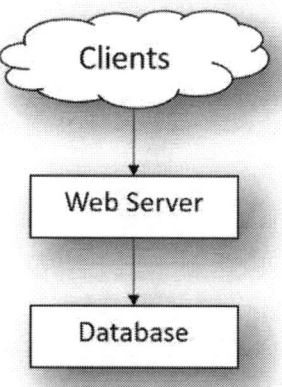

First, three requirements creating a profile, adding friends, posting messages are written some information to the database. While the last operation is reading data from the database.

The system will look like this:
1. Each user will have a profile.
2. There will be a list of friends in each user profile.
3. Each user will have their own homepage where his posts will be visible.

A user can like any post of their friend and that likes will reflect on the actual message shared by his friend.
If a user shares some post, then this post will be added to the user home page and all the other friends of the user will see this post as a new post.

Bottleneck

A number of requests posted per day is 100 million. Approximate some 1000 request are posted per second. There will be an uneven distribution of load so the system that we will design should be able to handle a few thousand requests per seconds.

Scalability

Since there is, a heavy load we need horizontal scaling many web servers will be handling the requests. In doing this we need to have a load balancer, which will distribute the request among the servers.

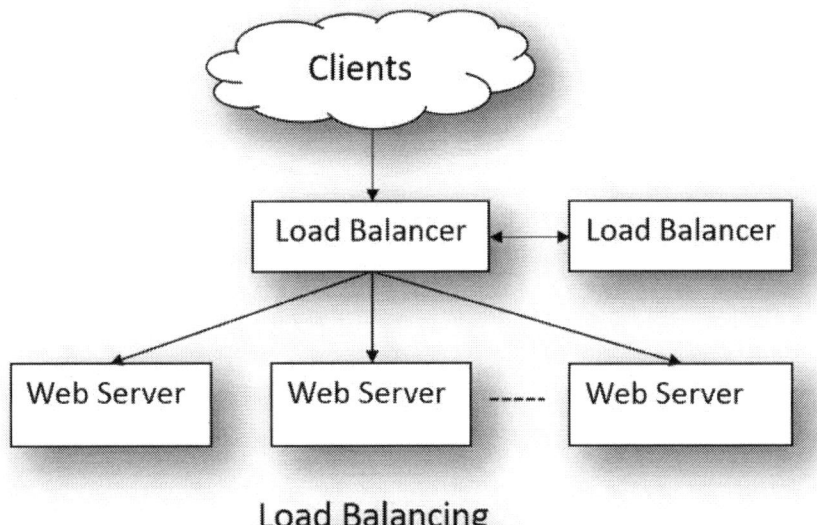

Load Balancing

This approach gives us a flexibility that when the load increases, we can add more web servers to handle the increased load.

These web servers are responsible for handling new post added by the user. They are responsible for generating various user homepage and timeline pages. In our diagram, the client is the web browser, which is rendering the page for the user.

We need to store data about user profile, Users friend list, User-generated posts, User like statues to the posts.

Let us find out how much storage we need to store all this data. The total number of users 10 million. Let us suppose each user is using Facebook for 5 to 6 years, so the total number of posts that a user had produced in this whole time is approximately 20,000 million or 20 billion. Let us suppose each message consists of 100 words or 500 characters. Let us assume each character take 2 bytes.

Total memory required = 20 * 500 * 2 billion bytes.
                              = 20,000 billion bytes
                              = 20, 000 GB
                              = 20 TB

1 gigabyte (GB) = 1 billion bytes
1000 gigabytes (GB) = 1 Terabytes

Most of the memory is taken from the posts and the user profile and friend list will take nominal as compared with the posts. We can use a relational database like SQL to store this data. Facebook and twitter are using a relational database to store their data.

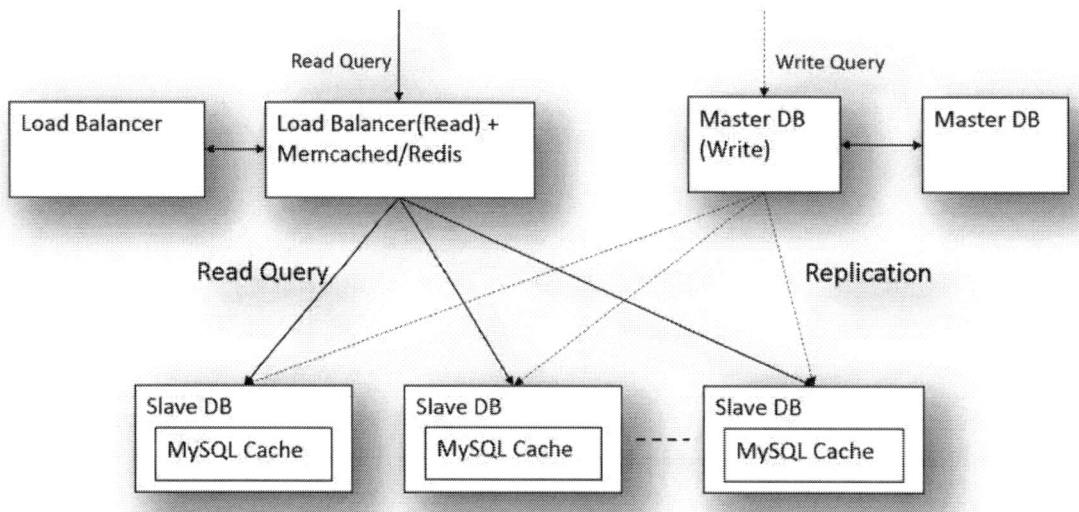

Responsiveness is key for social networking site. Databases have their own cache to increase their performance. Still database access is slow as databases are stored on hard drives and they are slower than RAM. Database performance can be increased by replication of the database. Requests can be distributed between the various copies of the databases.

Also, there will be more reads then writes in the database so there can be multiple slave DB which are used for reading and there can be few master DB for writing. Still database access is slow to we will use some caching mechanism like Memcached in between application server and database. Highly popular users and their home page will always remain in the cache.

There may be the case when the replication no longer solves the performance problem. In addition, we need to do some Geo-location based optimization in our solution.

Again, look for a complete diagram in the scalability theory section.

If it were asked in the interview how you would store the data in the database. The schema of the database can look like:

Table Users	Table Posts
<ul><li>User Id</li><li>First Name</li><li>Last Name</li><li>Email</li><li>Password</li><li>Gender</li><li>Birthday</li><li>Relationship</li></ul>	<ul><li>Post Id</li><li>Author Id</li><li>Date of Creation</li><li>Content</li></ul>
Table Friends	Table Likes
<ul><li>Relation Id</li><li>First Friend Id</li><li>Second Friend Id</li></ul>	<ul><li>Id</li><li>Post Id</li><li>User Id</li></ul>

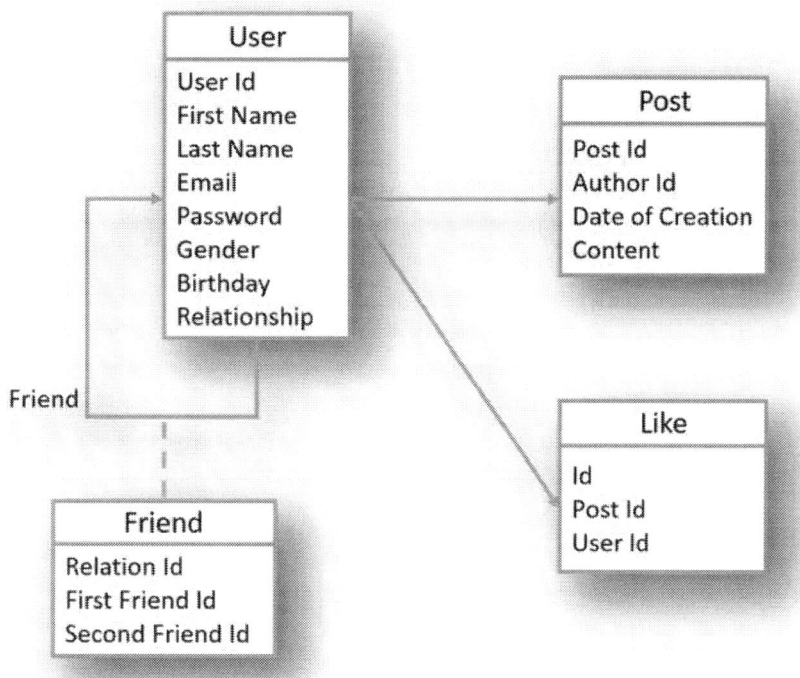

Design Facebook Friends suggestion function

Design a system to implement a friend suggestion functionality of Facebook, with millions of users. The algorithm should suggest all the friends of the immediate friends as a proposed list to add as friends.

Use Case

The system should suggest friends of the friends as suggested new friends.

Constraint

Millions of user's lot of data with billions of relations.

Basic Design

Forget about millions of users. Just consider there are only a few persons and they are connected with each other as friends.

Consider that people are represented by vertices of graphs and their friendship relation is represented by edges.

Since there are only a few people, then we can keep everything in memory and find the friend suggestion using Breadth First Traversal.
We just need to find the nodes, which are just 2 degrees apart from the starting node.

Bottleneck

Since there are millions of users, we cannot have everything in memory. Since there are millions of users, we cannot keep the data on one machine. One friends' profiles may lie on many different machines.

Scalability

Since there are millions of users, their user profile is distributed among many different database servers. User profiles can be distributed depending upon Geo-Location. The Indian users profile will lie in a server located in India and US citizen's profile lie in the server located in the US.

Each user will have corresponding User Id associated with them. Some portion of ID can be used to get Geo location of the user. Another portion of user id can find the user profile on that server.

The user profile is not that frequently updated so there is more read than write. So single master writer - multiple slave reader architecture is most suitable for this application.

The application server can process the data; it can do the optimization to query less from the database by accumulating user list to be processed.

```
public class system {
    private map<int, int> personIdToMachineIdMap;
    private map<int, Machine> machineIdToMachineMap;

    Machine getMachine(int machineId);
    Person getPerson(int personId)
    {
        int machienId = personIdToMachienIdMap[personId];
        Machine m = machineIdToMachineMap[machienId];
        return m.getPersonWithId(personId);
    }
}
```

Optimization: Reduced the number of jumps by first finding the list of friends whose profile is on the same machine. Then send the find next degree friends query which will return the list of next level friends. By doing, this work is distributed among various machines. Finally, the result of the various queries will be merged and then suggested the friends list.

Better result: You can calculate the degree of the friends with the friend list. The person who is a friend of many of my friends is more likely to be my friend than the person who is just a friend of one of my friends. We need to keep track of the friend reference counts by keeping Hash-Table for the friend list and make the count 1 whenever we find a new person otherwise increase the count by 1.

If we want to take advantage of caching, then we need to add some database cache in between.

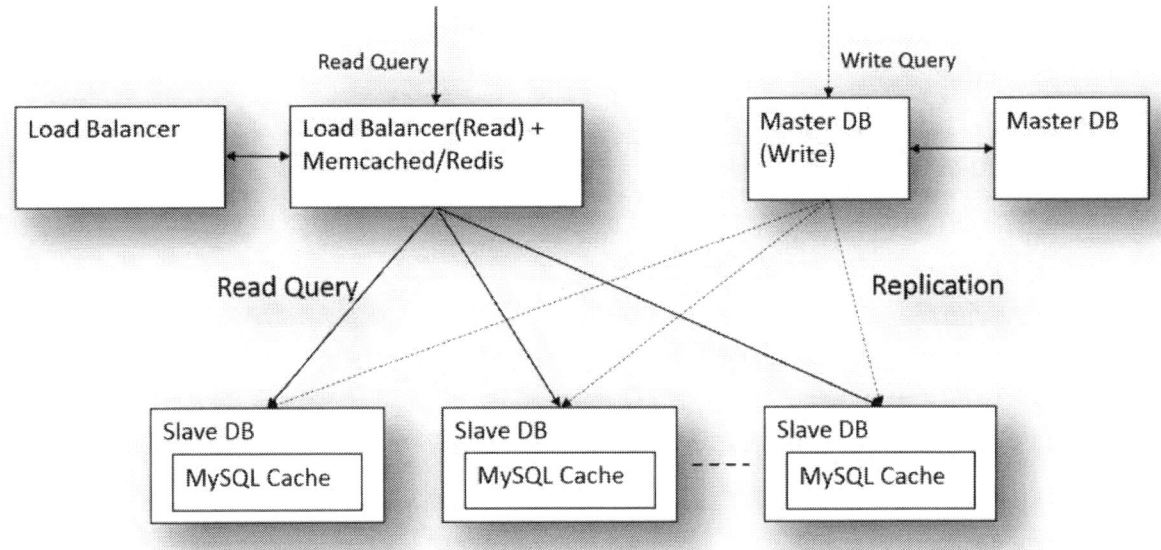

There can be multiple web servers, which will be querying the databases, and there will be multiple users who are accessing their Facebook profile and each one of them is proposed with new friends list so the final architecture is again same as the one proposed in the complete web server implemented in scalability theory.

Design a shortening service like Bitly

Use Case

Basic use case:
1. Shortening takes a URL and returns a short URL.
2. Redirection takes a short URL and redirects to the original URL.
3. Custom URL.
4. High availability of the system.

Additional use cases:
1. Analytics
2. Automatic link expiration.
3. Manual link removal.
4. Specific company URL.
5. UI or just API

Requirement Analysis/ Math
Firs we need to find the usage pattern.
You can directly ask this data from the interviewer or you can derive it using some data that the interviewer provides. Let us suppose that the interviewer tells that there will be 1 billion requests per month. In addition, out of these 10% times, it is a new request and 90% of the time, it is a redirection of the already shortened URL. Let us write down the data that we get.

1. 1BN requests per month
2. 10% are for new URL/shortening and 90% are for redirection.
3. New URLs per month is 100MLN
4. Requests per second 1BN/ (30*24*3600) = 385. Roughly, you can assume it 400 requests per seconds.
5. Total number of URLs stored in 5 years.
5* 12* 100 MLN = 6BN URLs in 5 years.
6. Let us suppose the space required by each URL is 500bytes.
7. Let us suppose the space required by each Hash code for corresponding URLs is 6byte long.
8. Total data we need to store in five years. 3TBs for all the URLs and 36gb for hashes
    6,000,000,000 * 500 bytes = 3 terabytes
    6,000,000,000 * 6 bytes = 36 gigabytes
9. New data write requests per second: 40 * (500+6): 20k

Basic design

Web server: provide the website for the Bitly service where users can generate the short URL.
Application Server: provides the following services:
1. Shortening service
2. Redirection service
3. Key = Hash Function (URL)

Database Server:
1. Keep track of hash to URL mapping.
2. Works like a huge Hash-Table stores the new mapping and retrieves old mapping given key.

Bottleneck
1. Traffic is not much
2. Data storage can be a problem.

Scalability

Application Server:
1. Start with the single machine.
2. Test how far it takes up.
3. Do a vertical scaling for some time?
4. Add load balancer and a cluster of machines to handle spikes and to increase availability.

Data Storage:
1. Billions of objects
2. Each object is fairly small
3. There is no relationship between objects
4. Reads are more than write.
5. 3TBs of URLs and 36GB of hash.

MySQL:
1. Widely used
2. A mature technology
3. Clean scaling paradigms (master/slave, master/ master)
4. Used by Facebook, google, twitter etc.
5. Index lookup is very fast.

Mappings: <Hash, URL>
1. Use only MySQL table with two fields.
2. Create a unique index on the hash we want to hold it in memory to speed up lookups.
3. Vertical scaling of MySQL for a while
4. Partition of data into many partitions
5. Master-slave (read from slave and write to master.)
6. Eventually, partition the data by taking the first character of the hash mod the number of partitions.

Stock Query Server

Implement a stock query service that can provide an interface to get stock price information like open price, close price, highest price, lowest price etc. You should provide an interface that will be used to enter these data and interface to read this data.

Use Case

There will be two interfaces to this system.
1) First interface to add daily stock price information to the system.
2) Second interface to read stock price information giving the date and stock id as input.

Constraints

Let us suppose the system will be used by thousands of users.
For each stock, there will be only one write operation per day. However, there will be any number of read operations that can happen per stock so the application is more read heavy then write heavy.
The solution should be flexible enough so that if new data fields need to be added to the stock they can easily be added to the system.
The solution provided should be secure.

Basic Design

We can use a database like SQL to store stock data. Client can access the database using the web server interface. Below diagram will show the basic architecture.

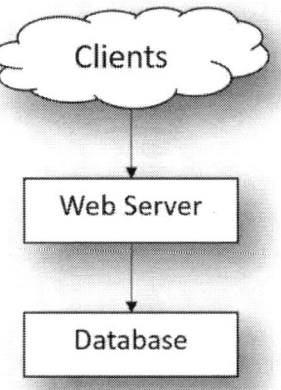

In the above architecture, the user can access the database using web service. Any number of flexibility can be provided for the use. For example, what is the max price of some stock in 6 months etc. At the same time, the user does not have access to the data they should not have. We can provide different access of read and write depending on the normal users or administrator. Well-defined rolling back, backing up data and security features are provided by the SQL database. The above architecture is easily extendable to use with a website or some mobile application.

Scalability

Since we have 1000's of users, then having a single web server and a single database is not extendable. We need to distribute data among N number of Databases, which sit behind some load balancer. In addition, multiple N number of web server which will sit behind some load balancer. Each of the load balancers needs to be provided with some redundancy as they will be a single point of failure. Finally, the solution will look like below diagram. (For details, see scalability theory explained earlier in this chapter)

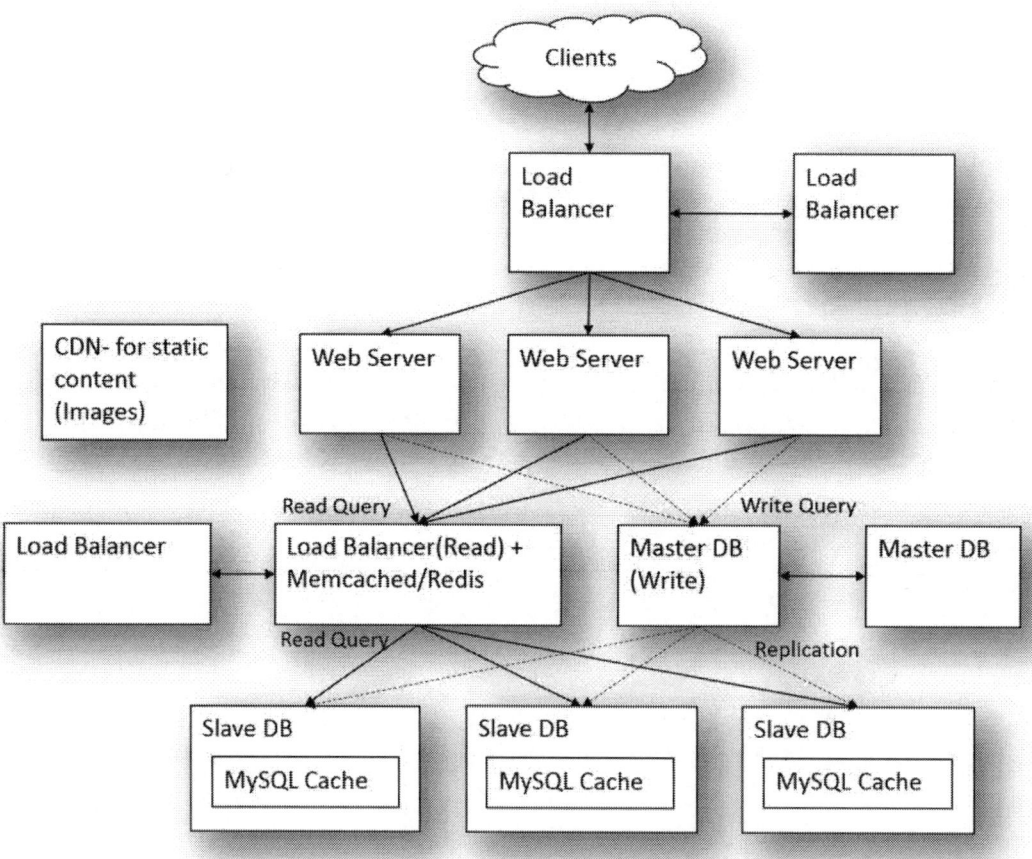

Design a basic search engine Database

You are given millions of URLs; how would you create your database. So that given a query string of words, how to find the URLs, which contain all the words of the query string. The words can come in any order.

Use Case

1) We are given a list of millions of URLs.
2) The user of the system will provide query string. And we need to return the URLs, which contain all the words of the query string.
3) It is some kind of search engine so we can pre-process the data and make our database.

Requirement Analysis

In the requirement step, you need to find out how many users are going to use this search engine. In our case, let us suppose there are not many users who are going to use our system so the sour main concern is a database.

Maybe we have N number of machines that can be used to fast our data pre-processing.

Basic Design

In this step, we will make the basic design so let us make a working system with just a few URLs. How would you find the required URL from the given URLs, which contains all the words of the input query string?

We can make a Hash-Table in which words are the keys and document ids are values.

"Hello" -> {url1, url2, url3}

"World" -> {url2, url4, url5}

To search the document, which contains "hello world", we can find the intersection of the two lists. In addition, url2 is the result.

Bottleneck

In this step, we will look back to our original problem in which there are millions of URLs that we need to pre-process. There may be a number of different words so it may not be possible to keep the whole Hash-Table on a single machine. Therefore, we need to divide the Hash-Table and keep it on a separate machine.

We need to retrieve the URLs that match a given word efficiently. So that we can find the intersection. Pre-processing all the millions, URLs by single machine will be slow. We need to find a way to parallel process pre-processing step.

Scalability

Let us look into the problem of keeping the Hash-Table in different databases. One solution is to divide the words alphabetically. We can make tables corresponding to each word. Each database contains tables of words under some range. For example, DB1 contains all the words, which start with alphabet "a", and DB2 contains all the words, which starts with the alphabets "b" and so on. Data is stored in the database and when a database reaches its maximum capacity, then the data is stored to next machine and a tree kind of structure can be made. Finding the list of URLs corresponding to some word is easy, we can go to the corresponding database and find the table and get all the data of that table. Finally, we can take the intersection of the result of various words. In addition, the result will be given as output.

Processing of the millions of URLs with a single machine is slow. Therefore, we can divide a bunch of URL processing among an N number of machines, each URL processing is independent of each other and the final Hash-Table of the URLs can be finally combined. This approach of processing independent data and finally combining their result is used in MapReduce.

MapReduce: A MapReduce divides the input dataset into independent chunks, which are processed in parallel. Then their output is combined to get the result.

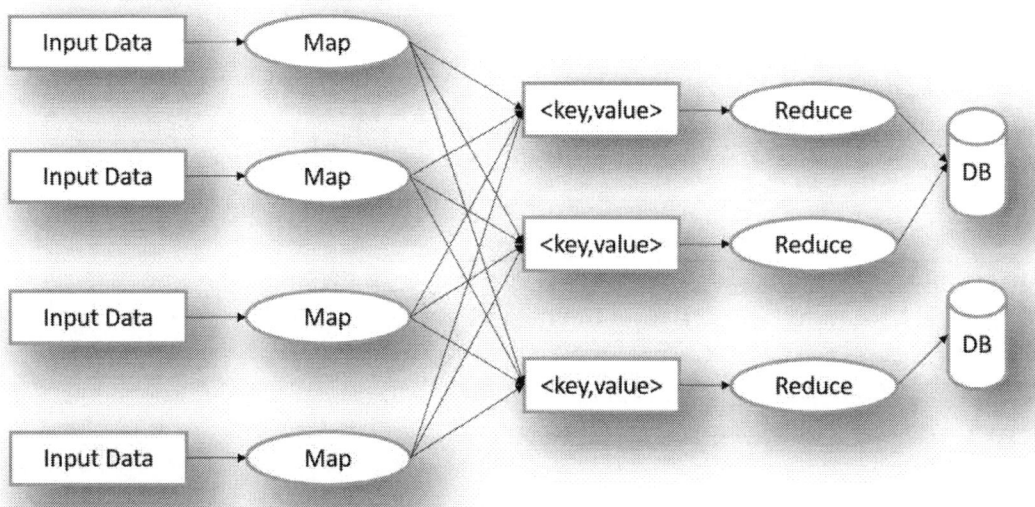

Duplicate integer in millions of documents

Given millions of documents with all distinct numbers, find the number, which occurs multiple times.

Basic Design

Consider there are just a few numbers, and we want to find the duplicate numbers.

The first approach is to find keep a sorted list of the numbers and see if the next read number matches with some number in the list.

Another better approach is to find a hash value corresponding to number and add that the number to a Hash-Table.

Constraint

Millions of documents and there is no range of number so we cannot keep everything in memory.

Scalability

We can find the hash value for all the integers and then add that integer to its corresponding hash value file or database. If there is some duplicate, then they will fall in the same file. In the first pass, various files are created and integers are distributed.

In the second pass all, the data of the individual files can be loaded into memory and sorted to find if there is some duplicate value.

We can use the same technique explained above to process the various documents of integer in parallel by different machines and then combine their output to get our result faster.

Design a basic search engine Caching

Given a search engine database implementation which supports QuerySearch() function which will return the best list of URLs based on the words of the query. This time you need to design the web server implementation of this such that there are N number of the web servers which are responding to the user queries. Any web server can be picked at random. QuerySearch() is a heavy operation so you need to design, caching for this system so that database access is reduced.

Use Case

1) The user of the system will provide query string and system will respond with the proper list of URLs corresponding to his request.
2) Given that, the database operations are very heavy we need to minimize them by caching the queries at the web server.
3) We need to keep the frequent queries in the cache and stale queries need to be removed from the cache.
4) We need to have some proper refresh mechanism for each query.

Basic Design

Let us lust forget about an N number of machines and assume that QuerySearch() operation happens on a single machine. Now we would like to cache queries. Each query will consist of some string. And the result of a query is a list of URLs.

We need to have a quick cache lookup so that we can get the result of the Query from the cache if it is present there. Also, need to have some proper refresh mechanism for each query.
The Hash-Table is most effective to keep the cache. By using a hash, a table lookup is fast. However, if the cache is filled how you would remove the least used data from the cache.

A linked list can be used to remove the old data. You can keep a double linked list to manage the old data removal. Whenever a data is accessed, it can be moved to the front of the linked list and the removal can happen from the end of the linked list.

Taking advantage of both the solutions, we can keep the cache in a linked list and add its reference to the Hash-Table.

Now the last problem of how to remove the data upon the expiry of it. For example, most frequently accessed query result will always remain in the linked list even though that result is changed and if it is accessed again from the database then it will give some updated result. For this, we need to have some TTL (time to live) associated with each query depending upon the result of URLs we get from each query. For example, some weather or current news related queries should have a TTL of days, on the other hand, some historic data should have a long TTL. The TTL can be derived from how frequently the URLs are changing in the query result.

Bottleneck

There are N different web servers. And any particular query can be served by any server.
Data access should be fast.

Scalability

The various solutions that we can think about are:

Approach 1:
Servers can have their own cache. If some query is sent to Machine1 it will catch it in its cache when the same query is sent to it again it will return it from its cache. However, if some query is sent to Machine1 first, it will cache it and if the same query is sent to Machine2, it will again do a database lookup and cache it to its own cache. This implementation is suboptimal as it is doing more number of database lookups than what is actually required.

Approach 2:
Another approach is that each machine stores identical cache. Whenever some database access happened then the same cache is updated by all the web server. This approach has a drawback that whenever a data is updated in cache same cache update is fired to all the N web servers. Another disadvantage is that all the cache stores the same data so we are wasting precious cache space.

Approach 3:
In this approach, we will divide our cache such that each web server holds a different part of the cache. When a query reach to some web server it knows which webserver actually holds the cache for this query or at least knows that which server is supposed to keep a cache of the particular query. To do this we need a hash-based approach. We find the server, which serves the query, by just finding the hash (query) percentage N.

When a query request come to some web server, it will find the webserver corresponding to this query by applying the formula. It will ask the QuerySearch() function to that particular server. That server will in turn will query the database if required or provide the result from its own cache.
Now, regarding the cache expiration and old cache removal. We are keeping the TTL corresponding to each query so there can be a thread running which looks for the expired data and remove it from the cache.
In addition, combination of linked list and Hash-Table is used to keep the cache to get rid of less accessed data when the cache is almost full.

As a further improvement, we can think of some sort of Geo location aware webserver selection and cache policy so that query related to India is more supposed to be done in India and query related to china is supposed to come more from china.

Zomato

Use Case

1. Given a location, the list of hotels in that locality needs to be displayed.
2. Given a hotel name that hotel's rating, review, and menu need to be displayed
3. There should be some option to find if a delivery option is there in the hotel.
4. There should be some option to select a hotel on veg/non-veg category.
5. There should be some option to select hotels, which serve alcohol.
6. The user should be able to add reviews, add personal ratings to the hotels.
7. The user has some account or can access as guest.
8. Users/Admin should be able to add a new hotel to the system.

Constraints

1. A number of queries per second, suppose 100 queries per second.
2. There are more reads than writes.
3. 90% of the time there is read operation and only 10% of the time there is a write operation.
4. 100 * 60 * 60 * 24 = 8,640,000

Basic Design

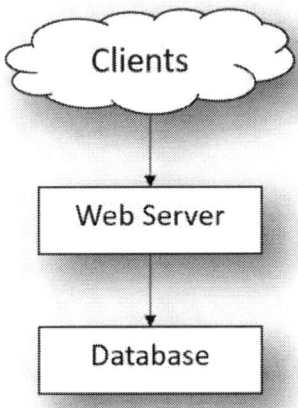

The scalability will be same as that of the examples explained above in the case of basic Facebook. Same concept of redundancy, load balancing, scalability etc.

Abstract Design

Each hotel has some hotel id associated with it.
1. Data of the hotel can be Name, Address, Rating, Review List, Veg-Nonveg, and Alcohol etc.
2. The region is a field in the address.
3. Search: When a user does a region based query all the entries of the hotels in that region need to be displayed to the user.
4. Search: User should be able to search specific hotel.

5. Add Review/ Rating: When users assess a hotel, then he should be able to add reviews and rating for the hotel.
6. Obviously the images are stored in CDN

Application Service layer
1. Start with fewer machines
2. All load balancer + a cluster of machines over time.
3. Traffic spike handling.
4. High availability.

Data Storage Layer:
1. Thousands of hotels.
2. There are no relationships between the object.
3. Reads are more than writing.
4. Relational database option is MySQL
5. Widely used.
6. Clear scaling paradigms. (Master-Master replication, Master-Slave replication)
7. Index lookups are very fast.

One optimization that we can assign an id to hotels the id can be derived from the locality so that it would be easy to find hotels in that locality.

YouTube

Scenarios

1. Users have some profile according to which content is shown.
2. Content thumbnails are shown when the user opens the YouTube web page.
3. When the user clicks on some thumbnail, then that video is played on flash player.

Constraints

1. Millions of users are going to use this service.
2. 200 million video requests served per day.
3. More reads than writes.

Design

1. YouTube is supposed to serve huge number of videos for which it has video serving clusters. A single video can be served from multiple servers in clusters and from multiple clusters thereby distributing the disk read which increases the performance of the system.
2. The most popular videos are served from CDN, CDN is more close to the user which reduce the response time. And reduce the load to the video serving clusters.
3. The rest of the metadata of the video is served from other servers as the user is not much interested in the metadata.
4. The rest of the application will be same as it will have an application server, database servers, load balancer, caching etc.

5. There is more read than write so the master server topology will be used. Therefore, there can be a single master for writing and multiple slaves for reading.
6. Master data is replicated to slaves. Since slaves are same as master then the master is down, then slaves can be promoted to make as master.
7. Page to be displayed to the user depends on his subscribed pages, History etc.
8. Information can be cached in the Memcached implemented near the database load balancer.

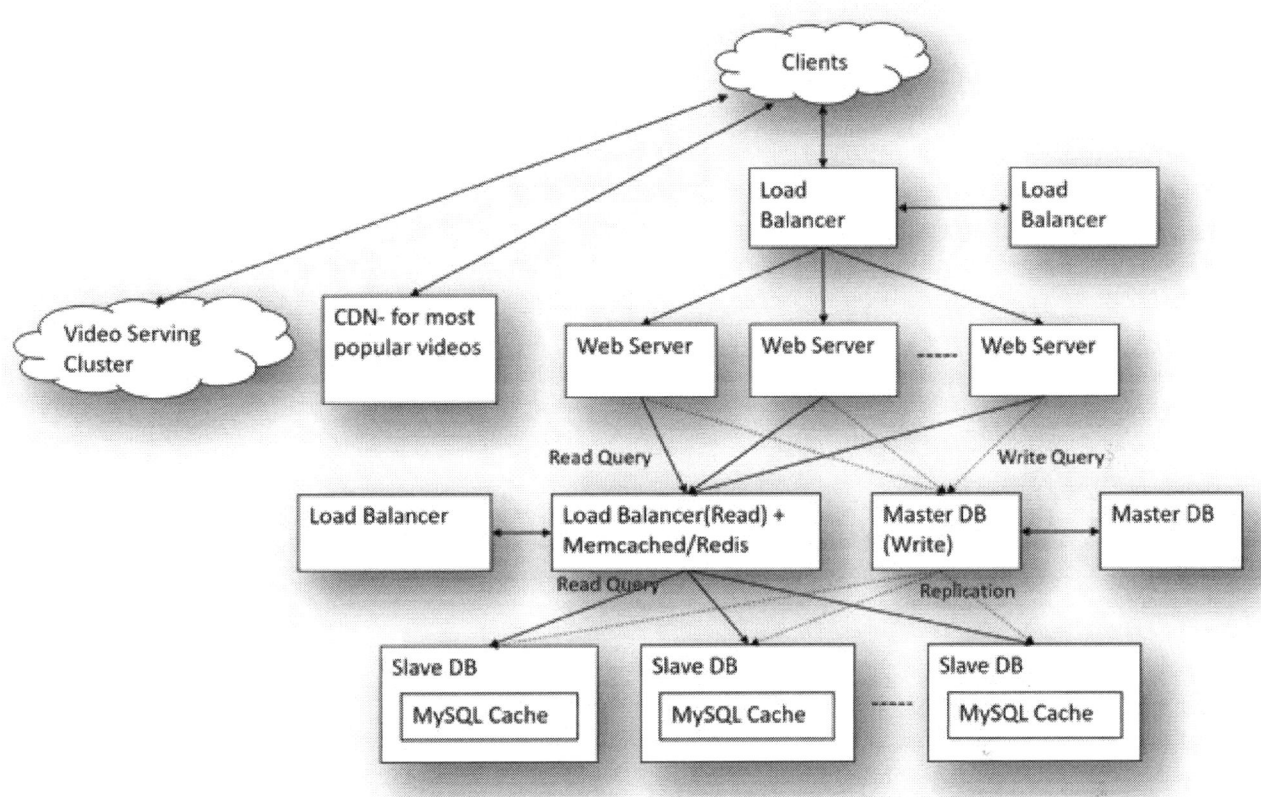

Design IRCTC

Scenario

1. The user should be able to query trains between two stations.
2. The query should be based on the date, quota, from and to the station.
3. The user should be able to see the availability in the train list retrieved from the above query.
4. The user should be able to book tickets for the available seats.

Constraints

1. There will be a huge number of people requesting the service. Let us suppose 0.01 percent of the population use the service daily once.
2. Geo-Redundancy should be provided.
3. More read query/ request then writes requests.

Design

1. The basic architecture from the scalability theory topic can be used here too.
2. There should be a huge number of servers which are serving the users.
3. There should be multiple servers at multiple geo locations to provide geo-redundancy. The database should be replicated at these multiple geo locations. There may be multiple servers in one particular zone too.
4. There is a huge number of read query, the user generally does a large number of query to find the seat he wants to book. There will be more reads then write so master-slave.
5. Queries can be cached; little old data is ok.
6. All the search will be served by slave servers.
7. When we book a ticket then transaction goes directly to the master server. Locks on train number can be taken to prevent race conditions. Once a lock is acquired then only you can book a ticket. Some counts can be used to avoid unnecessary locking and some counter can be used for this.
8. Each station has a quota in train and seats are allocated from that quota.
9. Each physical train will have two train ids one when trans go from source to destination station and one when it comes back. So in the system, there will be two train ids. Keeping these two separate ids will make the query easier to implement.
10. When final charting is done then each seat is swapped for the empty slots and we try to find the request from source station whose destination is also in that slot. The first fit is allotted that seat.
11. A load balancer is used to distribute traffic.
12. There may be multiple booking servers which ask for a booking token to the master server. Master server allocate a token for that server and reserve it for some time. When all the user information is filled and payment is done then only it allocates real seat depending upon user preference.
13. Slave server will handle user request till the end. Final booking request with the user payment and his complete information will go to the master server and the corresponding ticket will be booked.

Alarm Clock

How would you design an alarm clock?

Use Case

Alarm clock should have all the functions of clock. Should be able to show time.
The User can set the alarm time
The User can reset the alarm.
The User can set the alarm.

Constrains

The Granularity of alarm can be 15 min.

Test Case

Set the alarm at some time 6:00AM and set it.
The Alarm should work at 6:00AM

Stop the alarm, then alarm should stop ringing.

Design

There can be a clock class, which manages time and shoe time to the screen.
It has functions getTime() and setTime()
Alarm Clock extends Clock, and have some more function like startAlarm(), stopAlarm(), setAlarmTime(), ring()

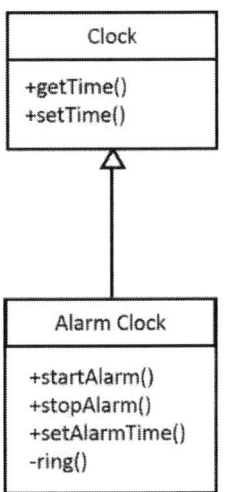

Implementation

A timer entry will run for granularity of 15 min. Or one min depends on customer requirement.
It will check if current time == alarm time if true then ring. And the alarm is on,
If start Alarm is called, it will set alarm on
If stop alarm is called, it will set alarm off.

Design for Elevator of a building

Scenarios

A typical lift has buttons (Elevator buttons) inside the cabin to let the user who got the lift to select his/her desired floor. Similarly, each floor has buttons (Floor buttons) to call the lift to go floors above and a floor below respectively. The buttons illuminate indicating the request is accepted. In addition, the button stops illuminating when the lift reaches the requested floor.

Use cases:
User
- Presses the floor button to call the lift
- Presses the elevator button to move to the desired floor

Floor Button & Elevator Button
- Illuminates when pressed by user
- Places an elevator request when pressed

Elevator
- Moves up/down as per instruction
- Opens/closes the door

Design

Each button press results in an elevator request which has to be served. Each of these requests is tracked at a centralized place. Elevator Requests, the class that stores, elevator requests can use different algo to schedule the elevator requests. The elevator is managed by a controller class, which we call Elevator Controller. Elevator controller class provide instructions to the elevator. Elevator controller reads the next elevator request to be processed and served.

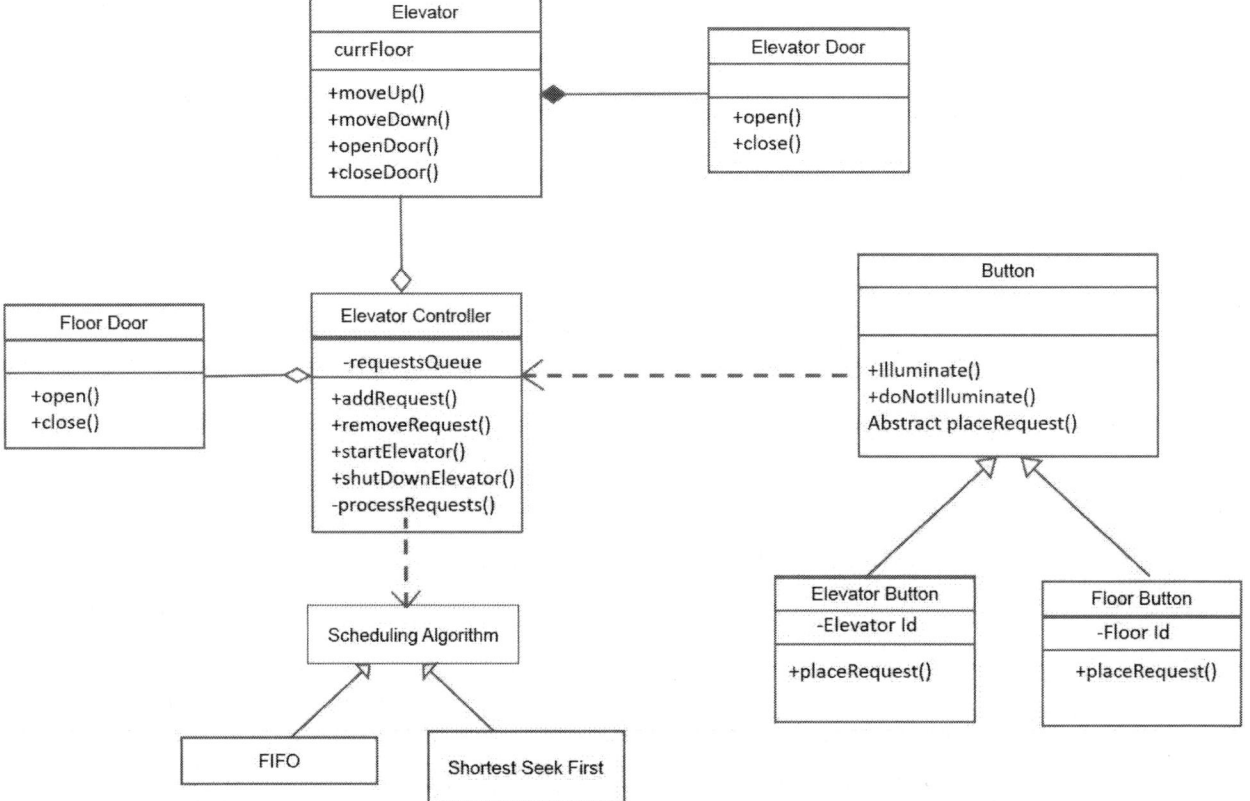

The button is an abstract class defining common behavior like illuminate, doNotIlluminate. FloorButton, Elevator Button extends Button type and define placeRequest () method which is invoked when a button is pressed. When a floor button or elevator button is presses a requests is added to a common queue.
ElevatorController reads the next request and instruct next action to the elevator.

How can we extend this to multiple elevators?

In the single elevator scenario, there is a single elevator and an elevator controller and a common server where the floor requests and the elevator button request are stored. Which are processed as per the scheduling algorithm.

To extend this to multiple elevator scenarios there will still be single elevator controller. Floor based requests can be served by any elevator whereas elevator button requests will be served only by the elevator to whom the button belongs.

FloorButton's placeRequest() adds a request to the common queue, which is accessed by the elevator controller thereby assigning the request to one of the elevators. ElevatorButton's placeRequest adds a request to the elevator directly as it is supposed to serve it. Elevator controller will be running various algorithms like shortest seek etc. to decide which lift is supposed to handle which request.

Valet parking system

Design a valet parking system.

Use Case

The requirements of the valet parking system should be:
1. Given a Parking lot having a fixed number of slots
2. Where a car can enter the slot if there is a free slot and then it will be given the direction of the free slot.
3. When exiting the car has to pay the fees for the duration of the time the car is in the slot.

Constraints

1. Parking slots come in multiple sizes- small, mid and large
2. Three types of vehicles, small, mid, large
3. A small vehicle can park in a small, medium, or large spot
4. A medium vehicle can park in a medium or large spot
5. A large vehicle can park only in a large spot

Design & Implementation

The parking lot will have the following interface

```java
public class parkingLot{
    private Map<int,Space> unreservedMap;
    private Map<int,Space> reservedMap;

    public boolean reserveSpace(Space)
    {
    // It will find if there is space in the unreserved map
    // If yes, then we will pick that element and
    // put into the reserved map with the current time value.
    }

    public int unreserveSpace(Space)
    {
    // It will find the entry in reserve map. If value found then
    // we will pick that Element and put into the unreserved map.
    // And return the charge units with the current time value.
    }
}
```

OO design for a McDonalds shop

Let's start with the description of how the McDonalds shop works.

1. In a McDonalds shop, the Customer selects the burger and directly places the order with the cashier.
2. In a McDonalds shop, the Customer waits for the order ready notification. Customer upon being notified that the order is ready collects the burger himself.

There are three different actors in our scenario and below is the list of actions they do.
Customer
1. Pays the cash to the cashier and places his order, get a token number and receipt
2. Waits for the intimation that order for his token is ready
3. Upon intimation/ notification, he collects the burger and enjoys his drink

Cashier
1. Takes an order and payment from the customer
2. Upon payment, creates an order and places it into the order queue
3. Provide token and receipt to the customer

Cook
1. Gets the next order from the queue
2. Prepares the burger
3. Places the burger in the completed order queue
4. Places a notification that order for token is ready

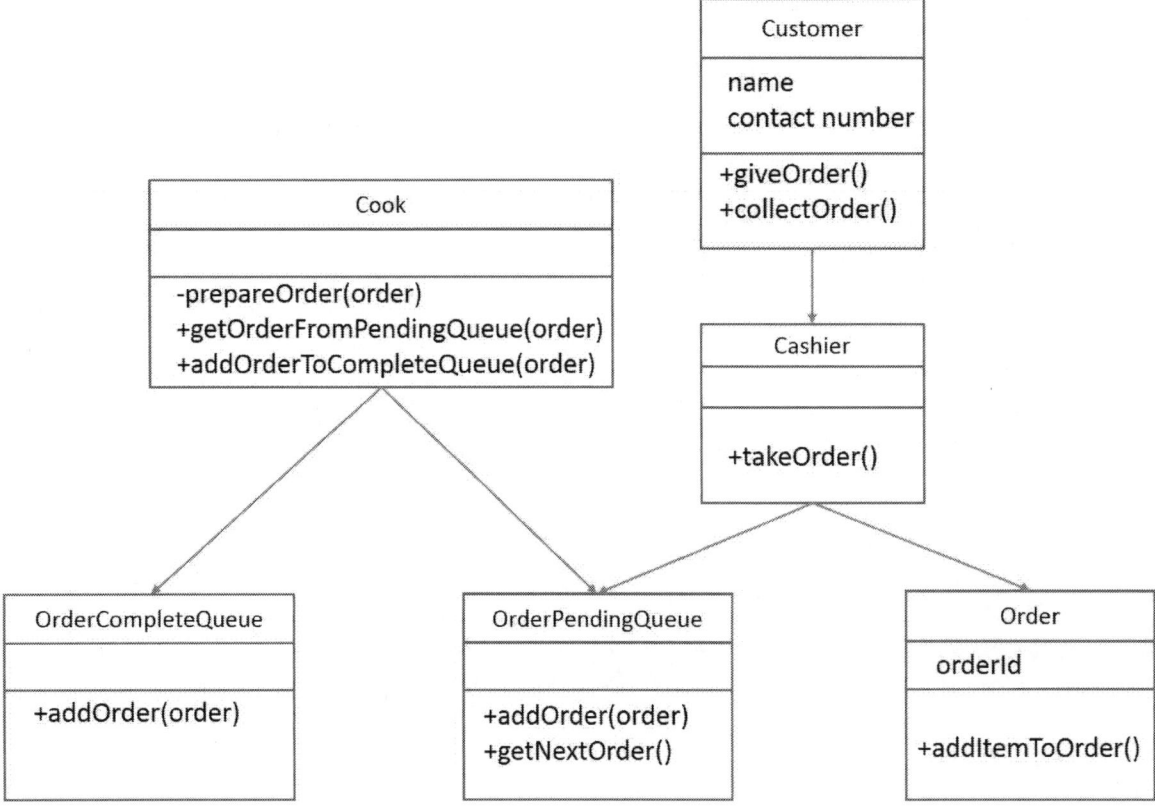

Object oriented design for a Restaurant

Let's describe how the restaurant works.
1. In a restaurant, the waiter takes order from the customer.
2. The waiter waits for the order to be ready and once ready serves the dishes to the customer.

These are the different actors in the model and I have listed the different actions against each actor

Customer
1. Selects the dish from the menu and call upon a waiter
2. Places the order
3. Enjoys his meal once the dish is served on his plate
4. Ask for the bill
5. Pays for the services

Waiter
1. Responds to the customers call on the tables he is waiting
2. Takes the customer's order
3. Places the order in the pending order queue
4. Waits for the order ready notifications
5. Once notification is received, collects the dish and serves the dish to the corresponding customer
6. Receives the bill request from customer
7. Asks the Cashier to prepare the bill
8. Gives the bill to the customer and accepts the payment

Cashier
1. Accepts the prepared bill request from the waiter for the given order details
2. Prepares the bills and hands it over to the waiter
3. Accepts the cash from the waiter towards the order

Cook
1. Gets the next order from the pending order queue
2. Prepares the dish and push the order to finished order queue
3. Sends a notification that the order is ready

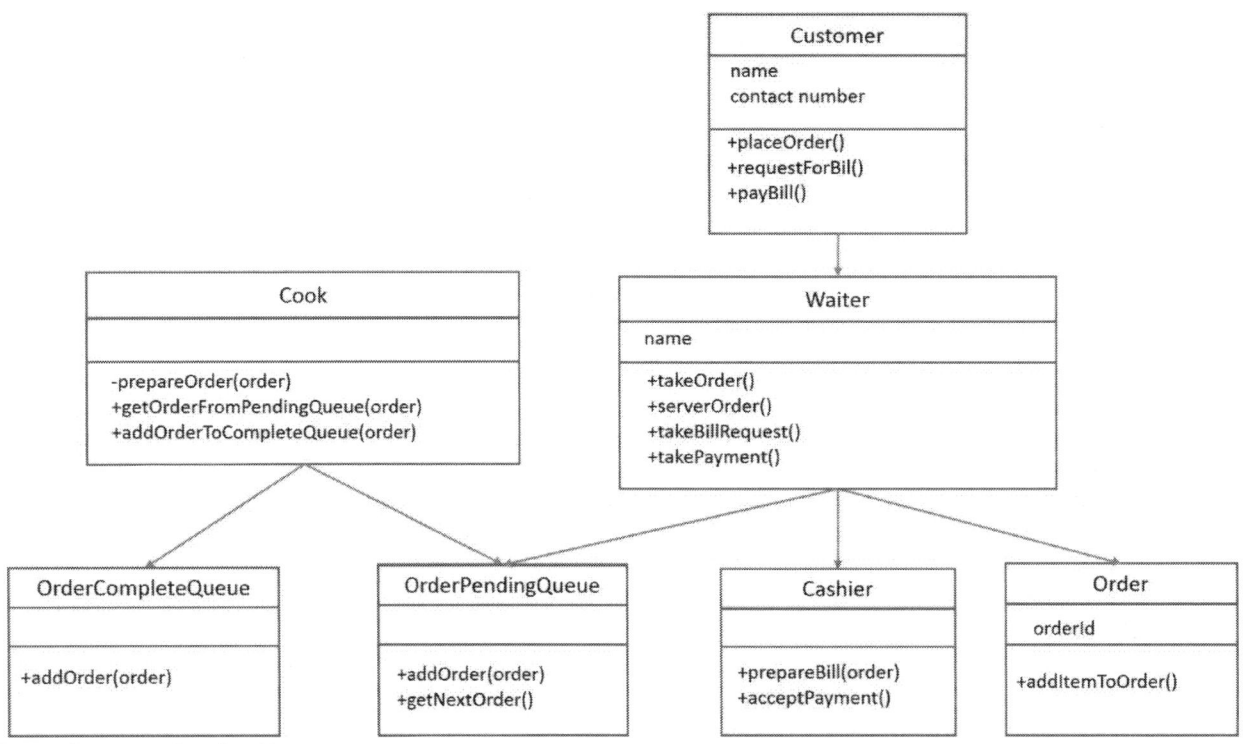

Class diagram for the Restaurant.

Object oriented design for a Library system

A library has a set of books, which the users can borrow for a certain period of time and return back. Users may choose to renew the return date if they feel they need more time to read the book.

The typical user actions with this online library would be
- Sign in/register
- Search books
- Borrow books
- Renew books
- Return books
- View his profile

The online library must keep track of the different books in the library currently available for users to borrow and the books already borrowed by users. Put it simply the inventory should be managed.

The various components of the system:
1. User
2. Librarian
3. Library
4. Book
5. Transection
6. Event Manager

The below class diagram, which depicts how these components inter-operates.

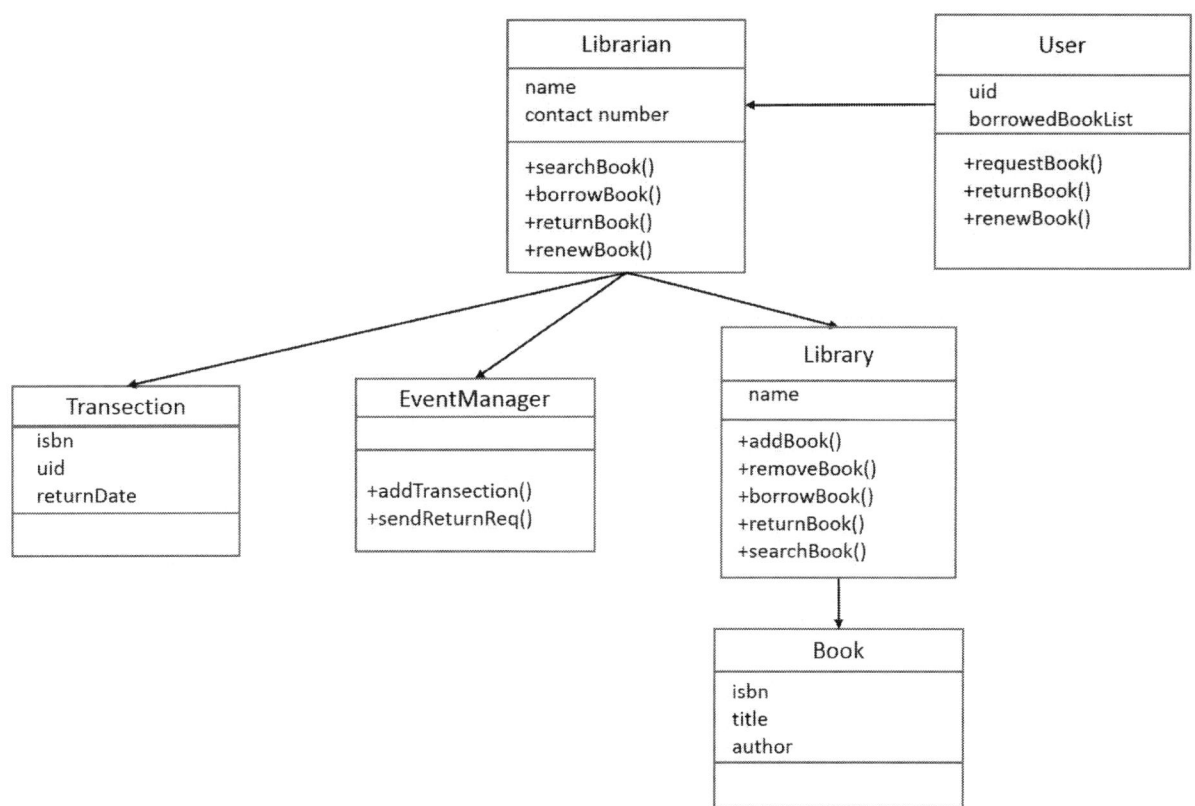

The User interacts with the Librarian, the user either request, return or renews a book. The Librarian will search for the book if the book is available in the Library then issue it to the user. A Transection will be created and added to the Event Manager. Event Manager will support add transaction and send return request interface. Once the book is overdue then the event manager will send an indication to the student that the book needs to be returned. When the book is renewed then the library state is not changed but the Transection detail is renewed at the Event Manager.

Suggest a shortest path

Use Case

The user had some coordinate by searching the coordinate from the name.
Show the whole map considering the coordinate as its centre.
Suggest the shortest path between two points.

Constraints

All paths are positive in cost.
For simplicity, I am considering all paths are for vehicle only, no pedestrian (pedestrian can walk in either direction even in one-way road.)

Design

The whole city map is stored as a graph in google.
We need to find the map by looking into the objects, which are in the distance shown by the browser.

The same path is stored as directed graph. And the graph that needs to be rendered depends on the zoom level. The preferred algorithm is a* for this application to get the shortest path.
Weight = h(x, y) + g(x, y)

Exercise

1. Design a system to implement social networking like Facebook, with millions of users? How would you find the connection between two people?

2. Autocomplete in www.booking.com. Design autocomplete feature for www.booking.com.

3. Instagram, Instagram is an online mobile-based photo sharing, video sharing service, which enables users to take pictures, and video upload them to the server and share them on social networking sites like Facebook or Twitter.
Note: - CDN is used to store active images.

4. Monolithic Website, assume you have a monolithic website and you are asked to rearchitect the website
Hint: - Discuss whole scalability theory section here.

5. Trip Advisor
URL's are parsed; content is collected from various services, and then applied to a template.

6. Cinchcast
Live audio streaming for business to do conferences.

7. BlogTalkRadio
Audio social network

8. Client based recommendation feature
How would you design a client based recommendation feature (based on customer history) on the product detail page? Design Customers who viewed item A also view item B and item C in an online shopping portal.

9. Car renting system
Design a car renting system, including reserving a car, checking in and checking out. Consider all the cases: reserve a car, then check out successfully; reserve a car, but the car is sold out before you check out...
Test Cases:
    1. Try to reserve a car for more than one person
    2. Try to reserve a car that is sold out
    3. Verify the checkout process. After checking out a particular, you should be able to reserve it for another customer.
    4. Try to reserve the same car for different customers in different dates

10. Online cab booking system (like Uber)

Admin Module
1. Admin should be able to add new driver / taxi details.
2. Should be able to calculate the amount that needs to be paid to the drivers. Monthly, weekly or daily.

User Module
1. Should be able to choose from and to location.
2. Available Taxies type, along with fare details.
3. Select a Taxi type
4. Book the taxi.
5. A confirmation message for the booking.

Driver Module
1. A driver should be able to register as a driver to Uber.
2. When a job is displayed to the driver he should be able to accept the job.
3. When the driver reaches to the customer then he should be able to start a trip.
4. When the driver had taken the customer to the desired location then he should stop the trip.
5. The driver should collect the fare based on the amount displayed in the app.
6. The driver should be able to give customer feedback.

Note: Just assume 2 minutes is equal to 1 KM.

11. Online teaching system

In an online teaching system, there are n number of teachers and each one teaches only one subject to any number of students.
And a student can join to any number of teachers to learn those subjects.
And each student can give one preference through which he can get updates about the subject or class timings etc.
Those preferences can be through SMS or Twitter/Facebook or Email etc.
Design above system and draw the diagram for above.

12. Customer Order Booking System

Admin Module
1. Should be able to add/edit/delete item, along with quantity, price, and unit.
2. Should be able to see all orders.

Customer Module
1. Should be able to enter his/her details for shipping, along will basic information like name, email, contact etc.
2. Can choose item, quantity
3. automatically payable price should be generated as per selected item and quantity.
4. Should be able to confirm the order.
5. After confirmation can see order confirmation report along with order number, which will be, system generated.

13. Online Movie Booking System

Admin Module

    1. Should be able to enter all movies, which have been released, and about to release in next week with all possible details like theatre location, price, show timings and seats.

    2. Should be able to delete movies, which are no longer in the theatre.

    3. Can see a number of booked tickets and remaining tickets for single theatre or for all theatre.

User Module

    1. User should be able to check all ongoing movies in theatre along with locations, availability of seats, price, and show timings

    2. The user should be able to check all upcoming movies for next week too.

    3. All movies those are running on theatre should be available for booking (one ticket or more than one ticket can be booked).

    4. After booking user should see the confirmation message of booking.

14. Design an online Auction system (similar to e-bay)

Functionalities include enlisting a product for auction by bid owner, placing the bid for a product by bidders, Bid winner selection, Notification of bid winner etc.).

APPENDIX

Appendix A

Algorithms	Time Complexity
Binary Search in a sorted array of N elements	$O(\log N)$
Reversing a string of N elements	$O(N)$
Linear search in an unsorted array of N elements	$O(N)$
Compare two strings with lengths L1 and L2	$O(min(L1, L2))$
Computing the Nth Fibonacci number using dynamic programming	$O(N)$
Checking if a string of N characters is a palindrome	$O(N)$
Finding a string in another string using the Aho-Corasick algorithm	$O(N)$
Sorting an array of N elements using Merge-Sort/Quick-Sort/Heap-Sort	$O(N * \log N)$
Sorting an array of N elements using Bubble-Sort	$O(N!)$
Two nested loops from 1 to N	$O(N!)$
The Knapsack problem of N elements with capacity M	$O(N * M)$
Finding a string in another string – the naive approach	$O(L1 * L2)$
Three nested loops from 1 to N	$O(N^3)$
Twenty-eight nested loops … you get the idea	$O(N^{28})$
Stack	
Adding a value to the top of a stack	$O(1)$
Removing the value at the top of a stack	$O(1)$
Reversing a stack	$O(N)$
Queue	
Adding a value to end of the queue	$O(1)$
Removing the value at the front of the queue	$O(1)$
Reversing a queue	$O(N)$
Heap	
Adding a value to the heap	$O \log N$
Removing the value at the top of the heap	$O(\log N)$
Hash	
Adding a value to a hash	$O(1)$
Checking if a value is in a hash	$O(1)$

INDEX

A

Abstract data type (ADT) 65
Adjacency List 299
Adjacency Matrix 298
Algorithm 48
All Pairs Shortest Paths 315
Array 35, 66
Array ADT Operations 67
Array Interview Questions 37
Asymptotic analysis 48

B

Backtracking 346, 384
Balanced Binary search tree 74
Bellman Ford Shortest Path 314
Big-O Notation 48
Binary Search 38, 90
Binary Search Tree (BST) for Strings 82
Binary Search Tree ADT Operations 74
Binary Search Trees (BST) 73
Binary Tree 73, 215
Breadth First Traversal 303
Breadth-First Search (BFS) 87
Brute Force Algorithm 342, 348
Bubble-Sort 119, 348
Bucket Sort 131

C

Caching 403
Call by Reference 23
Call by value 22
Circular Linked List 140, 169
Class co-NP 391
Class NP problems 388
Class P problems 387
Collision Resolution Techniques 283
Collisions 283
Comparisons of the various sorting algorithms 136
Complete binary tree 217
Constant Time 51
Counting Sort 88

D

Data-Structure 65
Decision problem 387
Depth First Traversal 301
Depth-First Search (DFS) 86
Dictionary 82
Digraph 296
Dijkstra's algorithm 311, 357
Directed Acyclic Graph 306
Divide-and-Conquer 343, 363
Doubly Circular list 176
Doubly Linked list 139
Doubly Linked List 159
Dynamic Programming 344, 375

E

External Sort 135

F

Fibonacci Number 44
Forest 298

G

Graph Algorithms 86
Graphs 85, 296
Greatest common divisor 44
Greedy Algorithm 343, 354

H

Hash Function 282
Hashing and Symbol Tables 91
Hashing with Open Addressing 283
Hashing with Separate chaining 287
Hash-Table 78, 82, 282, 325
Hash-Table Abstract Data Type 79, 282
Heap ADT Operations 77, 259
Heap-Sort 134, 268
Height-balanced Binary Tree 219

I

Insertion-Sort 122

K

Knuth-Morris-Pratt algorithm 320
Kruskal's Algorithm 310, 356

L

Left skewed binary tree 219
Linear Probing 283
Linear Search 89
Linear Time 51
Linked List 68, 139
Linked List ADT Operations 69
Load Balancer 401
Load Factor 283
Logarithmic Time 51

M

Master Theorem 56
Max Heap 258
Merge-Sort 126, 366
Min Heap 259
Minimum Spanning Trees (MST) 307

N

N-LogN Time 52
NP–Complete Problems 392
NP–Hard: 391

O

Omega-Ω Notation 49

P

Perfect binary tree 218
Preparation Plans 14
Prim's Algorithm 307, 355
Priority Queue 76, 257

Q

Quadratic Probing 284
Quadratic Time 52
Queue 71
Queue Abstract Data Type 203
Queue ADT Operations 72
Queue Using Array 203
Queue Using linked list 205
Quick Select 130
Quick-Sort 128, 367

R

Recursive Function 42
Redundancy 403
Right skewed binary tree 218
Robin-Karp algorithm 318

S

Selection of Best Sorting Algorithm 136
Selection-Sort 124, 349
Sequential Search 37, 349
Single Source Shortest Path 310
Singly Linked List 139, 140
Sorting Algorithms 87
Spanning tree 298
Stack 40, 70
Stack Abstract Data Type 182
Stack ADT Operations 70
Stack using Array 184
Stack using linked list 187
Strictly binary tree 218
Symbol Table 82, 322
System stack and Function Calls 40

T

Ternary Search Tree 84, 329
Ternary Search Trie 84, 329
Theta-Θ Notation 50
Time Complexity Order 51
Topological Sort 306
Tower of Hanoi 43
Tree Sorting 134
Trees 73
Trie 83, 326

U

Uses of Heap 275

V

Variable 18, 20

Made in the USA
San Bernardino, CA
26 February 2017